Methods in Molecular Biology

Series Editor
John M. Walker
School of Life and Medical Sciences
University of Hertfordshire
Hatfield, Hertfordshire, AL10 9AB, UK

For further volumes:
http://www.springer.com/series/7651

Microbial Carotenoids

Methods and Protocols

Edited by

Carlos Barreiro

Instituto de Biotecnología de León (INBIOTEC), Parque Científico de León, León, Spain;
Area of Microbiology, School of Health Studies, Campus de Ponferrada,
University of León, Ponferrada, Spain

José-Luis Barredo

Department of Biotechnology, Crystal Pharma, A Subsidiary of Albany
Molecular Research Inc. (AMRI), Parque Tecnológico de León, León, Spain

Humana Press

Editors
Carlos Barreiro
Instituto de Biotecnología de León (INBIOTEC)
Parque Científico de León
León, Spain

Area of Microbiology
School of Health Studies
Campus de Ponferrada
University of León
Ponferrada, Spain

José-Luis Barredo
Department of Biotechnology
Crystal Pharma, A Subsidiary of Albany
Molecular Research Inc. (AMRI)
Parque Tecnológico de León
León, Spain

ISSN 1064-3745 ISSN 1940-6029 (electronic)
Methods in Molecular Biology
ISBN 978-1-4939-8741-2 ISBN 978-1-4939-8742-9 (eBook)
https://doi.org/10.1007/978-1-4939-8742-9

Library of Congress Control Number: 2018951063

This Humana Press imprint is published by the registered company Springer Science+Business Media, LLC part of Springer Nature.
The registered company address is: 233 Spring Street, New York, NY 10013, U.S.A.

Preface

Plants, algae, humans, birds, crustaceans, insects, fungi, yeast and bacteria, in addition of their role as a part of life on Earth, have a much more colourful common link, the carotenoids. These natural pigments, which contain a chromophore responsible for the characteristic yellow to orange and reddish colours, are one of the most diverse classes of secondary metabolites. Carotenoid functions can be clustered in concordance to their photosynthetic or non-photosynthetic behaviour in the carrier organism, since they play different natural roles. Thus, they belong to light-harvesting complexes in photosynthesizing species but also play a photoprotective role against reactive oxygen species, which justify their presence in those non-photosynthetic organisms acting as (photo)protectors. Nowadays, they are commercialized as colorants, feed supplements and nutraceuticals and for medical, cosmetic and biotechnological purposes. This is the reason that this book focused on the upgraded methodologies and procedures scientifically and industrially used for carotenoid production, analysis, characterizations and discovery applied to fungi, algae and bacteria.

Currently, almost 80 years after the structural elucidation of β-carotene, more than 750 types of carotenoids have been described from natural sources, including the marine origin. About 50 carotenoids are incorporated in the human nutrition and 20 of them have been traditionally detected in human body due to their anti-inflammatory and antioxidant effects, potentially preventing human diseases and maintaining good health. Besides, in nature, different animal tissues are pigmented due to carotenoids: egg yolk, pink flesh of salmonids, pink plumage of flamingos, red exoskeleton of crustaceans, etc. Thus, feed supplementation of carotenoids in intensive farming and aquaculture is another relevant use to maintain this pigmentation. Even though the estimation of their economical profit result confuse, a range from 1.5 to 1.7 US$ billion on 2020 seems to be realistic, which highlights the industrial relevance of these compounds.

The three opening chapters of this book act as an introductory colourful funnel, which, on the one hand, present a wide overview of the different colours of the microbial biotechnology and, on the other hand, gradually focus the book on the origin, health utilities and economical relevance of these yellowish to reddish compounds. Afterwards, four well-defined thematic blocks present the specific methodological chapters so typical of this Springer Protocols series. Algae, Bacteria, Yeast and Fungi are the four chapter clusters that comprise the whole book, which has been written by outstanding experts in their field and provides a reference source for laboratory and industrial professionals, as well as for graduate students in a number of bio-sanitary disciplines, including biotechnology, microbiology, genetics, molecular biology, health, nutrition, farming, veterinary, etc.

We are indebted to the authors who, in spite of their professional activities, agreed to participate in this book; to Dr. J. Walker, series editor, for his encouragement and advice in reviewing the manuscripts; and to the staff of Springer, for their assistance in assembling this volume and their efforts in keeping this project on schedule. Last but, of course, not least, we warmly acknowledge our families, friends and colleagues of INBIOTEC and Crystal Pharma for their patience and kind support.

León, Spain

Carlos Barreiro
José-Luis Barredo

Contents

Contributors

JENNIFER ALCAÍNO • *Laboratorio de Genética, Departamento de Ciencias Ecológicas, Facultad de Ciencias, Universidad de Chile, Santiago, Chile*

MARÍA ISABEL ALVAREZ • *Área de Genética, Departamento de Microbiología y Genética, University of Salamanca, Salamanca, Spain*

JOSÉ MIGUEL ARAYA-GARAY • *Department of Microbiology, University of Santiago de Compostela, Santiago de Compostela, Spain*

DALAL ASKER • *Faculty of Agriculture, Food Science and Technology Department, Alexandria University, Alexandria, Egypt; Department of Materials Science and Engineering, University of Toronto, Toronto, ON, Canada*

JAVIER AVALOS • *Departamento de Genética, Facultad de Biología, Universidad de Sevilla, Sevilla, Spain*

TAREK S. AWAD • *Department of Materials Science and Engineering, University of Toronto, Toronto, ON, Canada*

MARCELO BAEZA • *Laboratorio de Genética, Departamento de Ciencias Ecológicas, Facultad de Ciencias, Universidad de Chile, Santiago, Chile*

JOSÉ-LUIS BARREDO • *Department of Biotechnology, Crystal Pharma, A Subsidiary of Albany Molecular Research Inc. (AMRI), Parque Tecnológico de León, León, Spain*

CARLOS BARREIRO • *Instituto de Biotecnología de León (INBIOTEC), Parque Científico de León, León, Spain; Area of Microbiology, School of Health Studies, Campus de Ponferrada, University of León, Ponferrada, Spain*

TERUHIKO BEPPU • *Life Science Research Center, College of Bioresource Sciences, Nihon University, Fujisawa, Japan*

SNÆDÍS HULD BJÖRNSDÓTTIR • *Center for Systems Biology and Faculty of Industrial Engineering, Mechanical Engineering and Computer Science, School of Engineering and Natural Sciences, University of Iceland, Reykjavík, Iceland*

SIGURDUR BRYNJOLFSSON • *Center for Systems Biology and Faculty of Industrial Engineering, Mechanical Engineering and Computer Science, School of Engineering and Natural Sciences, University of Iceland, Reykjavík, Iceland*

MARÍA DEL CARMEN CERÓN GARCÍA • *Department of Chemical Engineering, University of Almería, Almería, Spain*

CLÉMENCE CHAINTREUIL • *Laboratoire des Symbioses Tropicales et Méditerranéennes, IRD, UMR-IRD/SupAgro/INRA/UM/CIRAD, Montpellier Cedex 5, France*

VÍCTOR CIFUENTES • *Laboratorio de Genética, Departamento de Ciencias Ecológicas, Facultad de Ciencias, Universidad de Chile, Santiago, Chile*

FERNANDO COLABELLA • *Laboratorio de Microbiología Aplicada, Biotecnología y Bioinformática, Instituto Andino Patagónico de Tecnologías Biológicas y Geoambientales, IPATEC, UNComahue-CONICET, Bariloche, Argentina*

ARNOLD L. DEMAIN • *RISE Institute, Drew University, Madison, NJ, USA*

ARTURO P. ESLAVA • *Área de Genética, Departamento de Microbiología y Genética, University of Salamanca, Salamanca, Spain*

JOËL FARDOUX • *Laboratoire des Symbioses Tropicales et Méditerranéennes, IRD, UMR-IRD/SupAgro/INRA/UM/CIRAD, Montpellier Cedex 5, France*
JOSÉ MARÍA FERNÁNDEZ SEVILLA • *Department of Chemical Engineering, University of Almería, Almería, Spain*
MARÍA FERNÁNDEZ-LOBATO • *Centro de Biología Molecular Severo Ochoa, Departamento de Biología Molecular (UAM-CSIC), Universidad Autónoma de Madrid, Madrid, Spain*
JONAS FROHWITTER • *Genetics of Prokaryotes, Faculty of Biology and CeBiTec, Bielefeld University, Bielefeld, Germany*
WEIQI FU • *Center for Systems Biology and Faculty of Industrial Engineering, Mechanical Engineering and Computer Science, School of Engineering and Natural Sciences, University of Iceland, Reykjavík, Iceland*
CARLOS GARCÍA-ESTRADA • *INBIOTEC, Instituto de Biotecnología de León, León, Spain; Área de Toxicología, Departamento de Ciencias Biomédicas, Universidad de León, León, Spain*
VICTORIANO GARRE • *Departamento de Genética y Microbiología, Facultad de Biología, Universidad de Murcia, Murcia, Spain*
ERIC GIRAUD • *Laboratoire des Symbioses Tropicales et Méditerranéennes, IRD, UMR-IRD/SupAgro/INRA/UM/CIRAD, Montpellier Cedex 5, France*
CYNTHIA VICTORIA GONZÁLEZ LÓPEZ • *Department of Chemical Engineering, University of Almería, Almería, Spain*
LAURE HANNIBAL • *Laboratoire des Symbioses Tropicales et Méditerranéennes, IRD, UMR-IRD/SupAgro/INRA/UM/CIRAD, Montpellier Cedex 5, France*
NADJA A. HENKE • *Genetics of Prokaryotes, Faculty of Biology and CeBiTec, Bielefeld University, Bielefeld, Germany*
DÁMASO HORNERO-MÉNDEZ • *Departamento de Fitoquímica de los Alimentos, Instituto de la Grasa (CSIC), Campus Universidad Pablo de Olavide, Sevilla, Spain*
ENRIQUE A. ITURRIAGA • *Área de Genética, Departamento de Microbiología y Genética, University of Salamanca, Salamanca, Spain*
SOTIRIOS KIOKIAS • *Department of Food Technology, School of Agricultural Technology, Food Technology and Nutrition, Technological Educational Institute of Peloponnese, Kalamata, Greece*
KATARINA KOSALKOVÁ • *INBIOTEC, Instituto de Biotecnología de León, León, Spain*
PANAGIOTA LANGI • *Laboratory of Food Chemistry, Department of Chemistry, National and Kapodistrian University of Athens, Athens, Greece*
DIEGO LIBKIND • *Laboratorio de Microbiología Aplicada, Biotecnología y Bioinformática, Instituto Andino Patagónico de Tecnologías Biológicas y Geoambientales, IPATEC, UNComahue-CONICET, Bariloche, Argentina*
M. CARMEN LIMÓN • *Departamento de Genética, Facultad de Biología, Universidad de Sevilla, Sevilla, Spain*
JIN LIU • *Laboratory for Algae Biotechnology & Innovation, College of Engineering, Peking University, Beijing, China*
SERGIO LÓPEZ-GARCÍA • *Departamento de Genética y Microbiología, Facultad de Biología, Universidad de Murcia, Murcia, Spain*
SONIA MARTÍNEZ-CÁMARA • *Department of Biotechnology, Crystal Pharma, A Subsidiary of Albany Molecular Research Inc. (AMRI), Parque Tecnológico de León, León, Spain*

GALINA S. MINYUK • *The A.O. Kovalevsky Institute of Marine Biological Research, Russian Academy of Sciences, Sevastopol, Russia*
EMILIO MOLINA GRIMA • *Department of Chemical Engineering, University of Almería, Almería, Spain*
MARTÍN MOLINÉ • *Laboratorio de Microbiología Aplicada, Biotecnología y Bioinformática, Instituto Andino Patagónico de Tecnologías Biológicas y Geoambientales, IPATEC, UNComahue-CONICET, Bariloche, Argentina*
EUSEBIO NAVARRO • *Departamento de Genética y Microbiología, Facultad de Biología, Universidad de Murcia, Murcia, Spain*
MARÍA ISABEL NAVARRO-MENDOZA • *Departamento de Genética y Microbiología, Facultad de Biología, Universidad de Murcia, Murcia, Spain*
FRANCISCO E. NICOLÁS • *Departamento de Genética y Microbiología, Facultad de Biología, Universidad de Murcia, Murcia, Spain*
TAMÁS PAPP • *Department of Microbiology, University of Szeged, Szeged, Hungary*
CARLOS PÉREZ-ARQUES • *Departamento de Genética y Microbiología, Facultad de Biología, Universidad de Murcia, Murcia, Spain*
PETRA PETERS-WENDISCH • *Genetics of Prokaryotes, Faculty of Biology and CeBiTec, Bielefeld University, Bielefeld, Germany*
CHARALAMPOS PROESTOS • *Laboratory of Food Chemistry, Department of Chemistry, National and Kapodistrian University of Athens, Athens, Greece*
HANNAH DEL RÍO • *Department of Biotechnology, Crystal Pharma, A Subsidiary of Albany Molecular Research Inc. (AMRI), Parque Tecnológico de León, León, Spain*
MARTA RODRÍGUEZ-SÁIZ • *Department of Biotechnology, Crystal Pharma, A Subsidiary of Albany Molecular Research Inc. (AMRI), Parque Tecnológico de León, León, Spain*
SARA RUBIO • *Department of Biotechnology, Crystal Pharma, A Subsidiary of Albany Molecular Research Inc. (AMRI), Parque Tecnológico de León, León, Spain*
ISABEL-CLARA SÁNCHEZ-OREJAS • *INBIOTEC, Instituto de Biotecnología de León, León, Spain*
SERGIO SÁNCHEZ • *Instituto de Investigaciones Biomédicas, Universidad Nacional Autónoma de México, Ciudad de México, CDMX, México*
ALEXEI E. SOLOVCHENKO • *Lomonosov Moscow State University, Moscow, Russia*
YIXI SU • *Center for Systems Biology and Faculty of Industrial Engineering, Mechanical Engineering and Computer Science, School of Engineering and Natural Sciences, University of Iceland, Reykjavík, Iceland*
SANTIAGO TORRES-MARTÍNEZ • *Departamento de Genética y Microbiología, Facultad de Biología, Universidad de Murcia, Murcia, Spain*
KENJI UEDA • *Life Science Research Center, College of Bioresource Sciences, Nihon University, Fujisawa, Japan*
THEODOROS VARZAKAS • *Department of Food Technology, School of Agricultural Technology, Food Technology and Nutrition, Technological Educational Institute of Peloponnese, Kalamata, Greece*
PATRICIA VEIGA-CRESPO • *Department of Microbiology, University of Santiago de Compostela, Santiago de Compostela, Spain*
ANDRÉ VERMÉGLIO • *CEA, CNRS, Aix-Marseille Université, UMR 7265, Biosciences and Biotechnology Institute of Aix-Marseille, Saint-Paul-lez-Durance, France*
TOMÁS G. VILLA • *Department of Microbiology, University of Santiago de Compostela, Santiago de Compostela, Spain*

VOLKER F. WENDISCH • *Genetics of Prokaryotes, Faculty of Biology and CeBiTec, Bielefeld University, Bielefeld, Germany*
ZHIQIAN YI • *Center for Systems Biology and Faculty of Industrial Engineering, Mechanical Engineering and Computer Science, School of Engineering and Natural Sciences, University of Iceland, Reykjavík, Iceland*

Chapter 1

Advancement of Biotechnology by Genetic Modifications

Arnold L. Demain and Sergio Sánchez

Abstract

One of the greatest sources of metabolic and enzymatic diversity are microorganisms. In recent years, emerging recombinant DNA and genomic techniques have facilitated the development of new efficient expression systems, modification of biosynthetic pathways leading to new metabolites by metabolic engineering, and enhancement of catalytic properties of enzymes by directed evolution. Complete sequencing of industrially important microbial genomes is taking place very rapidly, and there are already hundreds of genomes sequenced. Functional genomics and proteomics are major tools used in the search for new molecules and development of higher-producing strains.

Key words Agriculture, Bioconversions, Biopharmaceuticals, Enzymes, Genetic engineering, Hosts, Metabolic engineering, Polymers, Primary metabolites, Organic acids, Alcohols, Secondary metabolites, Bioinsecticides, Recombinant proteins

1 Introduction

Advantages of microorganisms in the production of compounds, as compared to isolation from plants and animals or synthesis by chemists, include (a) rapid uptake of nutrients that supports high rates of metabolism and biosynthesis; (b) capability of carrying out a wide variety of reactions; (c) facility to adapt to a large array of different environments; (d) ease of genetic manipulation, both in vivo and in vitro, to increase production, to modify structures and activities, and to make entirely new products; (e) simplicity of screening procedures; and (f) a wide diversity.

Products from microbes are very diverse, ranging from very large molecules, such as proteins, nucleic acids, carbohydrate polymers, or even cells, to small molecules that are usually divided into primary metabolites, that is, those essential for vegetative growth, and secondary metabolites, that is, those nonessential for growth.

1.1 Primary Metabolites

Synthesis of microbial products during the exponential phase of growth is an integral part of the normal growth process. They are intermediates or end products of the pathways, are building blocks

Carlos Barreiro and José-Luis Barredo (eds.), *Microbial Carotenoids: Methods and Protocols*, Methods in Molecular Biology, vol. 1852, https://doi.org/10.1007/978-1-4939-8742-9_1,

for essential macromolecules (e.g., amino acids and nucleotides), or are converted into coenzymes (e.g., vitamins). Other primary metabolites (e.g., organic acids and ethanol) result from catabolic metabolism; they are not used for building cellular constituents, but their production, which is related to energy production and substrate utilization, is essential for growth. Industrially, the most important primary metabolites are amino acids, nucleotides, vitamins, alcohols, and organic acids.

Production of a particular primary metabolite by deregulated organisms may inevitably be limited by the inherent capacity of the particular organism to make the appropriate biosynthetic enzymes. Recent approaches utilize the techniques of modern genetic engineering to correct such deficiencies and develop strains overproducing primary metabolites. There are two ways to accomplish after this (a) to increase the number of copies of structural genes coding for these enzymes and (b) to increase the frequency of transcription.

Novel genetic technologies are important for the development of overproducers. Ongoing genome-sequencing projects, involving hundreds of genomes, the availability of sequences corresponding to model organisms, new DNA microarray and proteomic tools, as well as the new techniques for mutagenesis and recombination described below, will no doubt accelerate strain improvement programs.

Genome-based strain reconstruction achieves the construction of a superior strain, which only contains mutations crucial to hyperproduction, but not other unknown mutations, which accumulate by brute-force mutagenesis and screening [1].

The directed improvement of product formation or cellular properties via modification of specific biochemical reactions or introduction of new ones with the use of recombinant DNA technology is known as metabolic engineering [2, 3]. Analytical methods are combined to quantify fluxes and to control them with molecular biological techniques to implement suggested genetic modifications. The overall flux through a metabolic pathway depends on several steps, not just a single rate-limiting reaction. Amino acid production is one of the fields with many examples of this approach [4]. Other processes improved by this technique include vitamins, organic acids, ethanol, 1,3-propanediol, and carotenoids.

Reverse (inverse) metabolic engineering is another approach that involves choosing a strain which has a favorable cellular phenotype, evaluating and determining the genetic and/or environmental factors that confer that phenotype, and finally transferring that phenotype to a second strain via direct modifications of the identified genetic and/or environmental factors [5, 6].

Molecular breeding techniques are based on mimicking natural recombination by in vitro homologous recombination [7]. DNA shuffling not only recombines DNA fragments but also introduces point mutations at a very low controlled rate [8]. Unlike site-directed mutagenesis, this method of pooling and recombining parts of similar genes from different species or strains has yielded remarkable improvements in a very short amount of time [9]. Whole-genome shuffling is another technique that combines the advantage of multiparental crossing allowed by DNA shuffling with the recombination of entire genomes through recursive protoplast fusion [10, 11].

Systems biology is an integrated, systemic approach to the analysis and optimization of cellular processes by introducing a variety of perturbations and measuring the system response [12]. Altered phenotypes are created by molecular biological techniques or by altering environments. Further characterization of the phenotype leading to maximal product formation is analyzed and quantified through the use of genome-wide high-throughput *omics* data and genome-scale computational analysis.

1.2 Amino Acids

Many of the techniques mentioned above have made a great impact on the production of amino acids. Different strategies include (a) amplification of a rate-limiting enzyme of pathway, (b) amplification of the first enzyme after a branchpoint, (c) cloning of a gene encoding an enzyme with more or less feedback regulation, (d) introduction of a gene encoding an enzyme with a functional or energetic advantage as a replacement for the normal enzyme, (e) amplification of the first enzyme leading from central metabolism to increase carbon flow into the pathway, and (f) isolation of a transport mutant decreasing amino acid uptake and intracellular feedback control while improving excretion.

Among the amino acids, L-glutamate and L-lysine, mostly used as feed and food additives, represent the largest products in this category. Produced by fermentation are 1.5 million tons of L-glutamate and 850,000 tons of L-lysine HCl. The total amino acid market was about 4.5 billion dollars in 2004 [13]. High titers of amino acids are shown in Table 1.

1.2.1 L-Glutamate

Glutamate was the first amino acid to be produced by fermentation because of its use as a flavoring agent (monosodium glutamate, MSG). The process employs various species of the genera *Corynebacterium* and *Brevibacterium*. Molar yields of glutamate from sugar are 50–60%, and broth concentrations have reached 150 g/L. Glutamic acid overproduction is feedback regulated. However, mutant strains with modifications in the cell membrane are able to pump glutamate out of the cell, thus allowing its biosynthesis to proceed unabated. Introduction of the *Vitreoscilla* hemoglobin gene *vgb* into *Corynebacterium glutamicum* increased cell growth

Table 1
High amino acid levels produced by fermentation

Amino acid	Titer (g/L)
L-Alanine	120
L-Arginine	96
L-Glutamate	150
L-Glutamine	49
L-Histidine	42
L-Hydroxyproline	41
L-Isoleucine	40
L-Leucine	34
L-Lysine HCl	170
L-Methionine	25
L-Phenylalanine	51
L-Proline	108
L-Serine	65
L-Threonine	124
L-Tryptophan	60
L-Tyrosine	55
L-Valine	51

and glutamic acid and glutamine production via increased oxygen uptake [14]. Workers at Ajinomoto Co., Inc. increased glutamate production from glucose by 9% by suppressing CO_2 liberation in the pyruvate dehydrogenase reaction [15]. They did this by cloning the *xfp* gene-encoding phosphoketolase from *Bifidobacterium animalis* into *C. glutamicum* and overexpressing it.

1.2.2 L-Lysine

Since the bulk of the cereals consumed in the world are deficient in L-lysine, this essential amino acid became an important industrial product. Although the lysine biosynthetic pathway is controlled very tightly in an organism such as *Escherichia coli* and in lysine-producing organisms (e.g., mutants of *C. glutamicum* and its relatives), there is only a single aspartate kinase (AK), which is regulated via concerted feedback inhibition by threonine plus lysine. Metabolic engineering has been used in *C. glutamicum* to improve L-lysine production [4]. A chimeric AK, composed of the N-terminal catalytic region from *Bacillus subtilis* AKII and the C-terminal region from *Thermus thermophilus*, was evolved through random mutagenesis and then screened using a high-throughput synthetic RNA device which comprises an L-lysine-sensing riboswitch and a selection module. Of the three evolved aspartate kinases, the best mutant (BT3) showed 160% increased in vitro activity compared to the wild-type enzyme [16].

Comparative genome analysis between a wild-type strain and an L-lysine-producing *C. glutamicum* strain identified three mutations that increased L-lysine production when introduced into the wild-type strain. Introduction of the 6-phosphogluconate-dehydrogenase *gnd* mutation increased yield by 15% [17]. Further improvement was achieved by introducing the *mqo* mutation (malate: quinone oxidoreductase) resulting in an increased titer of 95 g/L in a fed-batch culture [18]. Transcriptome analysis revealed that strain B-6, as compared to the wild type, is upregulated in both the pentose phosphate pathway genes and the amino acid biosynthetic genes and downregulated in TCA cycle genes. Lysine HCl titers have reached 170 g/L [19]. Metabolic flux studies of wild-type *C. glutamicum* and improved lysine-producing mutants showed that yield increased from 1.2% to 24.9% relative to the glucose flux. Different approaches have been used to increase lysine production by *C. glutamicum* mutants including (1) deletion of different genes from the glycolytic pathway [20], (2) improving the availability of pyruvate by eliminating pyruvate dehydrogenase activity [21], (3) overexpression of pyruvate carboxylase or DAP dehydrogenase genes, and (4) overexpression of gene NCgl0855 encoding a methyltransferase or the *amtA-ocd-soxA* operon [22]. A genetically defined *C. glutamicum* L-lysine overproducer strain was developed by system metabolic engineering of the wild type. Implementation of only 12 defined genome-based changes in genes encoding central metabolic enzymes redirected major carbon fluxes as desired toward the optimal pathway usage predicted by in silico modeling. The final engineered *C. glutamicum* strain was able to produce lysine with a high yield of 0.55 g per gram of glucose, a titer of 120 g/L of lysine, and a productivity of 4.0 g/L/h in fed-batch culture [23].

1.2.3 L-Threonine

Overproduction of L-threonine has been achieved in *Serratia marcescens* by transductional crosses, which combined several feedback control mutations in a single organism. This resulted in titers up to 25 g/L [24]. Combination in a single strain of another six regulatory mutations derived by resistance to amino acid analogs led to desensitization and derepression of the key enzymes of threonine synthesis. The resultant transductant produced 40 g/L of threonine [25]. The use of recombinant DNA technology led to strains that made 63 g/L threonine [26]. An *E. coli* strain was developed via mutation and genetic engineering and optimized by the inactivation of threonine dehydratase (TD) resulting in a process yielding 100 g/L of L-threonine in 36 h of fermentation [27].

In *C. glutamicum* ssp. *lactofermentum*, L-threonine production reached 58 g/L when a strain producing both threonine and lysine was transformed with a plasmid carrying its own *hom*, *thrB*, and *thrC* genes [28]. A recombinant isoleucine auxotrophic strain of *E. coli* (carrying extra copies of the *thrABC* operon, an inactivated *tdh* gene and mutated to resist high concentrations of L-threonine and L-homoserine) produced 80 g/L L-threonine in 1.5 days with a yield of 50% [29]. Cloning extra copies of threonine export genes into *E. coli* also increased threonine production [30].

Systems metabolic engineering was used to develop an L-threonine1-overproducing *E. coli* strain [31]. Feedback inhibition of aspartokinase I and III (encoded by *thrA* and *lysC*, respectively) and transcriptional attenuation regulation (*thrL*) were removed. Deletion of *tdh* and mutation of *ilvA* avoid threonine degradation. The *metA* and *lysA* genes were deleted to make more precursors available for threonine biosynthesis. Further target genes to be engineered were identified by transcriptome profiling combined with in silico flux response analysis. The final engineered *E. coli* strain was able to reach a high yield of 0.393 g of threonine per gram of glucose and 82.4 g/L in a fed-batch culture. With the use of an L-isoleucine leaky (ILE^L) and α-amino-β-hydroxyvaleric acid resistant (AHV^r) *E. coli* TRFC strain, carrying a pTHR101 plasmid containing a Thr operon, a threonine production of 118 g/L was achieved after 38 h with a productivity of 3.1 g/L/h (46% conversion ratio from glucose to threonine) [32]. A further increase in amino acid production to 124.57 g/L was obtained in TRFC strain by decreasing acetic acid production using combined feeding strategies [33]. Acetic acid is an inhibitor of growth and threonine production.

1.2.4 L-Valine

Combined rational modification, transcriptome profiling, and systems-level in silico analysis was used to develop an *E. coli* strain for the production of this amino acid [34]. The *ilvA*, *leuA*, and *panB* genes were deleted to make more precursors available for L-valine biosynthesis. This engineered strain, harboring a plasmid overexpressing the *ilvBN* genes, produced 1.3 g/L L-valine. Overexpression of the *lrp* and *ygaZH* genes (encoding a global regulator Lrp and L-valine exporter, respectively), and amplification of the *lrp*, *ygaZH*, and *lrp-ygaZH* genes enhanced production of L-valine by 21.6%, 47.1%, and 113%, respectively. Further improvement was achieved by using in silico gene knockout simulation, which identified the *aceF*, *mdh*, and *pfkA* genes as knockout targets. The VAMF strain (Val Δ*aceF* Δ*mdh* Δ*pfkA*) was able to produce 7.5 g/L L-valine from 20 g/L glucose in batch culture, resulting in a high yield of 0.378 g of L-valine per gram of glucose. Production by mutant strain VAL1 of *C. glutamicum* amounted to 31 g/L [35]. The mutant was constructed by overexpressing biosynthetic enzymes via a plasmid, eliminating *ilvA* encoding threonine dehydratase, and deleting two genes encoding enzymes of pantothenate biosynthesis. The culture was grown with limiting concentrations of isoleucine and pantothenate. By applying metabolic engineering, a *C. glutamicum* strain (WCC003 harboring pJYW-4-*ilvBNC1-lrp1-brnFE*) deleted in the genes *aceE, alaT,* and *ilvA* and overexpressing the *ilvB, ilvN, ilvC, lrp1, brnF,* and *brnE* genes produced 51 g/L L-valine in fed-batch fermentation, with almost no detectable amino acid by-products such as L-alanine and L-isoleucine [36].

1.2.5 L-Isoleucine

Isoleucine processes have been devised in various bacteria such as *S. marcescens*, *C. glutamicum* ssp. *flavum*, and *C. glutamicum*. In *S. marcescens*, resistance to isoleucine hydroxamate and α-aminobutyric acid led to derepressed L-threonine deaminase (TD) and acetohydroxyacid synthase (AHAS) and production of 12 g/L of isoleucine [37]. Further work involving transductional crosses into a threonine overproducer yielded isoleucine at 25 g/L [24].

Feedback regulation in *C. glutamicum* was eliminated [38] by replacing the native threonine dehydratase gene *ilvA* with the feedback-resistant gene from *E. coli*. By introducing additional copies of genes encoding branched amino acid biosynthetic enzymes, lysine- or threonine-producing strains were converted into L-isoleucine producers with improved titers [39–41]. Amplification of the wild-type threonine dehydratase gene *ilvA* in a threonine-producing strain of *Corynebacterium lactofermentum* led to isoleucine production [42].

A threonine-overproducing strain of *C. glutamicum* was sequentially mutated to resistance to thiaisoleucine, azaleucine, and aminobutyric acid; it produced 10 g/L of isoleucine [43]. Metabolic engineering studies involving overexpression of biosynthetic genes were useful in improving isoleucine production by this species. Colon et al. [42] obtained an isoleucine-producing strain by cloning multiple copies of *hom* (encoding HDI) and wild-type *ilvA* (encoding TD) into a lysine overproducer and by increasing HK (encoded by *thrB*); a titer of 15 g/L isoleucine was obtained. Independently, Morbach et al. [44] cloned three copies of the feedback-resistant HD gene (*hom*) and multicopies of the deregulated TD gene (*ilvA*) in a deregulated lysine producer of *C. glutamicum*, yielding an isoleucine producer (13 g/L) with no threonine production and reduced lysine production. Application of a closed loop control fed-batch strategy raised production to 18 g/L [45]. Further metabolic engineering work involving amplification of feedback inhibition-insensitive biosynthetic enzymes converted lysine overproducers and threonine overproducers into *C. glutamicum* strains yielding 30 g/L of isoleucine [46].

C. glutamicum ssp. *flavum* studies employed resistance to α-amino-β-hydroxyvaleric acid, and the resultant mutant produced

11 g/L [47]. D-Ethionine resistance was used by Ikeda et al. [48] to yield a mutant producing 33 g/L in a fermentation continuously fed with acetic acid. Accumulated L-isoleucine in the cell is excreted via a two-component export system BrnFE, which is regulated by the global regulator Lrp. Overexpression of Lrp and BrnFE in *C. glutamicum* led to 27 g/L production in fed-batch fermentation [49]. A *C. glutamicum* strain overexpressing the threonine dehydratase (ilvA1) and acetohydroxy acid synthase (ilvBN1) feedback-resistant genes, as well as the ppnk1 gene, encoding NAD kinase (Strain IWJ001/pDXW-8-ilvBN1-ilvA1-ppnk1), led to a 32 g/L isoleucine production in a 72 h fed-batch fermentation [50]. Recently in a *Brevibacterium flavum* strain KM011 (Met^-+Eth^r+α-AB^r+Lys^L+AEC^r), betaine, vitamin B12, and vitamin B5 concentrations were optimized for isoleucine production in shake flasks. Under these conditions, a maximum of 13.35 g/L isoleucine was produced. However, a production of 35.26 g/L was obtained with optimized conditions using a 5-L fermenter [51].

1.2.6 L-Alanine

Lee et al. [52] introduced into an *E. coli* double mutant (lacking genes encoding a protein of the pyruvate dehydrogenase complex [*aceF*] and lactate dehydrogenase [*ldhA*]) a plasmid containing the *Bacillus sphaericus* alanine dehydrogenase gene (*alaD*). The strain produced L-alanine in 27 h with a yield on glucose of 0.63 g/g and a maximum productivity of 2 g/L/h. Further work has raised the titer to 114 g/L. A genetically engineered *E. coli* W (strain, XZ132), with a D-lactate dehydrogenase replaced by an alanine dehydrogenase from *Geobacillus stearothermophilus* and deletions in the methylglyoxal synthase (mgsA) and the catabolic alanine racemase (dadX) genes, produced 1279 mmol alanine [53]. Deletion of both enzymes reduced low levels of lactate and conversion of L- to D-alanine. A thermoregulated genetic switch, designed to dynamically control the expression of L-alanine dehydrogenase (alaD) from *G. stearothermophilus* on the *E. coli* B0016-060BC chromosome, led to an L-alanine titer of 120.8 g/L with higher overall and oxygen-limited volumetric productivities of 3.09 and 4.18 g/L/h, respectively, using glucose as the sole carbon source [54].

1.2.7 L-Proline

Proline-hyperproducing strains of bacteria, exhibiting reduced proline-mediated feedback inhibition of γ-glutamyl kinase (GK) activity (a result of single-base pair substitutions in the bacterial *proB* gene-coding region), have been isolated based on their resistance to toxic proline analogs (L-azetidine-2-carboxylic acid and 3,4-dehydro-DL-proline), compounds which inhibit GK activity while not interfering with protein synthesis. Cloning of the three genes of proline biosynthesis in *E. coli* on multicopy plasmids and selection of mutants of such plasmid-containing strains to resistance to 3,4-dehydroproline led to a process producing 20 g/L proline [55].

A mutant of *S. marcescens* resistant to 3,4-dehydroproline, thiazolidine-4-carboxylate, and azetidine-2-carboxylate and unable to utilize proline produced 50–55 g/L L-proline [56]. Cloning of a gene bearing the dehydroproline-resistance locus on a plasmid yielded a recombinant strain of *S. marcescens* producing 75 g/L [57]. Further development work increased the production to over 100 g/L [58].

A sulfaguanidine-resistant mutant of *C. glutamicum* ssp. *flavum* produced 35 g/L L-proline [59]. When a glutamate-producing strain of *C. glutamicum* was grown under modified conditions, it made 48 g/L [60]. A strain of *Corynebacterium acetoacidophilum* produced 108 g/L proline when grown in the presence of glutamate [61].

1.2.8 L-Hydroxyproline

Introduction of the proline 4-hydroxylase gene from *Dactylosporangium* sp. into recombinant *E. coli* producing L-proline at 1.2 g/L led to a new strain producing 25 g/L of hydroxyproline (*trans*-4-hydroxy-L-proline) [62]. When proline was added, the hydroxyproline titer reached 41 g/L, with a yield of 87% based on the amount of proline added. Optimization of the gene codons and vectors of proline-4-4-hydroxylase (*P4H*) in a recombinant *E. coli* BL21 strain and chemo-physical combination mutagenesis allowed isolation of strain NA45, able to grow in glycerol as a sole carbon source. With further systemic optimization of nutritional elements, the strain produced 25.4 g/L trans-4-hydroxy-L-proline at 48 h in fed-batch mode in a 5 L fermentor [63].

1.3 Nucleotides

Nucleotide fermentations became commercially important due to the activity of two purine ribonucleoside 5′-monophosphates, namely, guanylic acid (GMP) and inosinic acid (IMP), as enhancers of flavor. Titers of IMP and GMP have reached 30 g/L [64]. The techniques used to achieve such production are similar to those used for amino acid fermentations. In Japan, 2500 tons of GMP and IMP are produced annually with a market of $350 million. GMP can also be made by bioconversion of xanthylic acid (XMP). Genetic modification of *Corynebacterium ammoniagenes* involving transketolase, an enzyme of the nonoxidative branch of the pentose phosphate pathway, resulted in the accumulation of 39 g/L of XMP [65].

1.4 Vitamins

Riboflavin. Production of riboflavin (vitamin B_2) reached over 6000 tons per year and a titer of 20 g/L by overproducers such as the yeast-like molds, *Eremothecium ashbyii* and *Ashbya gossypii*. A bacterial process using *C. ammoniagenes* (previously *Brevibacterium ammoniagenes*) was developed by cloning and overexpressing the organisms own riboflavin biosynthetic genes [66] and its own promoter sequences. The resulting culture produced 15 g/L riboflavin in 3 days.

Genetic engineering of a *Bacillus subtilis* strain, already containing purine analog-resistance mutations, led to improved production of riboflavin [67]. The industrial strain of *B. subtilis* was produced by [1] making purine analog-resistance mutations to increase guanosine triphosphate (GTP; a precursor) production and [2] using a riboflavin analog (roseoflavin)-resistance mutation in *ribC* that deregulated the entire pathway [68]. Resultant production was over 25 g/L. A genome-wide transcript expression analysis [69] was successfully used to discover new targets for further improvement of the fungus *A. gossypii* [70]. The authors identified 53 genes of known function, some of which could clearly be related to riboflavin production. A metabolic engineering approach overexpressing the *ribA* (encoding 3,4-dihydroxy-2-butanone 4-phosphate synthase) gene in *B. subtilis* RB50::[pRF69]n[pRF93] strain led to productions of 15 g/L riboflavin [71].

Biotin has traditionally been made by chemical synthesis, but recombinant microbes have approached a competitive economic position. Cloning of a biotin operon (*bioABFCD*) on a multicopy plasmid allowed *E. coli* to produce 10,000 times more biotin than did the wild-type strain [72]. Sequential mutation of *S. marcescens* to resistance to the biotin antimetabolite acedomycin (=actithiazic acid) led to mutant strain SB412, which produced 20 mg/L of biotin [73]. Further improvements were made by mutating selected strains to ethionine resistance (strain ET2, 25 mg/L), then mutating ET2 to S-2-aminoethylcysteine resistance (strain ETA23, 33 mg/L), and finally cloning in the resistant *bio* operon yielding a strain able to produce 500 mg/L in a fed-batch fermentation along with 600 mg/L of biotin vitamers. Later advances led to production by recombinant *S. marcescens* of 600 mg/L of biotin [74]. A process using an *E. coli* mutant resistant to β-hydroxynorvaline (a threonine antimetabolite) yielding 970 mg/L has been patented [75]. Biotin production was further increased to over 1 g/L by the use of a *B. subtilis* strain resistant to 5-(2-thenyl) pentanoic acid (a biotin analog) and overexpressing several *bio* genes.

Vitamin C (ascorbic acid) has traditionally been made in a five-step chemical process by first converting glucose to 2-keto-L-gulonic acid (2-KGA) with a yield of 50% and then converting the 2-KGA by acid or base to ascorbic acid (L-AA). Annual production is 110,000 tons generating revenue of over 600 million dollars. A novel process for vitamin C synthesis involved the use of a genetically engineered *Erwinia herbicola* strain containing a gene from *Corynebacterium* sp. The engineered organism converted glucose into 1 g/L of 2-KGA [76]. A better process was devised independently which converted 40 g/L of glucose into 20 g/L of 2-KGA [77]. This process involved cloning and expressing the gene encoding 2,5-diketo-D-gluconate reductase from *Corynebacterium* sp. into *Erwinia citreus*. Another process used a recombinant strain of *Gluconobacter oxydans* containing genes encoding L-sorbose dehydrogenase and L-sorbosone dehydrogenase from *G. oxydans* T-100. The new strain was an improved producer of 2-KGA [78]. Further mutation to suppress the L-idonate pathway and to improve the promoter led to the production of 130 g/L of 2-KGA from 150 g/L of sorbitol. Further improvement of the strain was possible by suppressing the L-idonate pathway [79]. An *Erwinia herbicola*, transformed with the 2,5-diketo-D-gluconate reductase gene from *Corynebacterium* sp., produced up to 120 g/L 2-KGA in less than 120 h with the help of glucose dehydrogenase, gluconate dehydrogenase, ketogluconate dehydrogenase, and 2,5-diketo-D-gluconate reductase [80]. Similar to the chemical conversion of 2-KGA to vitamin C by lactonization, *X. campestris* probably lactonizes 2-KGA under oxidative stress to form L-AA. Using this method, 20.4 g/L L-AA were found in the extracellular broth which corresponds to the 14-fold amount of that detected in yeast cells directly synthesizing L-AA from D-glucose [81].

Vitamin B_{12} production depends on avoidance of feedback repression by vitamin B_{12}. The vitamin is industrially produced by *Propionibacterium shermanii* or *Pseudomonas denitrificans*. The early stage of the *P. shermanii* fermentation is conducted under anaerobic conditions in the absence of the precursor 5,6-dimethylbenzimidazole. These conditions prevent vitamin B_{12} synthesis and allow for the accumulation of the intermediate, cobinamide. Then, the culture is aerated, and dimethylbenzimidazole is added, converting cobinamide to the vitamin. In the *P. denitrificans* fermentation, the entire process is carried out under low levels of oxygen. A high level of oxygen results in an oxidizing intracellular environment, which represses the formation of the early enzymes of the pathway. Production of vitamin B_{12} has reached levels of 150 mg/L, 10 tons per year, and a world market of $71 million.

Other vitamins. Recombinant *E. coli*, transformed with genes encoding pantothenic acid (vitamin B_5) biosynthesis, and resistant to salicylic and/or other acids, produce 65 g/L of D-pantothenic acid from glucose using alanine as precursor [82]. A total of 7000 tons per year are made chemically and microbiologically. Thiamine (vitamin B_1) is produced synthetically at 4000 tons per year. Pyridoxine (vitamin B_6) is made chemically at 2500 tons per year.

1.5 Carotenoids

Carotenoid production processes have been extensively studied [83], but they have had difficulty in economically challenging chemical methods. Of over 600 microbial carotenoids, only β-carotene and astaxanthin are produced industrially by fermentation [84]. A semi-industrial β-carotene process was developed using mated cultures of *Blakeslea trispora* plus and minus strains. β-Carotene was produced at 1 g/L in the early 1960s [85]. By the addition of carotogenic chemicals and antioxidants, the titer was raised to over 3 g/L [86]. Processes in development include those yielding β-carotene, lycopene, zeaxanthin, and astaxanthin. Some have been improved by metabolic engineering and directed evolution [87–89]. Metabolic engineering of *E. coli* has led to strains forming 0.2 g/L of lycopene [90]. Lutein, a xanthophyll carotenoid with antioxidant properties, had sales as a food colorant of $150 million in the USA [91]. It is thought that this carotenoid prevents age-related macular degeneration and cataracts. It is made from petals of marigold, but microalgae are a potential new source.

Chemical production of *trans*-astaxanthin has a selling price of $2,000 kg [92] and a market of over $100 million per year. It is mainly used for pigmentation of salmonids raised in aquaculture, a multibillion dollar industry [93]. Astaxanthin can be made by the yeast *Phaffia rhodozyma* (*Xanthophyllomyces dendrorhous*) and the microalga *Haematococcus pluvialis.* Genetically improved strains of *P. rhodozyma* produce 10 mg/g cells in industrial fermentors. Recent improvements in astaxanthin production have been published by de la Fuente et al. [94] and Rodríguez-Sáiz et al. [95] to get maximal astaxanthin titers of 420 mg/L when *X. dendrorhous* is fermented under continuous white light. Recently astaxanthin production has been reported with an engineered *C. glutamicum* strain. Volumetric productivities of up to about 0.4 mg/L/h reported in simple shaking flask cultures by the recombinant strain compare favorably with those reported for the commercially used production hosts such as the green microalga *H. pluvialis*, the red yeast *X. dendrorhous*, and the recombinant *E. coli* [96].

1.6 Organic Acids

Microbial production of organic acids is an excellent approach for obtaining building-block chemicals from renewable carbon sources [97]. Production of some organic acids started decades ago, and titers have been improved by classical mutation and screening/selection techniques as well as by metabolic engineering [98].

1.6.1 Citric Acid

Citric acid has a market of $2 billion. About 1 million tons per year are produced annually by *Aspergillus niger* and yeasts. The commercial process employs *A. niger* in media deficient in iron and manganese. Other factors contributing to high citric acid production are a high intracellular concentration of fructose 2,6-biphosphate, inhibition of isocitrate dehydrogenase, and low pH (1.7–2.0). In approximately 4–7 days, the major portion (80%) of the sugar (glucose or sucrose) provided is converted into citric acid. *A. niger* titers have reached over 200 g/L [99].

Alternative processes have been developed with *Candida* yeasts, especially from hydrocarbons. Such yeasts are able to convert *n*-paraffins to citric and isocitric acids in extremely high yields

(150–170% on a weight basis). Titers as high as 225 g/L have been reached [100]. Anastasiadis and Rehm [101] reported production levels of 250 g/L with *Candida oleophila* ATCC 20177 under submerged continuous fermentation, with glucose as carbon source.

1.6.2 Acetic Acid

Titers of acetic acid reached 53 g/L with genetically engineered *E. coli* [102] and 83 g/L with a *Clostridium thermoaceticum* mutant [103]. Cloning of the aldehyde dehydrogenase gene from *Acetobacter polyoxogenes* on a plasmid vector into *Acetobacter aceti* subsp. *xylinum* increased the rate of acetic acid production by over 100% (1.8–4 g/L/h) and the titer by 40% (68–97 g/L) [104].

1.6.3 Lactic Acid

Whole-genome shuffling was used to improve the acid tolerance of a commercial lactic acid-producing *Lactobacillus* sp. [105]. Further approaches using this recursive protoplast fusion technique yielded strains of *Lactobacillus rhamnosus* ATCC 11443 with improved glucose tolerance (160–200 g/L glucose) while simultaneously enhancing L-lactic acid production by 71% as compared to the wild type. Shuffling of a mutant strain of *Lactobacillus delbrueckii* NCIM 2025 and *Bacillus amyloliquefaciens* ATCC 23842 produced a fusant that could utilize liquefied cassava bagasse starch directly to yield a titer of 40 g/L of lactic acid with a 96% conversion of starch to lactic acid [106].

Although lactobacilli make more lactic acid than *Rhizopus oryzae*, they produce mixed isomers. The fungus, however, produces L-(+) lactic acid exclusively. The yield is about 60–80% of added glucose, the remainder going to ethanol. By increasing lactic dehydrogenase levels via plasmid transformation with *ldhA*, more lactate could be made from pyruvate, and production was increased to 78 g/L, whereas the undesirable coproduct ethanol was reduced from 10.6 to 8.7 g/L [107]. A recombinant *Saccharomyces cerevisiae* strain containing six copies of bovine L-lactate dehydrogenase produced 122 g/L from sugar cane with an optical purity of 99.9% or higher [108]. Expression of the same bovine enzyme and a deletion of the pyruvate decarboxylase gene in *Kluyveromyces lactis* produced 109 g/L [109].

A recombinant *E. coli* strain was constructed that produced optically active pure D-lactic acid from glucose at virtually the theoretical maximum yield, for example, two molecules from one molecule of glucose [110]. D-Lactic acid has also been produced at 61 g/L by a recombinant strain of *S. cerevisiae* containing the D-lactic dehydrogenase gene from *Leuconostoc mesenteroides* [111].

1.6.4 Succinic Acid

Sanchez et al. [112] used metabolic engineering to create an *E. coli* strain, which had three deactivated genes of the central metabolic pathway, that is, *adhE*, *ldhA*, and *act-pta*, and an inactivated *iclR* gene, which resulted in activation of the glyoxylate pathway. The strain produced 40 g/L of succinate. Metabolic engineering of *Mannheimia succiniciproducens* led to a strain producing 52 g/L of succinic acid at a yield of 1.16 mol/mol glucose and a productivity of 1.8 g/L/h in fed-batch culture [113]. A metabolically engineered succinate-producing strain of *E. coli* yielded 58 g/L succinate in a 59 h fed-batch fermentation under aerobic conditions [114]. The average succinate yield was 0.94 mol/mol of glucose, the average productivity was 1.08 g/L/h, and the average specific activity was 90 mg/g/h. A titer of 99 g/L with a productivity of 1.3 g/L/h has been reached with recombinant *E. coli* [115].

1.6.5 Other Organic Acids

Metabolic engineering of *Clostridium tyrobutyricum* created a fermentation strain yielding 80 g/L butyric acid and a yield on glucose of 0.45 g/g [116]. *S. cerevisiae* normally produces 2 g/L of malic acid from fumaric acid. However, a recombinant strain containing a cloned fumarase gene was able to produce 125 g/L with a yield of almost 90% [117]. Microbial fermentation titers of some other organic acids are 135 g/L pyruvic acid, 107 g/L fumaric acid, 90 g/L shikimic acid, 69 g/L dehydroshikimic acid, 85 g/L itaconic acid, 504 g/L gluconic acid, 106 g/L propionic acid, 68 g/L oxalic acid, and 136 g/L glyceric acid. An oxidative bioconversion of saturated and unsaturated linear aliphatic 12–22 carbon substrates to their terminal dicarboxylic acids was developed by gene disruption and gene amplification [118]. Product concentrations reached 200 g/L, and problematic side reactions such as unsaturation, hydroxylation, and chain-shortening did not occur.

1.7 Alcohols

Ethanol. Fermentation of sugars by *S. cerevisiae* in the case of hexoses, and *Kluyveromyces fragilis* or *Candida* species with lactose or a pentose, results in the production of ethanol. Under optimum conditions, approximately 120 g/L ethanol can be obtained. Such a high concentration slows down growth and the fermentation ceases.

A *S. cerevisiae* fusant library obtained by genome shuffling was screened for growth at 35, 40, 45, 50, and 55 °C on agar plates containing different concentrations of ethanol [119]. After three rounds of genome shuffling, a strain was obtained which was able to grow on plates up to 55 °C, completely utilized 20% (w/v) glucose at 45–48 °C, produced 99 g/L ethanol, and tolerated 25% (v/v) ethanol stress.

In silico metabolic models have been used to overcome the redox imbalance in *S. cerevisiae* engineered with the *Xyl1* and *Xyl2* genes from *Pichia stipitis* [120, 121]. Overexpression of both genes led to an accumulation of NADH and a shortage of NADPH. Deletion of $NADP^+$-dependent glutamate dehydrogenase (GGH1) and overexpression of NAD^+-dependent GDH2 led to an increase in ethanol production using xylose as fermentation substrate. An in silico genome-scale gene insertion strategy was used to improve ethanol production and decrease the production of by-products glycerol and xylitol [122]. Introduction of glyceraldehyde-3-phosphate dehydrogenase in *S. cerevisiae* led to a 58% reduction in glycerol, a 33% reduction in xylitol, and a 24% increase in ethanol production.

When biomass is used as a carbon source for ethanol production, its breakdown results in liberation of acetic acid. The acid interferes with ethanol production. Tolerance of *Candida krusei* GL560 to acetic acid was improved by genome shuffling [123]. A mutant, S4-3, which was isolated and selected after four rounds, had a higher viability in different media containing acetic acid than did the parent strain GL560. The mutant also improved its multiple stress tolerance to ethanol, H_2O_2, heat, and freeze-thawing.

E. coli was converted into an ethanol producer (43 g/L) by cloning the alcohol dehydrogenase II and pyruvate decarboxylase genes from *Zymomonas mobilis* [124]. By cloning and expressing the same two genes in *Klebsiella oxytoca*, the recombinant was able to convert crystalline cellulose to ethanol in high yield when fungal cellulase was added [125]. Maximum theoretical yield was 81–86%, and a titer of 47 g/L of ethanol was produced from 100 g/L of cellulose.

Recombinant strains of *E. coli*, *Zymomonas*, and *Saccharomyces* can convert corn fiber hydrolysate to 21–35 g/L ethanol with yields of 0.41–0.50 ethanol per gram of sugar consumed [126]. For a recombinant *E. coli* strain making 35 g/L, the time was 55 h, and the yield was 0.46 g ethanol per gram of available sugar, which is 90% of the attainable maximum.

1,3-Propanediol. A strain of *Clostridium butyricum* converts glycerol to 1,3-propanediol (PDO) at a yield of 0.55 g/g glycerol consumed [127]. A major metabolic engineering feat was carried out in *E. coli* leading to a culture growing on glucose and producing PDO at 135 g/L, with a yield of 51% and a rate of 3.5 g/L/h [128]. To do this, eight new genes were introduced to convert dihydroxyacetone phosphate (DHAP) into PDO. Production was further improved by modifying 18 *E. coli* genes, including regulatory genes. PDO is the monomer used to chemically synthesize industrial polymers such as polyurethanes and the polyester fiber Sorono™ by DuPont. This bioplastic is polytrimethylene terephthalate (3GT polyester) made by reacting terephthalic acid with PDO [129]. PDO is also used as a polyglycol-like lubricant and as a solvent.

D-*Mannitol* is a naturally occurring polyol, widely used in the food, chemical, and pharmaceutical industries. A whole-cell bioconversion of D-fructose to D-mannitol was developed by metabolic engineering of *E. coli* [130]. The *mdh* gene encoding mannitol dehydrogenase from *Leuconostoc pseudomesenteroides* and the *fdh* gene encoding formate dehydrogenase from *Mycobacterium vaccae* were coexpressed in *E. coli* along with the *glf* gene encoding the glucose facilitator protein of *Z. mobilis*. The process yielded 75–91 g/L of D-mannitol, a specific productivity of 3.1–4.1 g/g/h and a molar yield of 84–92% with no by-products. An improved bioconversion process was developed with a recombinant *E. coli* strain in the presence of added glucose isomerase yielding 145 g/L of D-mannitol from 180 g/L glucose [131]. Supplementation of the medium used for mannitol production by *Candida magnolia* with Ca^{2+} and Cu^{2+} increased production up to 223 g/L [132].

Sorbitol, also called D-glucitol, is 60% as sweet as sucrose and is used in the food, pharmaceutical, and other industries. Its worldwide production is estimated to be higher than 500,000 tons per year, and it is made chemically by catalytic hydrogenation of D-glucose or syrup with a 50:50 mixture of glucose and fructose. It is also produced by extraction from seaweed as a by-product of alginate and iodine manufacture. However, excellent microbial processes have been developed [133]. Toluenized (permeabilized) cells of *Z. mobilis* produce 290 g/L of sorbitol and 283 g/L of gluconic acid from a glucose and fructose mixture in 16 h with yields nearly 95% for both products [134]. Other leading organisms are recombinant *C. glutamicum* at 285 g/L, *Lactobacillus intermedius* at 227 g/L, *C. magnoliae* at 223 g/L, and many others producing between 100 and 200 g/L. Metabolic engineering of *Lactobacillus plantarum* for high sorbitol production was successfully achieved by a simple two-step strategy overexpressing the two sorbitol-6-phosphate dehydrogenase genes (*srlD1* and *srlD2*) identified in the genome sequence [135].

n-Butanol is a good alternative fuel additive as it has two more carbons than ethanol, which results in an energy content about 40% higher. Also, automobile engines do not require modification until the percentage of butanol reaches over 40% of the total automobile fuel. In contrast, modification is required when ethanol is added to gasoline at levels exceeding 15%. Butanol can be obtained from the acetone–butanol–ethanol fermentation of *Clostridium beijerinckii* or *Clostridium acetobutylicum*. Butanol-resistant mutants showed increased production of butanol and acetone [136]. Biochemical engineering modifications were able to increase total acetone, butanol, and ethanol production (ABE) to 69 g/L [137]. A mutant in the presence of added acetate was able to produce almost 21 g/L butanol and 10 g/L of acetone from glucose [138].

Because butanol's octane number is lower than that of ethanol and the octane number increases with methyl branching and double bonds, other higher alcohols are also being considered as biofuels, for example, branched C 4 and C 5 alcohols. They are also desirable because of their higher energy density, lower vapor pressure, and lower hygroscopicity as compared to ethanol [139]. They include isopropanol, 1-propanol, 1-butanol (*n*-butanol), isobutanol (2-methyl-propanol), 3-methyl-1-butanol,

2-methyl-1-butanol, isopentanol (3-methyl-1 butanol), and isopentenol (3-methyl-3-buten-1-ol). A novel screening method based on overcoming the toxicity associated with the accumulation of prenyl diphosphate was used to screen a library of 19,000 clones harboring fragments of *Bacillus* genomic DNA [140]. Two genes, *yhfR* and *nudF*, coding for proteins capable of overcoming the toxicity associated with accumulating IPP and DMAPP were isolated. Both protein products have an affinity for IPP and DMAPP, converting them into isopentenol. *Clostridium beijerinckii* (*Clostridium butylicum*) produces 20 g/L of 1-butanol and 2 g/L of isopropanol as part of a mixed product. Recombinant *E. coli* can produce 4.9 g/L of isopropanol. Recently, a new strategy for the production of these alcohols has been reported [141]. This approach is based on the diversion of 2-keto acid intermediates from the endogenous amino acid pathway to alcohol biosynthesis especially that of isobutanol. As a result, engineered *E. coli* can produce 22 g/L of isobutanol in 110 h with a yield of 86% of the theoretical maximum.

Other alcohols. The noncariogenic, noncaloric, and diabetic-safe sweetener erythritol has 70–80% of the sweetness of sucrose. It can be produced by *Aureobasidium* sp. (165 g/L), an acetate-negative mutant of *Yarrowia lipolytica* (170 g/L), a *C. magnoliae* osmophilic mutant (187 g/L), the osmophile *Trichosporon* sp. (88 g/L), *Torula* sp. (200 g/L), and the yeast *Pseudozyma tsukubaensis* (245 g/L) [142].

Xylitol is a naturally occurring sweetener with anticariogenic properties, which is used for some diabetes patients. It can be produced by chemical reduction of D-xylose or by fermentation. Xylitol production at 150 g/L was obtained with *Candida guilliermondii* 2581 at pH 6.0 and shaking at 60 rpm [143].

2 Secondary Metabolites

As a group that includes antibiotics, pesticides, pigments, toxins, pheromones, enzyme inhibitors, immunomodulating agents, receptor antagonists and agonists, pesticides, antitumor agents, immunosuppressants, cholesterol-lowering agents, plant protectants, and animal and plant growth factors, these metabolites have tremendous economic importance. This remarkable group of compounds is produced by certain restricted taxonomic groups of organisms and is usually formed as mixtures of closely related members of a chemical family.

2.1 Antibiotics

Antibiotics are the most well-known secondary metabolites and have a tremendous importance. The most well known are the β-lactams, tetracyclines, aminoglycosides, chloramphenicol, macrolides, and other polyketides, polyenes, and glycopeptides, among others. They have been crucial in the increase in average life expectancy in the USA from 47 years in 1900 to 74 for males and 80 for women in 2000 [144]. More than 350 agents have reached the market as antimicrobials. The global market for finished antibiotics has reached $35 billion.

2.1.1 β-Lactams

The β-lactams are the most important class of antibiotics in terms of use. Included are the penicillins, cephalosporins, cephamycins, clavulanic acid, and carbapenems. Many of the current penicillins and cephalosporins are semisynthetic. All of the above are of great importance in chemotherapy of bacterial infections.

β-Lactamases of pathogenic bacteria are the major cause of resistance development, and there are over 450 such enzymes. However, β-lactams are still very useful due to the discovery of β-lactamase inhibitors. Although clavulanic acid is a β-lactam compound, it has only low antibacterial activity but is used widely as an inhibitor of β-lactamase, in combination with β-lactam antibiotics. Conventional strain improvement by protoplast fusion of auxotrophic strains yielded a fusant producing 30-fold more clavulanic acid than the wild type [145]. Inactivation of two G-3-P dehydrogenases, encoded by *gap1* and *gap2* by targeted gene disruption, doubled clavulanic acid production [146]. Also, increased dosage of biosynthetic genes *ceas* and *cs2* [147] or overexpression of positive regulatory genes increased production two- to threefold [148, 149].

Yield improvements have been achieved through different strategies. Thus, protoplast fusion between strains of *Penicillium chrysogenum* yielded a higher-producing strain of penicillin G [150]. Metabolic engineering was used to replace the normal promoter with the ethanol dehydrogenase promoter [151], increasing penicillin production up to 30-fold.

Protoplast fusion was also been carried out with strains of *Acremonium chrysogenum* to obtain a strain that produced 40% more cephalosporin C than the parent [152]. Production was also improved by cloning multiple copies of cyclase [153] or the *pcbC* and the *cefEF* genes [154].

Two improved cephamycin C-producing strains from *Nocardia lactamdurans* were fused to obtain cultures, which produced 10–15% more antibiotic [155]. Overexpression of *lat*, encoding lysine-aminotransferase, also led to an overproducing strain [156]. High-level expression of *ccaR*, a positive regulatory gene in *Streptomyces clavuligerus* [157], led to a two- to threefold increase in antibiotic production.

Chemical methods had traditionally been used to produce 7-aminocephalosporanic acid (7-ACA) and 7-aminodeacetoxycepalosporanic acid (7-ADCA), chemical precursors of semisynthetic cephalosporins. These processes are being replaced by safer microbiological processes. Transformation of *P.*

chrysogenum with bacterial *cefD* and *cefE* genes allowed the production of deacetoxycephalosporin C (DAOC) [158], another key intermediate in the commercial production of semisynthetic cephalosporins. Also, cloning of *cefE* from *S. clavuligerus* or *cefEF* and *cefG* from *Acremonium chrysogenum* into *P. chrysogenum* fed with adipic acid as side-chain precursor [159] resulted in the formation of several adipyl-6/7-intermediates. Enzymatic removal of the adipoyl side chain led to the production of 7-ADCA.

Disruption and one-step replacement of the *cefEF* gene of an industrial strain of *A. chrysogenum* yielded strains accumulating up to 20 g/L of penicillin N. Cloning and expression of the *cefE* gene from *S. clavuligerus* into those high-producing strains yielded recombinant strains producing high titers of DAOC [160]. An *E. coli* strain containing the D-amino acid oxidase gene from *Trigonopsis variabilis* and the glutaryl-7-aminocephalosporanic acid acylase gene from *Pseudomonas* sp. was able to convert cephalosporin C directly to 7-ACA [161].

Natural carbapenems, such as thienamycin, are made by *Streptomyces cattleya*, *Erwinia carotovora* subsp. *carotovora*, *Serratia* sp., and *Photorhabdus luminescens* [162]. Carbapenems are resistant to attack by most β-lactamases. The commercial carbapenems are made synthetically and include imipenem, meropenem, and ertapenem. Thienamycin is one of the most potent and broadest in spectrum of all antibiotics known today. Although a β-lactam, it is not a member of the penicillins or cephalosporins. It is active against aerobic and anaerobic bacteria, both Gram positive and Gram negative, including *Pseudomonas*. This novel structure was isolated in Spain from a new soil species, which was named *S. cattleya* [163]. Interestingly, this culture also produces penicillin N and cephamycin C. Also used to combat β-lactamase containing pathogens are new carbapenems, such as the recently approved doripenem (S-4661) which has broad-spectrum activity against resistant bacteria including *Pseudomonas aeruginosa*.

2.1.2 Other Antibiotics

The biosynthesis of polyketide macrolides has been subjected to genetic engineering [164]. This group of compounds includes antibiotics such as erythromycin, oleandomycin, pikromycin, tylosin, and amphotericin B. Reverse metabolic engineering increased erythromycin production by *Aeromicrobium erythreum* [165]. The technique is also known as inverse metabolic engineering and as combinatorial engineering. Tylosin production was increased up to 60% in *Streptomyces fradiae* by transposing a second copy of *tylF*, encoding macrocin *O*-methyltransferase, into a neutral site on its chromosome [166].

Genetic engineering of the nystatin biosynthetic pathway yielded polyenes with high antifungal activity, which are less toxic than amphotericin B [167]. The production of antibiotics in heterologous hosts via combinatorial biosynthesis is becoming very popular in antibiotic production and discovery [168]. New derivatives of antibiotics have been obtained after the biosynthetic paths were elucidated and the biosynthetic genes isolated [169]. Over 200 new polyketides have been made by combinatorial biosynthesis [170, 171]. The discovery of new antibiotics has also been achieved by genetic recombination between producers of different or even the same antibiotics [172–174].

Combinatorial biosynthesis has been used to construct macrolides with new sugar moieties [175, 176]. Methymycin and pikromycin, produced by a gene cluster of *Streptomyces venezuelae* and normally containing the sugar desosamine, were modified by cloning of a gene from the calicheamicin producer, *Micromonospora echinospora* spp. *calichensis.* The gene encodes TDP-glycerohexulose aminotransferase. Transfer of a 12.6 kb DNA fragment from the tetracenomycin C-producing *Streptomyces glaucescens* to *Streptomyces lividans* resulted in tetracenomycin C production by the latter [177]. The fragment contains 12 genes of biosynthesis and resistance. Novel hybrid tetracenomycins were produced by introducing a 25 kb cosmid from the elloramycin biosynthetic pathway of *Streptomyces olivaceus* into the polyketide synthase (PKS)-deleted mutant of the urdamycin producer, *S. fradiae*, and into the mithramycin producer, *Streptomyces argillaceus* [178]. The cosmid contains a glycosyltransferase gene whose enzyme has broad substrate specificity and thus produces hybrid products containing different D- and L-sugars. For more than 35 years, vancomycin and teicoplanin were the only antibiotics active against multidrug-resistant Gram-positive bacteria. Their use became severely limited by an increase in multidrug resistance. One group of narrow-spectrum compounds are the streptogramins which are synergistic pairs of antibiotics made by a single microbial strain. The pairs are constituted by a (Group A) polyunsaturated macrolactone containing an unusual oxazole ring and a dienylamide fragment and a (Group B) cyclic hexadepsipeptide possessing a 3-hydroxypicolinoyl exocyclic fragment. Such streptogramins include virginiamycin and pristinamycin [179]. Pristinamycin, made by *Streptomyces pristinaespiralis*, is a mixture of a cyclodepsipeptide (pristinamycin I) and a polyunsaturated macrolactone (pristinamycin II). Increasing resistance to pristinamycin in the pristinamycin producer *S. pristinaespiralis* was combined with genome shuffling to increase pristinamycin production ninefold [180].

An important new strategy to improve the discovery of new antibiotics is genome mining, which has come about due to advances in microbial genomics [181]. Mining of whole-genome sequences and genome scanning allows the rapid identification of more than 450 clusters of genes in antibiotic-producing cultures encoding biosynthesis of new bioactive products and the prediction of structure based on gene sequences [182, 183]. These efforts include mining of whole-genome sequences, genome scanning, heterologous expression, and discovery of novel chemistry. Genomics will also provide a huge group of new targets against which natural products can be screened [184]. Up to February 2017, more than 16,000 complete microbial genome sequences were available in one or two scaffolds. Of them, 411 belong to fungi and 699 to actinobacteria. Currently, 69 streptomycete genomes have been sequenced [185]. The smaller genome corresponds to *Streptomyces xiamenensis* with 5.95 Mb size, and the biggest is *Streptomyces rapamycinicus* with 12.7 Mb.

Consensus sequences can be obtained through alignments from the enzyme domains and used to construct Hidden Markov Models (HMM) of the chemical structure to predict SMILES. For this purpose, many online and stand-alone sources are freely available. Some of them include antiSMASH 2.0 [186], which detects 24 types of secondary metabolites. CLUSEAN, NaPDoS, NP.search, and Bagel2 are used for lantipeptides, while PKS/NRPS analysis, SBSPKS [187], and SMURF pipeline were developed for fungal systems. AntiSMASH is one of the most popular softwares since it is friendly and easy going.

2.2 Antitumor Agents

Microorganisms have played a crucial role in identifying compounds with therapeutic benefit against cancer [188]. Most of the important compounds used for chemotherapy of tumors are microbially produced antibiotics mainly made by actinomycetes. Among the most well known are actinomycin D (dactinomycin), anthracyclines (including daunorubicin, doxorubicin, epirubicin, pirarubicin, idarubicin, valrubicin, and amrubicin), glycopeptolides (bleomycin and phleomycin), mitomycin C, anthracenones (mithramycin, streptozotocin, and pentostatin), the enediyne calicheamicin attached to a monoclonal antibody (Mylotarg®), and, recently, the epothilones.

Novel anthracyclines have been produced by metabolic engineering, that is, cloning genes from antitumor-producing species into other producing or nonproducing strains, or by blocking deoxysugar biosynthesis. A new anthracycline, 11-hydroxyaclacinomycin A, was produced by cloning the doxorubicin resistance gene and the aklavinone 11-hydroxylase gene *dnrF* from the doxorubicin producer, *Streptomyces peucetius* subsp. *caesius*, into the aclacinomycin A producer [189]. The hybrid molecule showed greater activity against leukemia and melanoma than aclacinomycin A. Another hybrid molecule produced was 2″-amino-11-hydroxyaclacinomycin Y, which was highly active against tumors [190]. Additional new anthracyclines have been made by introducing DNA from *Streptomyces purpurascens* into *Streptomyces galilaeus*, both of which normally produce known anthracyclines [191]. Novel anthracyclines were produced by cloning DNA from the nogalomycin producer, *Streptomyces nogalater*, into *S. lividans* and into an aclacinomycin-negative mutant of *S.*

galilaeus [192]. Cloning of the actI, actIV, and actVII genes from *Streptomyces coelicolor* into the 2-hydroxyaklavinone producer, *S. galilaeus* 31,671, yielded novel hybrid metabolites, desoxyerythrolaccin and 1-*O*-methyl-desoxyerythrolaccin [193]. Similar studies yielded the novel metabolite aloesaponarin II [194].

Epirubicin (4′-epidoxorubicin) is a semisynthetic anthracycline with less cardiotoxicity than doxorubicin [195]. Genetic engineering of a blocked *S. peucetius* strain provided a new method to produce it [196]. The gene introduced was *avrE* of the avermectin-producing *Streptomyces avermitilis* or the *eryBIV* genes of the erythromycin producer, *Saccharopolyspora erythrea*. These genes and the blocked gene in the recipient are involved in deoxysugar biosynthesis.

Taxol®, a diterpene alkaloid, is approved for breast and ovarian cancer and acts by blocking depolymerization of microtubules. In addition, Taxol promotes tubulin polymerization and inhibits rapidly dividing mammalian cancer cells [197]. Taxol was originally isolated from the bark of the Pacific yew tree (*Taxus brevifolia*), but it took six trees of 100 years of age to treat one cancer patient [198]. It is now produced by plant cell culture or by semisynthesis from taxoids made by *Taxus* species. Early genetic engineering of *S. cerevisiae* yielded no taxadiene (the taxol precursor) because too little of the intermediate, geranylgeranyl diphosphate, was formed. When the *Taxus canadensis* geranylgeranyl diphosphate synthase gene was introduced, 1 mg/L of taxadiene was obtained [199]. More recent metabolic engineering studies [200] yielded a *S. cerevisiae* strain producing over 8 mg/L taxadiene and 33 mg/L geranylgeraniol. Taxol has sales of $1.6 billion per year.

Epothilones are an important group of new anticancer agents produced by the myxobacterium, *Sorangium cellulosum* [201]. Rounds of classical mutation and screening followed by recursive protoplast fusion resulted in fusants able to produce 130-fold more epothilone B compared to the starting strain. Epothilones have a mode of action similar to Taxol and, very importantly, are active against Taxol-resistant tumors.

2.3 Cholesterol-Lowering Agents

The largest segment of the pharmaceutical business is that of cholesterol-lowering drugs, amounting to about 30% of global sales. The first member of the fungal statins, that is, compactin, was discovered in cultures of *Penicillium brevicompactum* [202] and *Penicillium citrinum* [203]. A few years later, the more active methylated form of compactin, known as lovastatin (monacolin K, mevinolin, Mevacor™), was isolated from broths of *Monascus ruber* and *Aspergillus terreus* [204, 205]. Simvastatin (Zocor™), a semisynthetic derivative of lovastatin, reached a market of over $7 billion. Pravastatin, a product of compactin bioconversion, attained sales of $5 billion. The synthetic statin, atorvastatin (Lipitor™), became the world's leading drug at $12 billion per year.

Recently, association analysis, a strategy that integrates transcriptional and metabolic profiles, led to an improvement in lovastatin production of over 50% [206]. Improvement was achieved by increasing the dosage of lovastatin biosynthetic genes and of regulatory genes for secondary metabolism.

2.4 Antihelmintics

Microbially produced polyethers such as monensin, lasalocid, and salinomycin dominate the coccidiostat market.

The avermectins, a family of secondary metabolites having both antihelmintic and insecticidal activities, produced by *S. avermitilis*, have a market of over 1 billion dollars per year. Despite their macrolide structure, avermectins lack antibiotic activity and do not inhibit protein synthesis, nor are they ionophores; instead, they interfere with neurotransmission in many invertebrates.

Although the Merck Laboratories had earlier developed a commercially useful synthetic product, thiobenzole, they had enough foresight to also examine microbial broths for antihelmintic activity and found a nontoxic fermentation broth which killed the intestinal nematode, *Nematospiroides dubius*, in mice. The *S. avermitilis* culture, which was isolated by Omura and coworkers at the Kitasato Institute in Japan [207], produced a family of secondary metabolites having both antihelmintic and insecticidal activities. These were discovered by Merck scientists and named "avermectins." They are disaccharide derivatives of macrocyclic lactones with exceptional activity against parasites, that is, at least ten times higher than any synthetic antihelmintic agent known. They have activity against both nematode and arthropod parasites in sheep, cattle, dogs, horses, and swine.

Incorporation of multiple copies of *afsR2*, a global regulatory gene, from *S. lividans* into wild-type *S. avermitilis* increased avermectin production by 2.3-fold [208]. Transposon mutagenesis was used to eliminate production of the troublesome toxic oligomycin in *S. avermitilis* [209]. DNA shuffling of the *ave C* gene of *S. avermitilis* gave an improved ratio of the undesirable CHC-B2 to the desirable CHC-B1 of 0.07:1. This was an improvement of 21-fold over the ratio with the starting strain [210].

A semisynthetic derivative, 22,23-dihydroavermectin B1 (ivermectin), is 1000 times more active than thiobenzole and is a commercial veterinary product. Ivermectin is made by hydrogenation at C22–C23 of avermectin B1a and B1b with rhodium chloride acting as catalyst. By genetic engineering of *S. avermitilis*, in which certain PKS genes were replaced by genes from the PKS of *S. venezuelae* (the pikromycin producer), ivermectin could be made directly by fermentation, thus avoiding semisynthesis [211].

A new avermectin, called Doramectin (=cyclohexyl avermectin B1), was developed at Pfizer by the technique of mutational biosynthesis [212]. Indeed, it was the first commercially successful example of mutational biosynthesis.

2.5 Immunosuppressive Agents

Cyclosporin A was originally discovered as a narrow-spectrum antifungal peptide produced by the mold, *Tolypocladium niveum* (previously *Tolypocladium inflatum*). Discovery of immunosuppressive activity led to its use in heart, liver, and kidney transplants and to the overwhelming success of the organ transplant field. Sales of cyclosporin A have reached $1.5 billion. Although cyclosporine A had been the only product on the market for many years, two other products, produced by actinomycetes, provided new opportunities. These are rapamycin (=sirolimus) [213] and the independently discovered FK-506 (tacrolimus) [214]. They are both narrow-spectrum polyketide macrolide antifungal agents, which are 100-fold more potent than cyclosporin as immunosuppressants and less toxic. FK-506 and rapamycin have been used clinically for many years. FK-506 had a market of $2 billion in 2007. Mutants developed by increasing resistance to FK-506 produce higher titers [215].

Genome shuffling using mutants of the rapamycin producer, *Streptomyces hygroscopicus*, as well as interspecies fusion of protoplasts of *S. hygroscopicus* D7-804 and *Streptomyces erythreus* ZJU325, generated improved rapamycin-producing strains [216]. Two genes of the rapamycin biosynthetic cluster in *S. hygroscopicus*, that is, *rap G* and *rap H*, encode positive regulatory proteins for rapamycin production [217]. Overexpression of either gene increased rapamycin formation, whereas their deletions eliminated rapamycin biosynthesis. They act by affecting the promoter of the operon.

2.6 Bioinsecticides

The insecticidal bacterium, *Bacillus thuringiensis* (BT), owes its activity to its crystal protein produced during sporulation. Crystals plus spores had been applied to plants for years to protect them against lepidopteran insects. BT preparations are highly potent, some 300 times more active on a molar basis than synthetic pyrethroids and 80,000 times more active than organophosphate insecticides. In 1993, BT represented 90% of the biopesticide market and had annual sales of $125 million.

A very important insecticide is Spinosad (Naturalyte®) produced by *Saccharopolyspora spinosa* and used for protection of crops and feedstock animals. Spinosad is a mixture of two tetracyclic macrolides containing forosamine and tri-*O*-methyl rhamnose with different levels of methylation on the polyketide moiety. The two components are spinosyns A and D, which differ by a methyl group at position 6 of the polyketide. Spinosad is an excellent nontoxic agricultural insecticide. Genome shuffling has been used for strain improvement [218].

3 Recombinant Proteins: Biopharmaceuticals

Biopharmaceutical proteins can be categorized into four major groups: (1) protein therapeutics with enzymatic activity (e.g., insulin), (2) protein vaccines, (3) protein therapeutics with special targeting activity (e.g., monoclonal antibodies), and (4) protein diagnostics (e.g., biomarkers) [219]. Biologics accounted for over $80 billion in sales in 2008. Six of these therapeutic proteins were among the best selling drugs in the USA in that year. Monoclonal antibodies and Fc-fusion proteins made up 43% of this market value.

By means of genetic engineering, desired proteins are massively generated to meet the copious demands of industry [220]. Protein quality, functionality, production speed, and yield are the most important factors to consider when choosing the right expression system for recombinant protein production.

Non-glycosylated proteins are usually made in *E. coli* or yeasts, and they constitute 55% of the therapeutic protein market (39% by *E. coli*, 1% by other bacteria, and 15% by yeasts) [221]. On the other hand, N-glycosylated proteins are usually made in mammalian cells, which mimic human glycosylation. Chinese hamster ovary (CHO) cells provide about 35% of the therapeutic protein market, but the process is very expensive, and the glycoproteins made are not exactly of the human type. Although yeasts, molds, and insect cells are generally unable to provide mammalian glycosylation, the methylotrophic yeast, *Pichia pastoris*, has been genetically engineered to produce a human type of glycosylation [222].

Directed evolution of proteins has been reviewed by Yuan et al. [223]. Strategies include DNA shuffling, whole-genome shuffling, heteroduplex, random chimeragenesis of transient templates, assembly of designed oligonucleotides, mutagenic and unidirectional reassembly, exon shuffling, Y-ligation-based block shuffling, nonhomologous recombination, and the combining of rational design with directed evolution.

3.1 Bacteria

Bacterial systems are used to make somatostatin, insulin, bovine growth hormone for veterinary applications, α-1 antitrypsin, interleukin-2, tumor necrosis factor, β-interferon, and γ-interferon. *E. coli* is one of the earliest and most widely used hosts for the production of heterologous proteins [224]. As early as 1993, recombinant processes of *E. coli* were responsible for almost $5 billion worth of products, that is, insulin, human growth hormone, interferons, and G-CSF. Advantages of *E. coli* include rapid growth, rapid expression, ease of culture and genome modifications, low cost, and high product yields [225]. It is used for massive production of many commercialized proteins. This system is excellent for functional expression of non-glycosylated proteins.

E. coli genetics are far better understood than those of any other microorganism. Recent progress in the fundamental understanding of transcription, translation, and protein folding in *E. coli*, together with the availability of improved genetic tools, is making this bacterium more valuable than ever for the expression of complex eukaryotic proteins. Its genome can be quickly and precisely modified with ease, promotor control is not difficult, and plasmid copy number can be readily altered. This system also features alteration of metabolic carbon flow, avoidance of incorporation of amino acid analogs, formation of intracellular disulfide bonds, and reproducible performance with computer control. *E. coli* can accumulate recombinant proteins up to 80% of its dry weight and survives a variety of environmental conditions. Recombinant protein production in *E. coli* can be increased by mutations which eliminate acetate production [226]. Avecia Biologics achieved a titer of 14 g/L of recombinant protein using *E. coli* [227].

The value of transcriptome analysis in process improvement was shown by Choi et al. [228]. They analyzed an *E. coli* process yielding human insulin-like growth factor 1 fusion protein (IGF-If) in a high-density culture. Of 200 or so genes whose expression was downregulated after induction, the ones involved in biosynthesis of amino acids or nucleotides were studied. Amplification of two of these, *prsA* (encoding PRPP synthetase) and *glpF* (encoding the glycerol transporter), raised product formation from 1.8 to 4.3 g/L.

Bacilli have yields as high as 3 g/L. The organisms used are usually *Bacillus megaterium*, *B. subtilis*, and *Bacillus brevis*. *Staphylococcus carnosus* can produce 2 g/L of secreted mammalian protein.

An improved Gram-negative host for recombinant protein production has been developed using *Ralstonia eutropha* [229]. The system appears superior to *E. coli* with respect to inclusion body formation. Organophosphohydrolase, a protein prone to inclusion body formation with a production of less than 100 mg/L in *E. coli*, was produced at 10 g/L in *R. eutropha*. Another useful bacterium is *Pseudomonas fluorescens* MB101 developed by Dowpharma [230]. This system has produced 4 g/L of TNF-α.

3.2 Yeasts

Yeasts, the single-celled eukaryotic fungal organisms, are often used to produce recombinant proteins that are not produced well in *E. coli* because of problems dealing with folding or the need for glycosylation. The major advantages of yeast expression systems are listed in Table 2. The yeast strains are genetically well characterized and are known to perform many posttranslational modifications. They are easier and less expensive to work with than insect

Table 2
Advantages of yeast expression systems

Stable strains
High density of growth
Durability
High production titers and yields
Protein glycosylation
Reasonable cost
Product processing similar to that of mammalian cells
Suitable for isotopically labeled proteins
Suitable for S–S-rich proteins
Protein folding assisted
Chemically defined media supporting rapid growth

or mammalian cells and are easily adapted to fermentation processes.

The two most utilized yeast strains are *S. cerevisiae* and the methylotrophic yeast *P. pastoris*. Glucose oxidase from *A. niger* is produced by *S. cerevisiae* at 9 g/L. Recombinant products on the market which are made in *S. cerevisiae* are insulin, hepatitis B surface antigen, urate oxidase, glucagons, granulocyte macrophage colony-stimulating factor (GM-CSF), hirudin, and platelet-derived growth factor.

P. pastoris has the desirable qualities of dense growth and methanol-induced expression and secretion of recombinant proteins. It is used for the commercial production of non-glycosylated human serum albumin and glycosylated vaccines. Strains have been developed which are capable of human-type N-glycosylation, and such products are already in clinical testing. High recombinant protein yields can be obtained with *P. pastoris*, for example, 10 g/L of tumor necrosis factor [231], 14.8 g/L of gelatin, 15 g/L of mouse collagen [232], and *E. coli* phytase and *Candida parapsilosis* lipase/acetyltransferase at 6 g/L. Bacterial proteins such as intracellular tetanus fragment C were produced as 27% of protein with a titer of 12 g/L [233]. Production of serum albumin in *S. cerevisiae* amounted to 0.15 g/L, whereas in *P. pastoris*, the titer was 10 g/L [234]. Indeed, claims have been made that *P. pastoris* can make 20–30 g/L of recombinant proteins [235].

Heterologous gene expression in another methylotroph *Hansenula polymorpha* yielded 13.5 g/L of phytase, and other proteins were made at levels over 10 g/L. Among the advantages of methylotrophic yeasts over *S. cerevisiae* as a cloning host are the following: (1) higher protein productivity, (2) avoidance of hyperglycosylation, (3) growth in reasonably strong methanol solutions that would kill most other microorganisms, (4) a system that is cheap to set up and maintain, and (5) integration of multicopies of foreign DNA into chromosomal DNA yielding stable transformants [236].

3.3 Filamentous Fungi (Molds)

Filamentous fungi are attractive hosts for recombinant DNA technology because of their ability to secrete high levels of bioactive proteins with posttranslational processing such as glycosylation. *A. niger* excretes 25 g/L of native glucoamylase [99, 237]. Foreign genes can be incorporated via plasmids into chromosomes of the filamentous fungi where they integrate stably into the chromosome as tandem repeats providing superior long-term genetic stability. An excellent comprehensive review of heterologous expression in *Aspergillus* has been published by Lubertozzi and Keasling [238]. Like mammalian cells, fungi possess the cellular machinery for translation of proteins, protein folding, and posttranslational modification. Like bacteria, they are easy to culture. Genetic development of aspergilli has included (1) transformation systems; (2) expression constructs; (3) targeted integration and copy number manipulation; (4) promoters; (5) improved gene design; (6) engineering of proteases, secretion, and glycosylation; and (7) tools for tagging, targeting, and silencing of genes.

A 1000-fold increase in phytase production was achieved in *A. niger* by the use of recombinant technology [239]. Recombinant *Aspergillus oryzae* can produce 2 g/L of human lactoferrin [240] and 3.3 g/L of *Mucor* rennin [241]. Production of human lactoferrin [242] by *Aspergillus awamori* via rDNA technology and classical strain improvement amounted to 2 g/L of extracellular protein [240]. *A. niger* glucoamylase was made by *A. awamori* at 4.6 g/L.

The fungus *Chrysosporium lucknowense* has been genetically converted into a nonfilamentous, less viscous, low-protease-producing strain that is capable of producing very high yields of heterologous proteins [243]. Dyadic International, Inc., the company responsible for the development of the *C. lucknowense* system, claims protein production levels of up to 100 g/L of native extracellular protein.

3.4 Mammalian Cells

CHO cells constitute the preferred system for producing monoclonal antibodies and some other recombinant proteins. Other cell types include (1) various mouse myelomas such as NS0 murine myeloma cells [244], (2) baby hamster kidney (BHK) cells for production of cattle foot-and-mouth disease vaccine, (3) green monkey kidney cells for polio vaccine [245], and (4) human cell lines such as human embryonic kidney (HEK) cells. NSO is a nonsecreting subclone of the NS-1 mouse melanoma cell line. By 2006, production of therapeutic proteins by mammalian systems reached $20 billion [246].

Animal-free, protein-free, and even chemically defined media with good support of production have been developed [247]. Protein production by mammalian cells (CHO) went from 5–50 mg/L in 1985 to 50–500 mg/L in 1995 and 1–5 g/L in 2005 [248]. A number of mammalian processes are producing 3–5 g/L of recombinant protein [249], and in some cases, protein titers have reached 10 g/L in industry [250], including antibodies [251]. A rather new system is that of a human cell line known as PER.C6 of Crucell Holland BV, which, in cooperation with DSM Biologics, was reported to produce 15 g/L [252] and, then later, 27 g/L of a monoclonal antibody [253]. Protein production of over 20 g/L has been achieved in serum-free medium, but the production of 2–3 g/L in such media is more usual.

3.5 Insect Cells

Insect cells are able to carry out more complex posttranslational modifications than can be accomplished with fungi. They also have the best machinery for the folding of mammalian proteins and are therefore quite suitable for making soluble proteins of mammalian origin [254]. The most commonly used vector system for recombinant protein expression in insects is the baculovirus, especially the nuclear polyhedrosis virus (*Autographa californica*), which contains circular double-stranded DNA, is naturally pathogenic for lepidopteran cells, and can be grown easily in vitro. The usual host is the fall armyworm (*Spodoptera frugiperda*) in suspension culture. A larval culture can be used which is much cheaper than mammalian cell culture.

Baculovirus-assisted insect cell expression offers many advantages as follows:

1. Eukaryotic posttranslational modifications without complication, including phosphorylation, N- and O-glycosylation, correct signal peptide cleavage, proper proteolytic processing, acylation, palmitoylation, myristoylation, amidation, carboxymethylation, and prenylation [255, 256].
2. Proper protein folding and S–S bond formation, unlike the reducing environment of *E. coli* cytoplasm.
3. High expression levels. The virus contains a gene encoding the protein polyhedron, which is made at very high levels normally and is not necessary for virus replication. The gene to be cloned is placed under the strong control of the viral polyhedrin promoter, allowing expression of heterologous protein of up to 30% of cell protein.
4. Easy scale-up with high-density suspension culture.
5. Safety, expression vectors are prepared from the baculovirus which can attack invertebrates but not vertebrates or plants.
6. Lack of limit on protein size.

7. Efficient cleavage of signal peptides.
8. Simultaneous expression of multiple genes [257].

Production of recombinant proteins with the baculovirus expression vector system in insect cells reached 600 mg/L in 1988 [258]. Later information indicated that the baculovirus insect cell system can produce 11 g/L of recombinant protein [259]. Recombinant insect cell cultures have yielded over 200 proteins encoded by genes from viruses, bacteria, fungi, plants, and animals [260].

4 Enzymes

Over the years, high titers of enzymes were obtained using "brute-force" mutagenesis and random screening of microorganisms. Recombinant DNA technology acted as a boon for the enzyme industry in the following ways [261]: (1) plant and animal enzymes could be made by microbial fermentations, for example, chymosin; (2) enzymes from organisms difficult to grow or handle genetically were now produced by industrial organisms such as species of *Aspergillus* and *Trichoderma*, and *K. lactis*, *S. cerevisiae*, *Y. lipolytica*, and *Bacillus licheniformis* (e.g., thermophilic lipase was produced by *A. oryzae* and *Thermoanaerobacter* cyclodextrin glycosyl transferase by *Bacillus*); (3) enzyme productivity was increased by the use of multiple gene copies, strong promoters, and efficient signal sequences; (4) production of a useful enzyme from a pathogenic or toxin-producing species could now be done in a safe host; and (5) protein engineering was employed to improve the stability, activity, and/or specificity of an enzyme.

Genes encoding many microbial enzymes have been cloned, and the enzymes expressed at levels hundreds of times higher than those naturally produced. Over 60% of the enzymes used in different applications including detergent, food, and starch processing industry are recombinant proteins [262]. Recombinant molds are one of the main sources of enzymes for industrial applications. Yields as high as 4.6 g/L have been reached for several hosts including *A. niger*, *A. oryzae*, *A. awamori*, *C. lucknowense*, and *A. chrysogenum*.

Plant phytase [263], produced in recombinant *A. niger*, was used as a feed for 50% of all pigs in Holland. A 1000-fold increase in phytase production was achieved in *A. niger* by the use of recombinant technology [239]. Mammalian chymosin was cloned and produced by *A. niger* or *E. coli*, and recombinant chymosin was approved in the USA; its price was half that of natural calf chymosin.

Three fungal recombinant lipases are currently used in the food industry. They are from *Rhizomucor miehei*, *Thermomyces lanuginosus*, and *Fusarium oxysporum* and are produced in *A. oryzae*. They are used for laundry cleaning, interesterification of lipids, and esterification of glucosides producing glycolipids which have applications as biodegradable nonionic surfactants for detergents, skin care products, and contact lenses and as food emulsifiers. Washing powders have been improved in activity, and low-temperature operation has been achieved by the application of recombinant DNA technology and site-directed mutagenesis of genes encoding proteases and lipases [264, 265]. The first commercial recombinant lipase used in a detergent was from *Humicola lanuginosa*. The gene was cloned into the *A. oryzae* genome.

A multicopy plasmid of *B. subtilis* was used to increase by 2500-fold the production of an α-amylase from *B. amyloliquefaciens* [266]. An exoglucanase from *Cellulomonas fimi* was overproduced in *E. coli* to a level of over 20% of cell protein [267]. The same host has also been used to clone the endo-β-glucanase components from *Thermomonospora* and *Clostridium thermocellum* as well as the cellobiohydrolase I gene of *Trichoderma reesei* [268]. *P. pastoris* was engineered to produce and excrete *S. cerevisiae* invertase into the medium at 100 mg/L [269]. Self-cloning of the xylanase gene in *S. lividans* resulted in a sixfold overproduction of the enzyme [270].

The properties of many enzymes have been modified by random mutagenesis and screening of microorganisms over the years leading to changes in substrate specificity, feedback inhibition, kinetic parameters (V_{max}, K_m or K_i), pH optimum, thermostability, and carbon source inhibition. Based on this information, more rational techniques such as site-directed mutagenesis were used to introduce single changes in amino acid sequences yielding similar types of changes in a large variety of enzymes. Modification of eight amino acids increased heat tolerance and temperature stability at 100 °C of a protease from *Bacillus stearothermophilus* [271]. Interestingly, all mutations were far from the active site of the enzyme.

Molecular breeding techniques (e.g., *DNA shuffling* and *DNA family shuffling*) are being currently used to generate enzymes with improved properties such as activity and stability at different pH values and temperatures [272], increased or modified enantioselectivity [273], altered substrate specificity [274], stability in organic solvents [275], novel substrate specificity and activity [276], increased biological activity of protein pharmaceuticals and biological molecules [277], as well as novel vaccines [278, 279]. Two proteins from directed evolution work were already on the market by year 2000 [280]. These were green fluorescent protein of Clontech [281] and Novo Nordisk's LipoPrime® lipase.

5 Closing Remarks

The fermentation industry developed slowly from the beginning of the twentieth century to the early 1970s using brute-force mutagenesis followed by screening or selection. However, the birth of the era of recombinant DNA (rDNA) in 1971–1973 catalyzed a major change in the way useful processes could be developed. Production of primary metabolites was markedly improved by modern genetic techniques. Environmentally friendly fermentations replaced chemical synthesis to a great extent. Of great interest has been the application of rDNA technology to the production of secondary metabolites and to the elucidation of their biosynthetic pathways. Tools include transposition mutagenesis, targeted deletions, genetic recombination via combinatorial biosynthesis, transcriptome analysis, proteomics, metabolomics, metabolic engineering, etc. Genes encoding many enzymes have been cloned, and the enzymes have been expressed at levels hundreds of times higher than those naturally produced. Random redesign techniques have generated enzymes with improved properties including activity, stability, increased or modified enantioselectivity, altered substrate specificity, etc. An entirely new field of industrial microbiology has arisen out of rDNA, that is, the biopharmaceutical industry which is the most rapidly expanding segment of the biological industry, especially that of monoclonal antibodies. The best is yet to come from the fantastic combination of industrial microbiology and rDNA technology. Many of the new techniques are carried out by small companies and academic groups who could play a major role in rescuing us from the antibiotic crisis that we are now experiencing. Furthermore, we look forward to its role in eventually replacing the environmentally dangerous energy sources that we live with today, that is, petroleum, coal, etc., with future biofuels made from agricultural and forest biomass.

References

1. Ohnishi J, Mitsuhashi S, Hayashi M, Ando S, Yokoi H, Ochiai K et al (2002) A novel methodology employing *Corynebacterium glutamicum* genome information to generate a new L-lysine-producing mutant. Appl Microbiol Biotechnol 58:217–223
2. Stephanopoulos G (1999) Metabolic fluxes and metabolic engineering. Metab Eng 1:1–11
3. Nielsen J (2001) Metabolic engineering. Appl Microbiol Biotechnol 55:263–283
4. Sahm H, Eggeling L, de Graaf AA (2000) Pathway analysis and metabolic engineering in *Corynebacterium glutamicum*. Biol Chem 381:899–910
5. Santos CSS, Stephanopoulos G (2008) Combinatorial engineering of microbes for optimizing cellular phenotype. Curr Opin Chem Biol 12:168–176
6. Bailey JE, Sburlati A, Hatzimanikatis V, Lee K, Renner WA, Tsai PS (1996) Inverse metabolic engineering: a strategy for directed genetic engineering of useful phenotypes. Biotechnol Bioeng 52:109–121
7. Ness JE, del Cardayre SB, Minshull J, Stemmer WP (2000) Molecular breeding: the natural approach to protein design. Adv Protein Chem 55:261–292
8. Zhao H, Arnold FH (1997) Optimization of DNA shuffling for high fidelity recombination. Nucleic Acids Res 25:1307–1308
9. Patten PA, Howard RJ, Stemmer WP (1997) Applications of DNA shuffling to pharmaceuticals and vaccines. Curr Opin Biotechnol 8:724–733
10. Zhang YX, Perry K, Vinci VA, Powell K, Stemmer WP, del Cardayre SB (2002) Genome shuffling leads to rapid phenotypic improvement in bacteria. Nature 415:644–646
11. Hou L (2009) Novel methods of genome shuffling in *Saccharomyces cerevisiae*. Biotechnol Lett 31:671–677
12. Stephanopoulos G, Alper H, Moxley J (2004) Exploiting biological complexity for strain

improvement through systems biology. Nat Biotechnol 22:1261–1267
13. Leuchtenberger W, Huthmacher K, Drauz K (2005) Biotechnological production of amino acids and derivatives: current status and prospects. Appl Microbiol Biotechnol 69:1–8
14. Liu Q, Zhang J, Wei X-X, Ouyang S-P, Wu Q, Chen G-Q (2008) Microbial production of L-glutamate and L-glutamine by recombinant *Corynebacterium glutamicum* harboring *Vitreoscilla* hemoglobin gene *vgb*. Appl Microbiol Biotechnol 77:1297–1304
15. Chinen A, Kozlov YI, Hara Y, Izui H, Yasueda H (2007) Innovative metabolic pathway design for efficient L-glutamate production by suppressing CO_2 emission. J Biosci Bioeng 103:262–269
16. Wang J, Gao D, Yu X, Li W, Qi Q (2015) Evolution of a chimeric aspartate kinase for L-lysine production using a synthetic RNA device. Appl Microbiol Biotechnol 99:8527–8536
17. Ohnishi J, Katahira R, Mitsuhashi S, Kakita S, Ikeda M (2005) A novel *gnd* mutation leading to increased L-lysine production in *Corynebacterium glutamicum*. FEMS Microbiol Lett 242:265–274
18. Ikeda M, Ohnishi J, Hayashi M, Mitsuhashi S (2006) A genome-based approach to create a minimally mutated *Corynebacterium glutamicum* strain for efficient L-lysine production. J Ind Microbiol Biotechnol 33:610–615
19. Hiyashi M, Ohnishi J, Mitsuhashi S, Yonetani Y, Hashimoto S, Ikeda M (2006) Transcriptome analysis reveals global expression changes in an industrial L-lysine producer of *Corynebacterium glutamicum*. Biosci Biotechnol Biochem 70:546–550
20. Radmacher E, Eggeling L (2007) The three tricarboxylate synthase activities of *Corynebacterium glutamicum* and increase of L-lysine synthesis. Appl Microbiol Biotechnol 76:587–595
21. Blombach B, Schreiner ME, Moch M, Oldiges M, Eikmanns BJ (2007) Effect of pyruvate dehydrogenase complex deficiency on L-lysine production with *Corynebacterium glutamicum*. Appl Microbiol Biotechnol 76:615–623
22. Sindelar G, Wendisch VF (2007) Improving lysine production by *Corynebacterium glutamicum* through DNA microarray-based identification of novel target genes. Appl Microbiol Biotechnol 76:677–689
23. Becker J, Zelder O, Häfner S, Schröder H, Wittmann C (2011) From zero to hero-design-based systems metabolic engineering of *Corynebacterium glutamicum* for L-lysine production. Metab Eng 13:159–168
24. Komatsubara S, Kisumi M, Chibata I (1979) Transductional construction of a threonine producing strain of *Serratia marcescens*. Appl Environ Microbiol 38:1045–1051
25. Komatsubara S, Kisumi M, Chibata I (1983) Transductional construction of a threonine hyperproducing strain of *Serratia marcescens*: lack of feedback controls of three aspartokinases and two homoserine dehydrogenases. Appl Environ Microbiol 45:1445–1452
26. Sugita T, Komatsubara S (1989) Construction of a threonine-hyperproducing strain of *Serratia marcescens* by amplifying the phosphoenolpyruvate carboxylase gene. Appl Microbiol Biotechnol 30:290–293
27. Debabov VG (2003) The threonine story. Adv Biochem Eng Biotechnol 79:113–136
28. Ishida M, Kawashima H, Sato K, Hashiguchi K, Ito H, Enei H et al (1994) Factors improving L-threonine production by a three L-threonine biosynthetic genes-amplified recombinant strain of *Brevibacterium lactofermentum*. Biosci Biotechnol Biochem 58:768–770
29. Eggeling L, Sahm H (1999) Amino acid production: principles of metabolic engineering. In: Lee SY, Papoutsakis ET (eds) Metabolic engineering. Marcel Dekker, New York, pp 153–176
30. Kruse D, Kraemer R, Eggeling L, Rieping M, Pfefferle W, Tchieu JH et al (2002) Influence of threonine exporters on threonine production in *Escherichia coli*. Appl Microbiol Biotechnol 59:205–210
31. Lee KH, Park JH, Kim TY, Kim HU, Lee SY (2007) Systems metabolic engineering of *Escherichia coli* for L-threonine production. Mol Syst Biol 3:149–157
32. Chen N, Huang J, Feng Z, Yu L, Q-y X, T-y W (2009) Optimization of fermentation conditions for the biosynthesis of L-threonine by *Escherichia coli*. Appl Biochem Biotechnol 158:595
33. Wang J, Cheng LK, Chen N (2014) High-level production of L-threonine by recombinant *Escherichia coli* with combined feeding strategies. Biotechnol Biotechnol Equip 4(28):495–501
34. Park JH, Lee KH, Kim TY, Lee SY (2007) Metabolic engineering of *Escherichia coli* for the production of L-valine based on transcriptome analysis and in silico gene knockout stimulation. Proc Natl Acad Sci U S A 104:7797–7802
35. Lange C, Rittmann D, Wendisch VF, Bott M, Sahm H (2003) Global expression profiling and physiological characterization of *Corynebacterium glutamicum* grown in the presence of L-valine. Appl Environ Microbiol 69:2521–2532
36. Chen C, Li Y, Hu J, Dong X, Wang X (2015) Metabolic engineering of *Corynebacterium*

glutamicum ATCC13869 for L-valine production. Metab Eng 29:66–75
37. Kisumi M, Komatsubara S, Chibata I (1977) Enhancement of isoleucine hydroxamate-mediated growth inhibition and improvement of isoleucine-producing strains of *Serratia marcescens*. Appl Environ Microbiol 34:647–653
38. Guillouet S, Rodal AA, An G-H, Lessard PA, Sinskey AJ (1999) Expression of the *Escherichia coli* catabolic threonine dehydratase in *Corynebacterium glutamicum* and its effect on isoleucine production. Appl Environ Microbiol 65:3100–3107
39. Morbach S, Sahm H, Eggeling L (1996) L-Isoleucine production with *Corynebacterium glutamicum*: further flux increase and limitation of export. Appl Environ Microbiol 62:4345–4351
40. Hashiguchi K, Takesada H, Suzuki E, Matsui H (1999) Construction of an L-isoleucine overproducing strain of *Escherichia coli* K-12. Biosci Biotechnol Biochem 63:672–679
41. Eggeling L, Morbach S, Sahm H (1977) The fruits of molecular physiology: engineering the L-isoleucine biosynthesis pathway in *Corynebacterium glutamicum*. J Biotechnol 56:167–182
42. Colon GE, Nguyen TT, Jetten MSM, Sinskey A, Stephanopoulos G (1995) Production of isoleucine by overexpression of *ilvA* in *Corynebacterium lactofermentum* threonine producer. Appl Microbiol Biotechnol 43:482–488
43. Kase H, Nakayama K (1977) L-Isoleucine induction by analog-resistant mutants derived from threonine-producing strain of *Corynebacterium glutamicum*. Agric Biol Chem 41:109–116
44. Morbach S, Sahm H, Eggeling L (1995) Use of feedback-resistant threonine dehydratases of *Corynebacterium glutamicum* to increase carbon flux towards L-isoleucine. Appl Environ Microbiol 61:4315–4320
45. Morbach S, Kelle R, Winkels S, Sahm H, Eggeling L (1996) Engineering the homoserine dehydrogenase and threonine dehydratase control points to analyse flux towards L-isoleucine in *Corynebacterium glutamicum*. Appl Microbiol Biotechnol 45:612–620
46. Sahm H, Eggeling L, Morbach S, Eikmanns B (1999) Construction of L-isoleucine overproducing strains of *Corynebacterium glutamicum*. Naturwissenschaften 86:33–38
47. Shiio I, Sasaki A, Nakamori S, Sano K (1973) Production of L-isoleucine by AHV resistant mutants of *Brevibacterium flavum*. Agric Biol Chem 37:2053–2061
48. Ikeda S, Fujita I, Hirose Y (1976) Culture conditions of L isoleucine fermentation from acetic acid. Agric Biol Chem 40:517–522
49. Yin L, Shi F, Hu X, Chen C, Wang X (2013) Increasing L-isoleucine production in *Corynebacterium glutamicum* by overexpressing global regulator Lrp and two-component export system BrnFE. J Appl Microbiol 114(5):1369–1377
50. Yin L, Zhao J, Chen C, Hu X, Chen C, Wang X (2014) Enhancing the carbon flux and NADPH supply to increase L-isoleucine production in *Corynebacterium glutamicum*. Biotechnol Biopro Eng 19:132–142
51. Peng L, Lijun L, Wang J (2015) The optimization of L-isoleucine fermentation conditions by *Brevibacterium flavum* KM011. Microb Microbial Tech 1(2):002
52. Lee M, Smith GM, Eiteman MA, Altman E (2004) Aerobic production of alanine by *Escherichia coli aceF IdhA* mutants expressing the *Bacillus sphaericus alaD* gene. Appl Microbiol Biotechnol 65:56–60
53. Zhang X, Jantama K, Moore JC, Shanmugam KT, Ingram LO (2007) Production of L-alanine by metabolically engineered *Escherichia coli*. Appl Microbiol Biotechnol 77(2):355–366
54. Zhou L, Deng C, Cui W-J, Liu Z-M, Zho Z-M (2015) Efficient L-alanine production by a thermo-regulated switch in *Escherichia coli*. Appl Biochem Biotechnol 178:324–337
55. Bloom F, Smith CJ, Jessee J, Veileux B, Deutch AH (1984) The use of genetically engineered strains of *Escherichia coli* for the overproduction of free amino acids: proline as a model system. In: Downey K, Voellmy RW (eds) Advances in gene technology: molecular genetics of plants and animals. Academic, New York, pp 383–394
56. Sugiura M, Takagi T, Kisumu M (1995) Proline production by regulatory mutants of *Serratia marcescens*. Appl Microbiol Biotechnol 21:213–239
57. Sugiura M, Imai Y, Takagi T, Kisumi M (1985) Improvement of a proline-producing strain of *Serratia marcescens* by subcloning of a mutant allele of the proline gene. J Biotechnol 3:47–58
58. Masuda M, Takamatu S, Nishimura N, Komatsubara S, Tosa T (1993) Improvement of culture conditions for L-proline production by a recombinant strain of *Serratia marcescens*. Appl Biochem Biotechnol 43:189–197
59. Tsuchida T, Kubota K, Yoshinaga F (1986) Improvement of L-proline production by sulfaguanidine resistant mutants derived from

L-glutamic acid-producing bacteria. Agric Biol Chem 50:2201–2207
60. Nakanishi T, Yokote Y, Taketsugu Y (1973) Conversion of L-glutamic acid fermentation to a L-proline fermentation by *Corynebacterium glutamicum*. J Ferment Technol 51:742–749
61. Nakanishi T, Hirao T, Azuma T, Sakurai M, Hagino H (1987) Application of L-glutamate to L-proline fermentation by *Corynebacterium acetoacidophilum*. J Ferment Technol 65:139–144
62. Shibasaki T, Hashimoto S, Mori H, Ozaki A (2000) Construction of a novel hydroxyproline-producing recombinant *Escherichia coli* by introducing a proline 4-hydroxylase gene. J Biosci Bioeng 90:522–525
63. Wang J, Zhang Z, Liu H, Sun FF, Yue C, Hu J, Wang C (2016) Construction and optimization of trans-4-hydroxy-L-proline production recombinant *Escherichia coli* strain taking the glycerol as carbon source. J Chem Technol Biotechnol 91:2389–2398
64. Vandamme EJ (2007) Microbial gems: microorganisms without frontiers. SIM News 57:81–90
65. Kamada N, Yasuhara A, Takano Y, Nakano T (2001) Effect of transketolase modifications on carbon flow to the purine-nucleotide pathway in *Corynebacterium ammoniagenes*. Appl Microbiol Biotechnol 56:710–717
66. Koizumi S, Yonetani Y, Maruyama A, Teshiba S (2000) Production of riboflavin by metabolically engineered *Corynebacterium ammoniagenes*. Appl Microbiol Biotechnol 51:674–679
67. Perkins JB, Pero J (1993) Biosynthesis of riboflavin, biotin, folic acid, and cobalamin. In: Sonenshein AL, Hoch JA, Losick R (eds) *Bacillus subtilis* and other Gram-positive bacteria: biochemistry, physiology and molecular genetics. American Society for Microbiology, Washington, DC, pp 319–334
68. Perkins JB, Sloma A, Hermann T, Theriault K, Zachgo E, Erdenberger T et al (1999) Genetic engineering of *Bacillus subtilis* for the commercial production of riboflavin. J Ind Microbiol Biotechnol 22:8–18
69. Brenner S, Johnson M, Bridgham J, Golda G, Lloyd DH, Johnson D et al (2000) Gene expression analysis by massively parallel signature sequencing (MPSS) on microbead arrays. Nat Biotechnol 18:630–634
70. Karos M, Vilarino C, Bollschweiler C, Revuelta JL (2004) A genome wide transcription analysis of a fungal riboflavin overproducer. J Biotechnol 113:69–76
71. Shi SB, Chen T, Zhang ZG, Chen X, Zhao XM (2009) Transcriptome analysis guided metabolic engineering of *Bacillus subtilis* for riboflavin production. Metab Eng 11:243–252
72. Levy-Schil S, Debussche L, Rigault S, Soubrier F, Bacchette F, Lagneaux D et al (1993) Biotin biosynthetic pathway in a recombinant strain of *Escherichia coli* overexpressing bio genes: evidence for a limiting step upstream from KAPA. Appl Microbiol Biotechnol 38:755–762
73. Sakurai N, Imai Y, Masuda M, Komatsubara S, Tosa T (1994) Improvement of a D-biotin hyperproducing recombinant strain of *Serratia marcescens*. J Biotechnol 36:63–73
74. Masuda M, Takahashi K, Sakurai N, Yanagiya K, Komatsubara S, Tosa T (1995) Further improvement of D-biotin production by a recombinant strain of *Serratia marcescens*. Process Biochem 30:553–562
75. Matsui J, Ifuku O, Kanzaki N, Kawamoto T, Nakahama K (2001) Microorganism resistant to threonine analogue and production of biotin. US patent 6284500
76. Anderson S, Marks CB, Lazarus R, Miller J, Stafford K, Seymour J et al (1985) Production of 2-keto-L-gulonate: an intermediate in L-ascorbate synthesis by a genetically modified *Erwinia herbicola*. Science 230:144–149
77. Grindley JF, Payton MA, van de Pol H, Hardy KG (1988) Conversion of glucose to 2-keto-L-gulonate, an intermediate in L-ascorbate synthesis, by a recombinant strain of *Erwinia citreus*. Appl Environ Microbiol 54:1770–1775
78. Saito Y, Ishii Y, Hayashi H, Imao Y, Akashi T, Yoshikawa K et al (1997) Cloning of genes coding for L-sorbose and L-sorbosone dehydrogenases from *Gluconobacter oxydans* and microbial production of 2-keto-gulonate, a precursor of L-ascorbic acid, in a recombinant *G. oxydans strain*. Appl Environ Microbiol 63:454–460
79. Saito Y, Ishii Y, Hayashi K, Yoshikawa K, Noguchi Y, Yoshida S, Soeda S, Yoshida M (1998) Direct fermentation of 2-keto-L-gulonic acid in recombinant *Gluconobacter oxydans*. Biotechnol Bioeng 58:309–315
80. Chotani G, Dodge T, Hsu A, Kumar M, LaDuca R, Trimbur D, Weyler W, Sanford K (2000) The commercial production of chemicals using pathway engineering. Biochim Biophys Acta 1543:434–455
81. Rao YM, Sureshkumar GK (2000) Direct biosynthesis of ascorbic acid from glucose by *Xanthomonas campestris* through induced free-radicals. Biotechnol Lett 22:407–411
82. DeBaets S, Vandedrinck S, Vandamme EJ (2000) Vitamins and related biofactors, microbial production. In: Lederberg J (ed)

Encyclopedia of microbiology, vol 4, 2nd edn. Academic, New York, pp 837–853
83. Johnson EA, Schroeder WA (1995) Microbial carotenoids. Adv Biochem Eng Biotechnol 53:119–178
84. López-Nieto MJ, Costa J, Peiro E, Méndez E, Rodríguez-Sáiz M, de la Fuente JL, Cabri W, Barredo JL (2004) Biotechnological lycopene production by mated fermentation of *Blakeslea trispora*. Appl Microbiol Biotechnol 66:153–159
85. Ciegler A (1965) Microbial carotenogenesis. Adv Appl Microbiol 7:1–34
86. Ninet L, Renaut J (1979) Carotenoids. In: Peppler HJ, Perlman D (eds) Microbial technology, vol 1, 2nd edn. Academic, New York, pp 529–544
87. Barkovich R, Liao JC (2001) Metabolic engineering of isoprenoids. Metab Eng 3:27–39
88. Lee PC, Schmidt-Dannert C (2002) Metabolic engineering towards biotechnological production of carotenoids in microorganisms. Appl Microbiol Biotechnol 60:1–11
89. Tao L, Jackson RE, Cheng Q (2005) Directed evolution of copy number of a broad host range plasmid for metabolic engineering. Metab Eng 7:10–17
90. Alper H, Miyaoku K, Stephanopoulos G (2006) Characterization of lycopene-overproducing *E. coli* strains in high cell density fermentations. Appl Microbiol Biotechnol 72:968–974
91. Fernández-Sevilla JM, Acien Fernández FG, Molina Grima E (2010) Biotechnological production of lutein and its applications. Appl Microbiol Biotechnol 86:27–40
92. Margalith PZ (1999) Production of ketocarotenoids by microalgae. Appl Microbiol Biotechnol 51:431–438
93. Johnson EA (2003) *Phaffia rhodozyma:* a colorful odyssey. Int Microbiol 6:169–174
94. de la Fuente JL, Rodríguez-Sáiz M, Schleissner C, Díez B, Peiro E, Barredo JL (2010) High titer production of astaxanthin by the semiindustrial fermentation of *Xanthophyllomyces dendrorhous.* J Biotechnol 148:144–146
95. Rodríguez-Sáiz M, de la Fuente JL, Barredo JL (2010) *Xanthophyllomyces dendrorhous* for the industrial production of astaxanthin. Appl Microbiol Biotechnol 88:645–658
96. Henke NA, Heider SAE, Peters-Wendisch P, Wendisch VF (2016) Production of the marine carotenoid astaxanthin by metabolically engineered *Corynebacterium glutamicum.* Mar Drugs 14:124
97. Sauer M, Porro D, Mattanovich D, Branduardi P (2008) Microbial production of organic acids: expanding the markets. Trends Biotechnol 26:100–108
98. Sánchez S, Demain AL (2008) Metabolic regulation and overproduction of primary metabolites. J Microbial Biotechnol 1:283–319
99. Ward OP, Qin WM, Hanjoon JD, Singh EJYA (2006) Physiology and biotechnology of *Aspergillus.* Adv Appl Microbiol 58:1–75
100. Kubicek CP, Roehr M (1986) Citric acid fermentation. Crit Rev Biotechnol 3:331–373
101. Anastasiadis S, Rehm HJ (2005) Continuous citric acid secretion by a high specific pH dependent active transport system in yeast *Candida oleophila* ATCC 20177. Electron J Biotechnol 8(2):26–42
102. Causey TB, Zhou S, Shanmugam KT, Ingram LO (2003) Engineering the metabolism of *Escherichia coli* W3110 for the conversion of sugar to redox-neutral and oxidized products: homoacetate production. Proc Natl Acad Sci U S A 100:825–832
103. Parekh SR, Cheryan M (1994) High concentrations of acetate with a mutant strain of *C. thermoaceticum.* Biotechnol Lett 16:139–142
104. Fukaya M, Tayama K, Tamaki T, Tagami H, Okumura H, Kawamura Y et al (1989) Cloning of the membrane-bound aldehyde dehydrogenase gene of *Acetobacter polyoxogenes* and improvement of acetic acid production by use of the cloned gene. Appl Environ Microbiol 55:171–176
105. Patnaik R, Louie S, Gavrilovic V, Perry K, Stemmer WPC, Ryan CM et al (2002) Genome shuffling of *Lactobacillus* for improved acid tolerance. Nat Biotechnol 20:707–712
106. John RP, Gangadharan D, Madhavan Nampoothiri K (2008) Genome shuffling of *Lactobacillus delbrueckii* mutant and *Bacillus amyloliquefaciens* through protoplastic fusion for L-latic acid production from starchy wastes. Bioresour Technol 99:8008–8015
107. Skory CD (2004) Lactic acid production by *Rhizopus oryzae* transformants with modified lactate dehydrogenase activity. Appl Microbiol Biotechnol 64:237–242
108. Saito S, Ishida N, Onishi T, Tokuhiro K, Nagamori E, Kitamoto K et al (2005) Genetically engineered wine yeast produces a high concentration of L-lactic acid of extremely high optical purity. Appl Environ Microbiol 71:2789–2792
109. Porro D, Bianchi MM, Brambilla L, Menghini R, Bolzani D, Carrera V et al (1999) Replacement of a metabolic pathway for large scale production of lactic acid from engineered yeasts. Appl Environ Microbiol 65:4211–4215
110. Zhou S, Yomano LP, Shanmugam KT, Ingram LO (2003) Fermentation of 10% (w/v) sugar

to D(−)-lactate by engineered *Escherichia coli* B. Biotechnol Lett 27:1891–1896
111. Ishida N, Suzuki T, Tokuhiro K, Nagamori E, Onishi T, Saitoh S et al (2006) D-Lactic acid production by metabolically engineered *Saccharomyces cerevisiae.* J Biosci Bioeng 101:172–177
112. Sánchez AM, Bennett GN, San KY (2005) Novel pathway engineering design of the anaerobic central metabolic pathway in *Escherichia coli.* Metab Eng 7:229–239
113. Lee SJ, Song H, Lee SY (2006) Genome-based metabolic engineering of *Mannheimia succiniciproducens* for succinic acid production. Appl Environ Microbiol 72:1939–1948
114. Lin H, Bennett GN, San KY (2005) Fedbatch culture of a metabolically engineered *Escherichia coli* strain designed for high-level succinate production and yield under aerobic conditions. Biotechnol Bioeng 90:775–779
115. Vemuri GN, Eiteman MA, Altman E (2002) Succinate production in dual-phase *Escherichia coli* fermentations depends on the time of transition from aerobic to anaerobic conditions. J Ind Microbiol Biotechnol 28:325–332
116. Liu X, Yang S-T (2005) Metabolic engineering of *Clostridium tyrobutyricum* for butyric acid fermentation. Proceedings of the 229th ACS National Meeting, San Diego, Abstract 70
117. Neufeld RJ, Peleg Y, Rokem JS, Pines O, Goldberg I (1991) L-Malic acid formation by immobilized *Saccharomyces cerevisiae* amplified for fumarase. Enzyme Microb Technol 13:991–996
118. Picataggio S, Rohver T, Deander K, Lanning D, Reynolds R, Mielenz J et al (1992) Metabolic engineering of *Candida tropicalis* for the production of long-chain dicarboxylic acids. Nat Biotechnol 10:894–898
119. Shi DJ, Wang CL, Wang KM (2009) Genome shuffling to improve thermotolerance, ethanol tolerance and ethanol productivity of *Saccharomyces cerevisiae.* J Ind Microbiol Biotechnol 36:139–147
120. Chu BCH, Lee H (2007) Genetic improvement of *Saccharomyces cerevisiae* for xylose fermentation. Biotechnol Adv 25:425–441
121. Jeffries TW (2006) Engineering yeast for xylose metabolism. Curr Opin Biotechnol 17:320–326
122. Bro C, Regenberg B, Forster J, Nielsen J (2006) In silico aided metabolic engineering of *Saccharomyces cerevisiae* for improved bioethanol production. Metab Eng 8:102–111
123. Wei P, Li Z, He P, Lin Y, Jiang N (2008) Genome shuffling in the ethanologenic yeast *Candida krusei* to improve acetic acid tolerance. Biotechnol Appl Biochem 49:113–120
124. Ingram LO, Conway T, Clark DP, Sewell GW, Preston JF (1987) Genetic engineering of ethanol production in *Escherichia coli.* Appl Environ Microbiol 53:2420–2425
125. Doran JB, Ingram LO (1993) Fermentation of crystalline cellulose to ethanol by *Klebsiella oxytoca* containing chromosomally integrated *Zymomonas mobilis* genes. Biotechnol Prog 9:533–538
126. Dien BS, Nichols NN, O'Bryan PJ, Bothast RJ (2000) Development of new ethanologenic *Escherichia coli* strains for fermentation of lignocellulosic biomass. Appl Biochem Biotechnol 84:181–196
127. Papanikolaou S, Ruiz-Sánchez P, Pariset B, Blanchard F, Fick M (2000) High production of 1,3-propanediol from industrial glycerol by a newly isolated *Clostridium butyricum* strain. J Biotechnol 77:191–208
128. Sanford K, Valle F, Ghirnikar R (2004) Pathway engineering through rational design. Tutorial: designing and building cell factories for biobased production. Genet Eng News 24:44–45
129. Nakamura C, Whited G (2003) Metabolic engineering for the microbial production of 1,3-propanediol. Curr Opin Biotechnol 14:454–459
130. Kaup B, Bringer-Meyer S, Sahm H (2004) Metabolic engineering of *Escherichia coli*: construction of an efficient biocatalyst for D-mannitol formation in a whole-cell biotransformation. Appl Microbiol Biotechnol 64:333–339
131. Kaup B, Bringer-Meyer S, Sahm H (2005) D-Mannitol formation from D-glucose in a whole-cell biotransformation with recombinant *Escherichia coli.* Appl Microbiol Biotechnol 69:397–403
132. Lee J-K, Oh D-K, Song H-Y, Kim I-W (2007) Ca^{2+} and Cu2+ supplementation increases mannitol production by *Candida magnoliae.* Biotechnol Lett 29:291–294
133. Song SH, Vieille C (2009) Recent advances in the biological production of mannitol. Appl Microbiol Biotechnol 84:55–62
134. Chun UH, Rogers PL (1988) The simultaneous production of sorbitol and gluconic acid by *Zymomonas mobilis.* Appl Microbiol Biotechnol 29:19–24
135. Ladero V, Ramos A, Wiersma A, Goffin P, Schanck A, Kleerbezem M (2007) High-level; production of the low-calorie sugar sorbitol by *Lactobacillus plantarum* through metabolic engineering. Appl Environ Microbiol 73:1864–1872
136. Hermann M, Fayolle F, Marchal R, Podvin L, Sebald M, Vandecasteele J-P (1985) Isolation and characterization of butanol-resistant

mutants of *Clostridium acetobutylicum*. Appl Environ Microbiol 50:1238–1243
137. Qureshi N, Maddox IS, Freidl A (1992) Application of continuous substrate feeding to the ABE fermentation: relief of product inhibition using extraction, perstraction, stripping and pervaporation. Biotechnol Prog 8:382–390
138. Chen C-K, Blaschek HP (1999) Acetate enhances solvent production and prevents degeneration in *Clostridium beijerinckii* BA101. Appl Microbiol Biotechnol 52:170–173
139. Connor MR, Liao JC (2009) Microbial production of advanced transportation fuels in non-natural hosts. Curr Opin Biotechnol 20:307–315
140. Whiters ST, Gottlieb SS, Lieu B, Newman JD, Keasling JD (2007) Identification of isopentenol biosynthetic genes from *Bacillus subtilis* by a screening method based on isoprenoid precursor toxicity. Appl Environ Microbiol 73:6277–6283
141. Atsumi S, Hanai T, Liao JC (2008) Nonfermentative pathways for synthesis of branched-chain higher alcohols as biofuels. Nature 451:86–89
142. Jeya M, Lee K-M, Tiwari MK, Kim J-S, Gunasekaran P, Kim S-Y, Kim I-W, Lee J-K (2009) Isolation of a novel high erythritol producing *Pseudomonas tsukubaensis* and scale-up of erythritol fermentation to industrial level. Appl Microbiol Biotechnol 83:225–231
143. Zagustina NA, Rodionova NA, Mestechkina NM, Shcherbukhin VD, Bezborodov AM (2001) Xylitol production by a culture of *Candida guilliermondii* 2581. Appl Biochem Microbiol 37:489–492
144. Lederberg J (2000) Pathways of discovery: infectious history. Science 288:287–293
145. Choi KP, Kim KH, Kim JW (1997) Strain improvement of clavulanic acid producing *Streptomyces clavuligerus*. Proc 10th Internat Symp Biol Actinomycetes (ISBA), Beijing, Abstr 12, 9
146. Li R, Townsend CA (2006) Rational strain improvement for enhanced clavulanic acid production by genetic engineering of the glycolytic pathway in *Streptomyces clavuligerus*. Metab Eng 8:240–252
147. Pérez-Redondo RA, Rodríguez-García A, Martín JF, Liras P (1999) Deletion of the *pyc* gene blocks clavulanic acid biosynthesis except in glycerol-containing medium: evidence for two different genes in formation of the C3 unit. J Bacteriol 181:6922–6928
148. Paradkar AS, Aiodoo KA, Jensen SE (1998) A pathway-specific transcriptional activator regulates late steps of clavulanic acid biosynthesis in *Streptomyces clavuligerus*. Mol Microbiol 27:831–843
149. Pérez-Llarena FJ, Liras P, Rodríguez-García A, Martín JF (1997) A regulatory gene (*ccaR*) required for cephamycin and clavulanic acid production in *Streptomyces clavuligerus*: amplification results in overproduction of both β-lactam compounds. J Bacteriol 179:2053–2059
150. Lein J (1986) The Panlabs penicillin strain improvement program. In: Vanek Z, Hostalek Z (eds) Overproduction of microbial metabolites; strain improvement and process control strategies. Butterworth, Boston, pp 105–139
151. Kennedy J, Turner G (1996) δ-(L-α-Aminoadipyl)-L-cysteinyl-D-valine synthetase is a rate limiting enzyme for penicillin production in *Aspergillus nidulans*. Mol Gen Genet 253:189–197
152. Hamlyn PF, Ball C (1979) Recombination studies with *Cephalosporium acremonium*. In: Sebek OK, Laskin AI (eds) Genetics of industrial microorganisms. American Society for Microbiology, Washington, DC, pp 185–191
153. Skatrud PL, Fisher DL, Ingolia TD, Queener SW (1987) Improved transformation of *Cephalosporium acremonium*. In: Alacevic M, Hranueli D, Toman Z (eds) Genetics of industrial microorganisms, part B. Pliva, Zagreb, pp 111–119
154. Skatrud PL, Tietz AJ, Ingolia TD, Cantwell CA, Fisher DL, Chapman JL et al (1989) Use of recombinant DNA to improve production of cephalosporin C by *Cephalosporium acremonium*. Nat Biotechnol 7:477–485
155. Wesseling AC, Lago B (1981) Strain improvement by genetic recombination of cephamycin producers, *Nocardia lactamdurans* and *Streptomyces griseus*. Dev Ind Microbiol 22:641–651
156. Chary VK, de la Fuente JL, Leitao AL, Liras P, Martín JF (2000) Overexpression of the *lat* gene in *Nocardia lactamdurans* from strong heterologous promoters results in very high levels of lysine-6-aminotransferase and up to a two-fold increase in cephamycin C production. Appl Microbiol Biotechnol 53:282–288
157. Bignell DRD, Tahlan K, Colvin KR, Jensen SE, Leskiw BK (2005) Expression of *ccaR*, encoding the positive activator of cephamycin C and clavulanic acid production in *Streptomyces clavuligerus*, is dependent on *bldG*. Antimicrob Agents Chemother 49:1529–1541
158. Cantwell C, Beckmann R, Whiteman P, Queener SW, Abraham EP (1992) Isolation of deacetoxycephalosporin C from fermentation broths of *Penicillium chrysogenum* transformants: construction of a new fungal biosynthetic pathway. Proc R Soc Lond (Biol) 248:283–289
159. Crawford L, Stepan AM, Mcada PC, Rambosek JA, Conder MJ, Vinci VA et al

(1995) Production of cephalosporin intermediates by feeding adipic acid to recombinant *Penicillium chrysogenum* strains expressing ring expansion activity. Nat Biotechnol 13:58–62
160. Velasco J, Adrio JL, Moreno MA, Díez B, Soler G, Barredo JL (2000) Environmentally safe production of 7-aminodeacetoxycephalosporanic acid (7ADCA) using recombinant strains of *Acremonium chrysogenum*. Nat Biotechnol 18:857–861
161. Luo H, Yu H, Qiang L, Shen Z (2004) Cloning and co-expression of D-amino acid oxidase and glutaryl-7-aminocephalosporanic acid acylase genes in *Escherichia coli*. Enzyme Microb Technol 35:514–518
162. Coulthurst SJ, Barnard AM, Salmond GP (2005) Regulation and biosynthesis of carbapenem antibiotics in bacteria. Nat Rev Microbiol 3:295–306
163. Kahan JS, Kahan FM, Goegelman R, Currie SA, Jackson M, Stapley EO, Miller TW, Miller AK, Hendlin D, Mochales S, Hernandez S, Woodruff HB, Birnbaum J (1979) Thienamycin, a new β-lactam antibiotic. 1. Discovery, taxonomy, isolation and physical properties. J Antibiot 32:1–12
164. Park SR, Han AR, Ban Y-H, Yoo YJ, Kim EJ et al (2010) Genetic engineering of macrolide biosynthesis: past advances, current state and future prospects. Appl Microbiol Biotechnol 85:1227–1239
165. Reeves AR, Cernota WH, Brikun IA, Wesley RK, Weber JM (2004) Engineering precursor flow for increased erythromycin production in *Aeromicrobium erythreum*. Metab Eng 6:300–312
166. Solenberg PJ, Cantwell CA, Tietz AJ, McGilvray D, Queener SW, Baltz RH (1996) Transposition mutagenesis in *Streptomyces fradiae*: identification of a neutral site for the stable insertion of DNA by transposon exchange. Gene 16:67–72
167. Brautaset T, Sletta H, Nedal A, Borgos SEF, Degnes KF, Bakke I, Volokhan O, Sekurova ON, Treshalin ID, Mirchink EP, Dikiy A, Ellingsen TE, Zotchev SB (2008) Improved antifungal polyene macrolides via engineering of the nystatin biosynthetic genes in *Streptomyces noursei*. Chem Biol 15:1198–1206
168. Galm U, Shen B (2006) Expression of biosynthetic gene clusters in heterologous hosts for natural product production and combinatorial biosynthesis. Expert Opin Drug Discovery 1:409–437
169. Méndez C, Salas JA (2003) On the generation of novel anticancer drugs by recombinant DNA technology: the use of combinatorial biosynthesis to produce novel drugs. Comb Chem High Throughput Screen 6:513–526
170. Rodríguez E, McDaniel R (2001) Combinatorial biosynthesis of antimicrobials and other natural products. Curr Opin Microbiol 4:526–534
171. Trefzer A, Blanco G, Remsing L, Kunzel E, Rix U, Lipata F et al (2002) Rationally designed glycosylated premithramycins: hybrid aromatic polyketides using genes from three different biosynthetic pathways. J Am Chem Soc 124:6056–6062
172. Gomi S, Ikeda D, Nakamura H, Naganawa H, Yamashita F, Hotta K et al (1984) Isolation and structure of a new antibiotic, indolizomycin, produced by a strain SK2-52 obtained by interspecies fusion treatment. J Antibiot 37:1491–1494
173. Traxler P, Schupp T, Wehrli W (1982) 16, 17-dihydrorifamycin S and 16,17-dihydro-17-hydroxyrifamycin S, two novel rifamycins from a recombinant strain C5/42 of *Nocardia mediterranei*. J Antibiot 35:594–601
174. Okanishi M, Suzuki N, Furuta T (1996) Variety of hybrid characters among recombinants obtained by interspecific protoplast fusion in streptomycetes. Biosci Biotechnol Biochem 60:1233–1238
175. Zhou L, Ahlert J, Xue Y, Thorson JS, Sherman DH, Liu H-W (1999) Engineering a methymycin/pikromycin-calicheamicin hybrid: construction of two new macrolides carrying a designed sugar moiety. J Am Chem Soc 121:9881–9882
176. Méndez C, Salas JA (2001) Altering the glycosylation pattern of bioactive compounds. Trends Biotechnol 19:449–456
177. Decker H, Hutchinson CR (1993) Transcriptional analysis of the *Streptomyces glaucescens* tetracenomycin biosynthesis gene cluster. J Bacteriol 175:3887–3892
178. Wohlert S-E, Blanco G, Lombo F, Fernández E, Brana AF, Reich S, Udvarnoki G, Méndez C, Decker H, Frevert J et al (1998) Novel hybrid tetracenomycins through combinatorial biosynthesis using a glycosyltransferase encoded by the *elm* genes in cosmid 16 F4 which shows a very broad sugar substrate specificity. J Am Chem Soc 120:10,596–10,601
179. Barriere JC, Berthaud N, Beyer D, Dutka-Malen S, Paris JM, Desnottes JF (1998) Recent developments in streptogramin research. Curr Pharm Des 4:155–180
180. Xu B, Jin Z, Wang H, Jin Q, Jin X et al (2008) Evolution of *Streptomyces pristinaespiralis* for resistance and production of pristinamycin by

genome shuffling. Appl Microbiol Biotechnol 80:261–267
181. Van Lanen SG, Shen B (2006) Microbial genomics for the improvement of natural product discovery. Curr Opin Microbiol 9:252–260
182. Jenke-Kodama H, Sandmann A, Müller R, Dittmann E (2005) Evolutionary implications of bacterial polyketide synthases. Mol Biol Evol 22:2027–2039
183. Zazopoulos E, Hwang K, Staffa A, Liu W, Bachmann BO, Nonaka K et al (2003) A genomics-guided approach for discovering and expressing cryptic metabolic pathways. Nat Biotechnol 21:187–190
184. Moir DT, Shaw KJ, Hare RS, Vovis GF (1999) Genomics and antimicrobial drug discovery. Antimicrob Agents Chemother 43:439–446
185. NCBI (National Center for Biotechnology Information) (2017) Microbial genomes. Complete Eukaryotic and Prokariotic genomes. http://www.ncbi.nlm.nih.gov/genome/browse/. Accessed 15 Feb 2017
186. Blin K, Medema MH, Kazempour D et al (2013) antiSMASH 2.0—a versatile platform for genome mining of secondary metabolite producers. Nucleic Acids Res 41:W204–W212
187. Anand S, Prasad MV (2010) Yadav G (2010) SBSPKS: structure based sequence analysis of polyketide synthases. Nucleic Acids Res 38(Web Server issue):W487–W496
188. Newman DJ, Shapiro S (2008) Microbial prescreens for anticancer activity. SIM News 58:132–150
189. Hwang CK, Kim HS, Hong YS, Kim YH, Hong SK, Kim SJ, Lee JJ (1995) Expression of *Streptomyces peucetius* genes for doxorubicin resistance and aklavinone 11-hydroxylase in *Streptomyces galilaeus* ATCC 31133 and production of a hybrid aclacinomycin. Antimicrob Agents Chemother 39:1616–1620
190. Kim HS, Hong YS, Kim YH, Yoo OJ, Lee JJ (1996) New anthracycline metabolites produced by the aklavinone 11-hydroxylase gene in *Streptomyces galilaeus* ATCC 3113. J Antibiot 49:355–360
191. Niemi J, Mäntäslä P (1995) Nucleotide sequences and expression of genes from *Streptomyces purpurascens* that cause the production of new anthracyclines. J Bacteriol 177:2942–2945
192. Ylihonko K, Hakala J, Kunnari T, Mäntsälä P (1996) Production of hybrid anthracycline antibiotics by heterologous expression of *Streptomyces nogalater* nogalamycin biosynthesis genes. Microbiology 142:1965–1972
193. Strohl WR, Bartel PL, Li Y, Connors NC, Woodman RH (1991) Expression of polyketide biosynthesis and regulatory genes in heterologous streptomycetes. J Ind Microbiol 7:163–174
194. Bartel PL, Zhu CB, Lampel JS, Dosch DC, Connors NC, Strohl WR, Beale JM Jr, Floss HG (1990) Biosynthesis of anthraquinones by interspecies cloning of actinorhodin genes in streptomycetes: clarification of actinorhodin gene functions. J Bacteriol 172:4816–4826
195. Arcamone F, Penco S, Vigevani A, Redaelli S, Franchi G, Di Marco A, Casazza AM, Dasdia T, Formelli F, Necco A, Soranzo C (1975) Synthesis and antitumor properties of new glycosides of daunomycinone and adriamycinone. J Med Chem 18:703–707
196. Madduri K, Kennedy J, Rivola G, Inventi-Solari A, Filippini S, Zanuso G, Colombo AL, Gewain KM, Occi JL, MacNeil DJ, Hutchinson CR (1998) Production of the antitumor drug epirubicin (4′-epidoxorubicin) and its precursor by a genetically engineered strain of *Streptomyces peucetius*. Nat Biotechnol 16:69–74
197. Manfredi JJ, Horowitz SB (1984) Taxol: an antimitotic agent with a new mechanism of action. Pharmacol Ther 25:83–125
198. Horwitz SB (1994) Taxol (paclitaxel): mechanisms of action. Ann Oncol 5(Suppl 6):S3–S6
199. Dejong JM, Liu Y, Bollon AP, Long RM, Jennewein S, Williams D, Croteau RB (2005) Genetic engineering of taxol biosynthetic genes in *Saccharomyces cerevisiae*. Biotechnol Bioeng 93:212–224
200. Engels B, Dahm P, Jennewein S (2008) Metabolic engineering of taxadiene biosynthesis in yeast as a first step towards Taxol (Paclitaxel) production. Metab Eng 10:201–206
201. Borzlleri RM, Vite GD (2002) Epothilones: new tubulin polymerization agents in preclinical and clinical development. Drugs Future 27:1149–1163
202. Brown AG, Smale TC, King TJ, Hasenkamp R, Thompson RH (1976) Crystal and molecular structure of compactin: a new antifungal metabolite from *Penicillium brevicompactum*. J Chem Soc Perkin Trans I (11):1165–1170
203. Endo A, Kuroda M, Tsujita Y (1976) ML-236A, ML-236B and ML-236C, new inhibitors of cholesterol genesis produced by *Penicillium citrinum*. J Antibiot 29:1346–1348
204. Endo A (1979) Monacolin K, a new hypocholesterolemic agent produced by *Monascus* species. J Antibiot 32:852–854

205. Alberts AW, Chen J, Kuron G, Hunt V, Huff J, Hoffman C et al (1980) Mevinolin: a highly potent competitive inhibitor of hydroxymethylglutaryl-coenzyme A reductase and a cholesterol-lowering agent. Proc Natl Acad Sci U S A 77:3957–3961
206. Askenazi M, Driggers EM, Holtzman DA, Norman TC, Iverson S, Zimmer DP et al (2003) Integrating transcriptional and metabolite profiles to direct the engineering of lovastatin-producing fungal strains. Nat Biotechnol 21:150–156
207. Stapley EO (1982) Avermectins, antiparasitic lactones produced by *Streptomyces avermitilis* isolated from a soil in Japan. In: Umezawa H, Demain AL, Hata R, Hutchinson CR (eds) Trends in antibiotic research. Japan Antibiotic Research Association, Tokyo, pp 154–170
208. Lee J-Y, Hwang Y-S, Kim S-S, Kim E-S, Choi C-Y (2000) Effect of a global regulatory gene, *afsR2*, from *Streptomyces lividans* on avermectin production in *Streptomyces avermitilis*. Biosci Bioeng 89:606–608
209. Ikeda H, Takada Y, Pang C-H, Tanaka H, Omura S (1993) Transposon mutagenesis by Tn4560 and applications with avermectin-producing *Streptomyces avermitilis*. J Bacteriol 175:2077–2082
210. Stutzman-Engwall K, Conlon S, Fedechko R, McArthur H, Pekrun K, Chen Y, Jenne S, La C, Trinh N, Kim S, Zhang Y-X, Fox R, Gustafsson C, Krebber A (2005) Semisynthetic DNA shuffling of *aveC* leads to improved industrial scale production of doramectin by *Streptomyces avermitilis*. Metab Eng 7:27–37
211. Zhang X, Chen Z, Li M, Wen Y, Song Y, Li J (2006) Construction of ivermectin producer by domain swaps of avermectin polyketide synthase in *Streptomyces avermitilis*. Appl Microbiol Biotechnol 72:986–994
212. McArthur HIA (1998) The novel avermectin, Doramectin-a successful application of mutasynthesis. In: Hutchinson CR, McAlpine J (eds) Developments in industrial microbiology-BMP 97. Society for Industrial Microbiology, Fairfax, pp 43–48
213. Vezina C, Kudelski A, Sehgal SN (1975) Rapamycin (AY 22,989), a new antifungal antibiotic. I. Taxonomy of the producing streptomycete and isolation of the active principle. J Antibiot 28:721–726
214. Kino T, Hatanaka H, Hashimoto M, Nishiyama M, Goto T, Okuhara M, Kohsaka M, Aoki H, Imanaka H (1987) FK-506, a novel immunosuppressant isolated from *Streptomyces*. I. Fermentation, isolation and physico-chemical and biological characteristics. J Antibiot 40:1249–1255
215. Jung S, Moon S, Lee K, Park Y-J, Yoon S et al (2009) Strain development of *Streptomyces* sp. for tacrolimus production using sequential adaptation. J Ind Microbiol Biotechnol 36:1467–1471
216. Chen X, Wei P, Fan L, Yang D, Zhu X, Shen W et al (2009) Generation of high-yield rapamycin-producing strains through protoplast-related techniques. Appl Microbiol Biotechnol 83:507–512
217. Kuscer E, Coates N, Challis I, Gregory M, Wilkinson B, Sheridan R, Petkovic H (2007) Roles of *rapH* and *rapG* in positive regulation of rapamycin biosynthesis in *Streptomyces hygroscopicus*. J Bacteriol 189:4756–4763
218. Jin ZH, Xu B, Lin SZ, Jin QC, Cen PL (2009) Enhanced production of spinosad in *Saccharopolyspora spinosa* by genome shuffling. Appl Biochem Biotechnol 159:655–663
219. Leader B, Baca QJ, Golan DE (2008) Protein therapeutics: a summary and pharmacological classification. Nat Rev Drug Discov 7:21–39
220. Demain AL, Vaishnav P (2009) Production of recombinant proteins by microbes and higher organisms. Biotechnol Adv 27:297–306
221. Rayder RA (2008) Expression systems for process and product improvement. Bioprocess Int 6:4–9
222. Choi BK, Bobrowicz P, Davidson RC, Hamilton SR, Kung DH, Li H et al (2003) Use of combinatorial genetic libraries to humanize N-linked glycosylation in the yeast *Pichia pastoris*. Proc Natl Acad Sci U S A 100:5022–5027
223. Yuan L, Kurek I, English J, Keenan R (2005) Laboratory-directed protein evolution. Microbiol Mol Biol Rev 69:373–392
224. Terpe K (1996) Overview of bacterial expression systems for heterologous protein production: from molecular and biochemical fundamentals to commercial systems. Appl Microbiol Biotechnol 72:211–223
225. Swartz J (1996) *Escherichia coli* recombinant DNA technology. In: Neidhardt FC (ed) *Esherichia coli* and *Salmonella*: cellular and molecular biology, 2nd edn. American Society of Microbiology, Washington, DC, pp 1693–1771
226. Wong MS, Wu S, Causey TB, Bennett GN, San K-Y (2008) Reduction of acetate accumulation in *Escherichia coli* cultures for increased recombinant protein production. Metab Eng 10:97–108
227. Morrow KJ (2009) Grappling with biologic manufacturing concerns. Genet Eng Biotechnol News 29(5):54–55

228. Choi JH, Lee SJ, Lee SJ, Lee SY (2003) Enhanced production of insulin-like growth factor I fusion protein in *Escherichia coli* by coexpression of the down-regulated genes identified by transcriptome profiling. Appl Environ Microbiol 69:4737–4742
229. Barnard GC, Henderson GE, Srinivasan S, Gerngross TU (2004) High level recombinantT7 RNA polymerase based amplification. Protein Expr Purif 38:264–271
230. Squires CH, Lucy P (2008) Vendor voice: a new paradigm for bacterial strain engineering. Bioprocess Int 6:22–27
231. Sreekrishana K, Nelles L, Potenz R, Cruze J, Mazzaferro P et al (1989) High level expression, purification, and characterization of recombinant human tumor necrosis factor synthesized in the methylotrophic yeast *Pichia pastoris*. Biochemistry 28:4117–4125
232. Werten MWT, van den Bosch TJ, Wind RD, Mooibroek H, De Wolf FA (1999) High yield secretion of recombinant gelatins by *Pichia pastoris*. Yeast 15:1087–1096
233. Clare JJ, Rayment FB, Ballantine SP, Sreekrishna K, Romanos MA (1991) High-level expression of tetanus toxin fragment C in *Pichia pastoris* strains containing multiple tandem integrations of the gene. Nat Biotechnol 9:455–460
234. Nevalainen KMH, Te'o VSJ, Bergquist PL (2005) Heterologous protein expression in filamentous fungi. Trends Biotechnol 23:468–474
235. Morrow KJ (2007) Strategic protein production. Genet Eng Biotechnol News 27:50–54
236. Gellison G, Janowicz ZA, Weydemann U, Melber K, Strasser AWM, Hollenberg CP (1992) High-level expression of foreign genes in *Hansenula polymorpha*. Biotechnol Adv 10:179–189
237. Meyer V (2008) Genetic engineering of filamentous fungi-progress, obstacles and future trends. Biotechnol Adv 26:177–185
238. Lubertozzi D, Keasling JD (2009) Developing *Aspergillus* as a host for heterologous expression. Biotechnol Adv 27:53–75
239. Van Hartinsveldt W, van Zeijl CM, Harteeld GM, Gouka RJ, Suykerbuyk M, Luiten RG et al (1993) Cloning, characterization and overexpression of the phytase-encoding gene (*phyA*) of *Aspergillus niger*. Gene 127:87–94
240. Ward PP, Piddlington CS, Cunningham GA, Zhou X, Wyatt RD, Conneely OM (1995) A system for production of commercial quantities of human lactoferrin: a broad spectrum natural antibiotic. Nat Biotechnol 13:498–503
241. Christensen T, Woeldike H, Boel E, Mortensen SB, Hjortshoej K, Thim L et al (1988) High level expression of recombinant genes in *Aspergillus oryzae*. Nat Biotechnol 6:1419–1422
242. Headon DR, Wyatt RD (1995) Human lactoferrin from *Aspergillus* spp. SIM News 45:113–117
243. Verdoes JC, Punt PJ, Burlingame R, Bartels J, van Dijk R, Slump E, Meens M, Joosten R, Emalfarb M (2007) A dedicated vector for efficient library construction and high throughput screening in the hyphal fungus *Chrysosporium lucknowense*. Ind Biotechnol 3:48–57
244. Andersen DC, Krummen L (2002) Recombinant protein expression for therapeutic applications. Curr Opin Biotechnol 13:117–123
245. Wrotnowski C (1998) Animal cell culture; novel systems for research and production. Genet Eng News 18(3):13–37
246. Griffin TJ, Seth G, Xie H, Bandhakavi S, Hu W-S (2007) Advancing mammalian cell culture engineering using genome-scale technologies. Trends Biotechnol 25:401–408
247. Decaria P, Smith A, Whitford W (2009) Many considerations in selecting bioproduction culture media. Bioprocess Int 7:44–51
248. Scott C, Montgomery SA, Rosin LJ (2007) Genetic engineering leads to microbial, animal cell, and transgenic expression systems. BIO Internat Convention:27–34
249. Morrow KJ (2007) Improving protein production processes. Genet Eng Biotechnol News 27:44–47
250. Ryll T (2008) Antibody production using mammalian cell culture—how high can we push productivity? SIM Annual Meeting Program & Abstract, San Diego, S146, p.101
251. Meyer HP, Biass J, Jungo C, Klein J, Wenger J, Mommers R (2008) An emerging star for therapeutic and catalytic protein production. Bioprocess Int 6:10–21
252. CocoMartin JM, Harmsen MM (2008) A review of therapeutic protein expression by mammalian cells. Bioprocess Int 6:28–33
253. Jarvis LM (2008) A technology bet. DSM's pharma product unit leverages its biotech strength to survive in a tough environment. Chem Eng News 86:30–31
254. Agathos SN (1991) Production scale insect cell culture. Biotechnol Adv 9:51–68
255. Luckow VA, Summers MD (1988) Trends in the development of baculovirus expression vectors. Nat Biotechnol 6:47–55
256. Miller LK (1988) Baculoviruses as gene expression vectors. Annu Rev Microbiol 42:177–199
257. Wilkinson BE, Cox M (1998) Baculovirus expression system: the production of proteins for diagnostic, human therapeutic or vaccine use. Genet Eng News 35(Nov):18

258. Maiorella B, Harano D (1988) Large scale insect cell culture for recombinant protein production. Nat Biotechnol 6:1406–1409
259. Morrow KJ Jr (2007) Improving protein production processes. Genet Eng News 27:50–54
260. Knight P (1991) Baculovirus vectors for making proteins in insect cells. ASM News 57:567–570
261. Falch E (1991) Industrial enzymes-developments in production and application. Biotechnol Adv 9:643–658
262. Cowan D (1996) Industrial enzyme technology. Trends Biotechnol 14:177–178
263. Vohra A, Satyanarayana T (2003) Phytases: microbial sources, production, purification, and potential biotechnological applications. Crit Rev Biotechnol 23:29–60
264. Vaishnav P, Demain AL (2009) Industrial biotechnology overview. In: Schaechter M, Lederberg J (eds) Encyclopedia of microbiology, 3rd edn. Elsevier, Oxford, p 335
265. Wackett LP (1997) Bacterial biocatalysis: stealing a page from nature's book. Nat Biotechnol 15:415–416
266. Palva I (1982) Molecular cloning of alpha-amylase gene from *Bacillus amyloliquefaciens* and its expression in *B. subtilis*. Gene 19:81–87
267. O'Neill GP, Kilburn DG, RAJ W, Miller RC (1986) Overproduction from a cellulose gene with a high guanosine-plus-cytosine content in *Escherichia coli*. Appl Environ Microbiol 52:737–743
268. Shoemaker S, Schweickart V, Ladner M, Gelfand D, Kwok S, Myambo K et al (1983) Molecular cloning of exo-cellobiohydrolase I derived from *Trichoderma reesei* strain L27. Nat Biotechnol 1:691–696
269. Van Brunt J (1986) Fungi: the perfect hosts? Biotechnology 4:1057–1062
270. Mondou F, Shareck F, Morosoli R, Kleupfel D (1986) Cloning of the xylanase gene of *Streptomyces lividans*. Gene 49:323–329
271. Van den Burg B, Vriend G, Veltman O, Venema G, Eijsink VGH (1998) Engineering an enzyme to resist boiling. Proc Natl Acad Sci USA 95:2056–2060
272. Ness JE, Welch M, Giver L, Bueno M, Cherry JR, Borchert TV et al (1999) DNA shuffling of subgenomic sequences of subtilisin. Nat Biotechnol 17:893–896
273. Jaeger KE, Reetz MT (2000) Directed evolution of enantioselective enzymes for organic chemistry. Curr Opin Chem Biol 4:68–73
274. Suenaga H, Mitsokua M, Ura Y, Watanabe T, Furukawa K (2001) Directed evolution of biphenyl dioxygenase: emergence of enhanced degradation capacity for benzene, toluene, and alkylbenzenes. J Bacteriol 183:5441–5444
275. Song JK, Rhee JS (2001) Enhancement of stability and activity of phospholipase A(1) in organic solvents by directed evolution. Biochim Biophys Acta 1547:370–378
276. Raillard S, Krebber A, Chen Y, Ness JE, Bermudez E, Trinidad R et al (2001) Novel enzyme activities and functional plasticity revealed by recombining highly homologous enzymes. Chem Biol 8:891–898
277. Kurtzman AL, Govindarajan S, Vahle K, Jones JT, Heinrichs V, Patten PA (2001) Advances in directed protein evolution by recursive genetic recombination: applications to therapeutic proteins. Curr Opin Biotechnol 12:361–370
278. Marshall SH (2002) DNA shuffling: induced molecular breeding to produce new generation long-lasting vaccines. Biotechnol Adv 20:229–238
279. Locher CP, Soong NW, Whalen RG, Punnonen J (2004) Development of novel vaccines using DNA shuffling and screening strategies. Curr Opin Mol Ther 6:34–39
280. Tobin MB, Gustafsson C, Huisman GW (2000) Directed evolution: the 'rational' basis for 'irrational' design. Curr Opin Struct Biol 10:421–427
281. Crameri A, Whitehorn A, Stemmer WPC (1996) Improved green fluorescent protein by molecular evolution using DNA shuffling. Nat Biotechnol 14:315–319

Chapter 2

Carotenoids Production: A Healthy and Profitable Industry

Carlos Barreiro and José-Luis Barredo

Abstract

Carotenoids relevance as natural pigments is mainly due to their uses as colorants, feed supplements, nutraceuticals and for medical, cosmetic, and biotechnological purposes. Since they have putative health beneficial effects, the demand and market of carotenoids are growing significantly. There is a diversity of natural and synthetic carotenoids, but only a few of them are commercially produced, including carotenes (β-carotene and lycopene) and xanthophylls (astaxanthin, canthaxanthin, lutein, zeaxanthin, and capsanthin). Some biotechnological processes for carotenoids production were established some years ago, but new strains and technologies are being developed nowadays for carotenoids widely in demand. This chapter shows a revision of the main carotenoids from a commercial point of view.

Key words Carotenoids, Carotene, Xanthophyll, Astaxanthin, Beta-carotene, Lutein, Lycopene, Zeaxanthin, Canthaxanthin

1 Introduction

Carotenoids are one of the most diverse classes of secondary metabolites, which define a subfamily of isoprenoids (also called "terpenoids"). This group includes tetraterpenoid organic pigments that mainly comprise eight isoprene units with a C40 skeleton. Carotenoids are lipophilic metabolites insoluble in water containing a chromophore that comprise a long polyene central chain of conjugated double bonds (maximal absorbance wavelength 400–500 nm), which is responsible for the characteristic yellow to orange and reddish colors of these compounds [1–3]. According to their oxygenation degree, these compounds can be chemically classified into carotenes and xanthophylls. On the one hand, carotenes (e.g., α-carotene, β-carotene, or lycopene) are strictly non-oxygenated molecules (hydrocarbons). On the other hand, xanthophylls (e.g., canthaxanthin, astaxanthin, zeaxanthin, or lutein) are oxygenated molecules (oxycarotenoids) with an epoxy, hydroxy, and/or oxo group [3, 4]. Nowadays, almost 80 years after the structural elucidation of

Carlos Barreiro and José-Luis Barredo (eds.), *Microbial Carotenoids: Methods and Protocols*, Methods in Molecular Biology, vol. 1852, https://doi.org/10.1007/978-1-4939-8742-9_2,

β-carotene by Kuhn and Karrer, more than 750 carotenoid types have been described from natural sources (over 250 from marine origin) [5–7]. About 50 carotenoids are components of the human nutrition and 20 of them have been traditionally detected in human blood and tissues [8–10].

Carotenoids play different natural roles, and these functions can be grouped in function of the photosynthetic or non-photosynthetic behavior of the carrier organism. Thus, their activities in photosynthesizing species are associated with the light-harvesting complexes, extending the range of light absorbed by the photosynthetic system. In those organisms, carotenoids also play a photoprotective role by quenching triplet-state chlorophyll molecules and scavenging singlet oxygen and other toxic oxygen species formed within the chloroplast. In example, zeaxanthin dissipates damaging excitation energy under stress conditions as light excess, which generates reactive oxygen species (ROS) like singlet oxygen and other free radicals. This particular ability as oxidative species quenchers justifies their presence in all non-photosynthetic organisms acting as (photo)protectors. Thus, carotenoids play a capital role as antioxidants [3, 8, 11–13].

2 Why Carotenoids Are Industrially Relevant Compounds?

The deep description of ROS-induced oxidative stress mechanisms, as well as the search for suitable strategies to decrease or avoid the oxidative stress, has focused a relevant number of medical research efforts. Although reactive oxygen species are involved in positive functions as regulation control (protein phosphorylation influence, intracellular and extracellular signalling processes, and gene expression interference), the unbalance between detoxification and ROS generation may lead to their accumulation resulting in abnormalities associated with chronic diseases (photosensitive disorders, immunity diseases, neurological disorders, cancer, diabetes, cardiovascular diseases). This fact is the core of the industrial relevance of carotenoids applied to human healthcare, since several studies linked oxidative stress and degenerative pathogenesis, as well as carotenoids acting, as palliative treatment of degenerative diseases such as Alzheimer or Parkinson [8, 14]. Additionally, carotenoids are generally recognized as nontoxic even though in high doses. Just canthaxanthin can cause a reversible retinopathy, and β-carotene is not recommended due to the increased risk of stomach and lung cancer for smokers [15]. Additionally, new trends in the lifestyle, the aim to achieve disease prevention bypassing the recommendations on dietary changes, or the increase of the health perception of the worldwide population have boosted the demand for nutritional supplements with health potential benefits (such as immunostimulants, proteins, amino acids, and vitamins). Currently, this market of supplements is also including carotenoids, which is likely to be increased over the next few years [16, 17]. Third human application area is cosmetics due to the beauty-enhancing effect since carotenoids block the effects of ROS and suppress cell disruption and damage, which naturally occurs in skin [18]. Finally, in nature different animal tissues are pigmented due to carotenoids (egg yolk, pink flesh of salmonids, flamingo's pink plumage, red exoskeleton of crustaceans), which should be kept similar in intensive farming and aquaculture, and then feed supplementation is another relevant use of carotenoids [14]. Nowadays, there are six main carotenoids industrially produced and involved in human health and feed supplementation: astaxanthin, β-carotene, canthaxanthin, lutein, lycopene, and zeaxanthin [14, 16].

3 Astaxanthin

Astaxanthin (3,3′-dihydroxy-β,β-carotene-4,4′-dione) is a red-orange pigment with a market value of $2500–7000/kg. Commercial astaxanthin is from either chemical synthesis or natural resources (microalgae or yeast fermentative production and crustacean by-products). However, 95% of the global market comes from synthetically obtained astaxanthin, although natural sources are being promoted due to the petrochemical origin of synthetic compounds. Three natural sources are a realistic alternative to synthetic astaxanthin: (1) bacteria such as *Paracoccus carotinifaciens*, (2) yeasts like *Xanthophyllomyces dendrorhous* (*Phaffia rhodozyma*), and (3) algae like *Haematococcus pluvialis* and also *Chlorella zofingiensis* or *Coelastrella striolata*. Nowadays, DSM (the Netherlands), BASF (France), and NHU (China) are the major manufacturers [3]. Astaxanthin is used as a feed additive in aquaculture (salmon and trout), poultry farming (chicken and quail), and egg production [4, 19, 20]. Besides, antioxidant properties have been demonstrated for astaxanthin in both experimental animals and clinical studies, which support its beneficial effects in cardiovascular, diabetes, carcinogenic, immune, and neurodegenerative diseases and as an antiaging and sun proofing agent, in addition to male fertility [3, 21–31].

4 β-Carotene

β-carotene is a red-orange carotenoid carrying β-rings at both ends, which acts as an intermediary molecule for the astaxanthin biosynthesis. This compound can be found in different fruits and vegetables such as carrots (65–210 μg/g fresh weight), pumpkin (0.6–148 μg/g fresh weight), mangos (4.45–32.67 μg/g fresh weight), and papaya (2.3–19.81 μg/g fresh weight). β-carotene is industrially produced due to its use in the feed, food, cosmetic, and pharmaceutical industries. Its antioxidant properties, relevant coloring traits, and its provitamin A activity [main dietary precursor for vertebrate animal species of vitamin A (retinol)] support its production and purification becoming the most important industrial carotenoid [32, 33]. Technical Services of the USDA (US Department of Agriculture) describes three β-carotene manufacturing methods: chemical synthesis, biological synthesis, and extraction from plant sources (https://www.ams.usda.gov/sites/default/files/media/Beta%20Carotene%20Synthetic%20Supplemental%20TR.pdf). Chemical synthesis was initiated by Roche (1954) and BASF (1960) using different condensation methods but starting from the same precursor (β-ionone). Biotechnological production of β-carotene is carried out by different microorganisms such as (1) the microalgae *Dunaliella salina*, which can accumulate up to 10% of β-carotene on dry cell weight at certain environmental conditions, or *Coccomyxa onubensis* and *Dunaliella salina*; (2) the fungus *Blakeslea trispora*; (3) and the yeasts *Sporobolomyces roseus* or *Rhodotorula glutinis* and genetically engineered strains of *Saccharomyces cerevisiae* [14, 34–37]. Vitamin A is essential for growth and cell and tissue differentiation; that is why the provitamin A carotenoid β-carotene is essential during pregnancy and throughout the breastfeeding period [38]. Besides, β-carotene prevents UV-induced erythema [39]; collaborates, in addition to other dietary supplements (e.g., vitamin C and vitamin E), to delay progression of age-related macular degeneration [40]; helps in the prevention of acute and chronic coronary syndromes [14, 41]; produces sustained remissions of oral leucoplakia [42]; and improves respiratory and pulmonary function [43].

5 Canthaxanthin

Canthaxanthin (4,4′-diketo-β-carotene) is an orange-red ketocarotenoid (xanthophyll) commonly recognized as an antioxidant. Its initial extraction came from an edible mushroom (*Cantharellus cinnabarinus*) in 1950 [44]. Canthaxanthin has been detected in several organisms as plants, trout, birds (flamingo), crayfish, fungi (*Aspergillus carbonarius*), algae (*Coelastrella striolata*, *Chlorella vulgaris*, *Chlorella zofingiensis*, and *Dictyococcus cinnabarinus*), and bacteria (*Bradyrhizobium* sp., *Dietzia* sp., *Gordonia jacobaea*, *Micrococcus roseus*, and *Rhodococcus maris*) [37, 45]. Nowadays, it can be obtained through microbial fermentation of algae (*Haematococcus pluvialis*, *Chlorella zofingiensis*, *Dactylococcus dissociates*) or bacteria (*Paracoccus* sp., *Dietzia* sp.) and by chemical synthesis [45, 46]. It is relevant in medical, chemical, cosmetic, food, and feed industries [45, 47]. DSM website describes the broiler and egg yolk pigmentation as the main commercial canthaxanthin used, in addition to colorant in beverages, food, and pharma (https://www.dsm.com/markets/foodandbeverages/en_US/products/carotenoids.html). Canthaxanthin has been reported as a growth inhibitor of the cancer cell lines and apoptosis inducer of tumor cell lines [16]. Besides, its immunomodulatory activity has been also studied [46].

6 Lutein

Lutein is a xanthophyll traditionally found in high quantities in green leafy vegetables such as kale and spinach, where the green chlorophyll hides its yellowish color. Its presence in human tissues comes totally from diet. Thus, in Spain its uptake is mainly achieved from spinach (775.0 μg/g fresh weight), lettuce (0.73–26.35 μg/g fresh weight), oranges, eggs (288.0 μg/100 g of dry matter), broccoli (6.8–12.8 μg/g fresh weight), peas, potatoes, tangerines, peppers, and leeks [17, 48, 49]. Marigold flowers' petals (*Tagetes erecta*) are traditionally used as natural lutein sources, since their dried flowers contain 0.1%–0.2% dry matter of carotenoids (80% lutein diesters). It is grown for economical purposes in Ecuador, Peru, Mexico, Spain, China, or India. Microalgae (*Scenedesmus obliquus*, *Chlorella minutissima*, and *Chlorella vulgaris*) are also suitable candidates due to their growth rate, lutein content, little handle requirements, and physical space. However, downstream extraction and separation, which means 50%–80% of the total production cost, is the main industrial drawback [49–52]. Industrial and medical interest of lutein comes from its use as a food colorant, in addition to human eye health since it decreases the risk of cataract and age-related macular degeneration [51, 53]. Besides, lutein and zeaxanthin are relevant for the reduction or prevention of cardiovascular diseases, stroke, neurodegenerative disorders, and several skin inflammatory responses and decrease and possibly diminish different types of cancer (e.g., breast and lung) [48, 54]. Few years ago, the European Food Safety Authority (EFSA) concluded as weak the evidence submitted for the approval of a health claim for lutein in concordance with the decision of the US FDA almost a decade before. In contrast, the Brazilian Health Surveillance Agency (ANVISA) and the Natural Health Product unit of Health Canada approved a lutein claim (eye health-maintaining antioxidant) in the same period of time. So, as Nwachukwu and coworkers (2016) said: "Even in the absence of formal health claims, it is difficult to completely ignore currently available data pointing to a link between lutein/zeaxanthin and the promotion of vision" [49].

7 Lycopene

Lycopene is a non-provitamin A natural fat-soluble red carotenoid synthesized by some plants and microorganisms, which is found in human plasma as a mixture of two isoforms [(Z)-lycopene:(all-E)-lycopene; approximated ratio 1:1]. Microbial production of lycopene by means of the fungus *Blakeslea trispora*, the yeasts *Rhodotorula* and *Yarrowia lipolytica*, the bacteria *Dietzia natronolimnaea*, or the microalga *Dunaliella salina* in fermentative processes has provided a metabolic engineering proof of concept, which suggests a future replacement of plants as the primary source of this compound. However, up to date, due to different factors, including concentration and toxicity of the enzyme inhibitor used in the production processes or type and time of fermentation, lycopene is only available in natural form. In the United States, the primary dietary lycopene source is the processed tomato (8.8–42.0 μg/g fresh weight), even though it is also present in pink grapefruit (33.6 μg/g fresh weight), watermelon (23.0–72.0 μg/g fresh weight), papaya (20.0–53.0 μg/g fresh weight), pink guava (54.0 μg/g fresh weight), and other fruits [14, 55–59]. Although new clinical studies are needed, cell culture and animal tests, in addition to epidemiological studies, suggested long time ago that lycopene and lycopene-containing food consumption may decrease cancer (prostatic hyperplasia and prostate cancer), cardiovascular disease risk (atherosclerosis and acute and chronic coronary syndromes), male fertility, or osteoporosis [14, 55, 59–63].

8 Zeaxanthin

Zeaxanthin (β,β-carotene-3,3′-diol) is a yellow xanthophyll alcohol traditionally described as important for vision [3, 64]. Up to date, different fruits and vegetables have been described as enriched in zeaxanthin in nature such as wolfberries, spinach (15.1–3.3 μg/g fresh weight), pepper (*Capsicum annuum*), yellow corn (5.28 μg/g fresh weight), butternut (2.80 μg/g fresh weight), and collards (2.66 μg/g fresh weight) [49, 56, 65]. Besides, it is naturally synthesized by many microorganisms (bacteria, yeast, filamentous fungi, and microalgae) [64]. Zhang and co-workers described in 2017 [64] three zeaxanthin production methods: (1) chemically synthesis, which presents poor biological activity [66]; (2) extraction from vegetables and fruits (corn, mangos, oranges, melons, peppers), which is limited by the extraction rate, waste release, and environmental concerns [67]; and (3) microbial production that is under evolution by means of natural producers (e.g., *Mesoflavibacter zeaxanthinifaciens*, *Muricauda* sp., *Siansivirga zeaxanthinifaciens*, *Sphingomonas natatoria*) and genetically modified strains (e.g.,

Synechocystis sp., *Mucor circinelloides*, *Escherichia coli*) [64, 68]. Zeaxanthin is highly valued in poultry and fish production since it enhances the pigmentation providing a true color of flesh and eggs [65]. In humans, zeaxanthin and lutein are specifically accumulated in the retinal macular region where zeaxanthin is located at the macula, whereas lutein is homogeneously distributed throughout the retina [65]. This fact is the link between age-related macular degeneration and putative beneficial properties of zeaxanthin, which in concordance with the incident blue light absorption by means of lutein (40%–90%) protecting the retina from light-related damage [69].

9 Carotenoids Market

Nowadays, the bulk of industry-produced carotenoids is chemically manufactured, and a lower amount is obtained from microorganisms (yeast, bacteria, and fungi), algae, and plant extraction. Vilchez and co-workers (2011) [14], based on the Business Communications report (March 2008), brought the global carotenoids market at 766 US$ million with clear expectations of rising to 919 US$ million in 2015. Later, Mordor Intelligence on December 2017 released the report "Feed Carotenoids Market—By Type, Animal Type and Geography-Trends and Forecasts (2017–2022)" where they estimate the global feed carotenoids market in 2013 at 1,742.5 US$ million (www.mordorintelligence.com). In turn, BBC Research (www.bccresearch.com) defined the carotenoid global market in 2014 at 1.5 US$ billion. Curiously, Sathasivam and co-workers (2018) show the global carotenoid market at 1.24 US$ billion in 2016 reaching 1.53 US$ billion by 2021, also based on BBC Research data (www.bccresearch.com). Grand View Research expects to reach the level of 1.74 US$ billion by 2025 (www.grandviewresearch.com). Even though the global market estimation seems difficult, it is clear the economical relevance of carotenoids due to their application in food, feed, pharma, and cosmetics.

Almost 60% of the total market value is shared by astaxanthin, β-carotene, and lutein [1]. Vilchez and co-workers (2011) defined the β-carotene market volume at 247 US$ million in 2007 expecting to reach 285 US$ million in 2015 [14]. In contrast, Cardoso and co-workers (2017) described β-carotene as the most profitable carotenoid due to its highest value, which reached 233 US$ million in 2010, and it is expected to be increased up to 309 US$ million by 2018 [37]. Same authors presented astaxanthin as the third carotenoid with a global market of 225 US$ million, which can reach 253 US$ million by 2018 [37], whereas Vilchez and co-workers (2011) [14] showed astaxanthin market volume just in aquaculture at 260 US$ million in 2009. However, PRNewswire (www.prnewswire.com) on August 2015, based on BCC Research study reviews, established the astaxanthin market at 369 US$ million in 2014 expecting to reach at 423 US$ million in 2019. Once again, this dancing around the figures describes the relevance of the business but also shows the difficulty of obtaining a real view of the global market.

10 Conclusions and Future Prospects

Carotenes (β-carotene and lycopene) and xanthophylls (the oxygenated carotenoids: astaxanthin, canthaxanthin, lutein, zeaxanthin, and capsanthin) play an important role as colorants, feed supplements, and nutraceuticals in the food, medical, and cosmetic industries. Between carotenes, β-carotene is by far the most demanded in the market, and among xanthophylls, astaxanthin has a large importance in aquaculture for salmonids and crustacean pigmentation and lutein to prevent macular degeneration.

Biotechnological processes are an alternative to the chemical synthesis, which at present is the major source in the market. Non-GMO strains are preferred for the biotechnological production of carotenoids due to the resistance of the consumers to incorporate GMO foods to their diet. *X. dendrorhous* and *H. pluvialis* are the microorganisms of choice for the industrial production of astaxanthin, as well as *Blakeslea trispora* is chosen for β-carotene and lycopene manufacture. New strains and technologies are being developed nowadays for carotenoids widely in demand, including *Paracoccus carotinifaciens* for the production of astaxanthin and other xanthophylls and GMO strains of *Yarrowia lipolytica* for lycopene production. Besides, an important effort is being focused on new producer strain description, which is allowing new carotenoid detection.

Even though market estimations vary among different data suppliers, it is clear the relevance of these compounds in addition to the promising trends of microbial production titers and consumption increase. Such increase will be expanded if FDA and EFSA institutions accept future health claims supporting the positive effects assigned to these colorful compounds.

Acknowledgments

We would like to thank for their collaboration to all the members of INBIOTEC and BIONICE. Special thanks to the European Union programs: i) ERA-IB 7th Joint Call [ProWood project (ERA-IB-16-040) through the APCIN call of the Spanish Ministry of Economy, Industry and Competitiveness and the State Research

Agency (AEI) (Project ID: PCIN-2016-081)] and ii) ERA CoBioTech Joint Call [Syntheroids project (ID: 48) through the Spanish Ministry of Economy, Industry and Competitiveness (MINECO), State Research Agency (AEI) and Centre for the Development of Industrial Technology (CDTI)].

References

1. Saini RK, Keum YS (2017) Progress in microbial carotenoids production. Indian J Microbiol 57:129–130. https://doi.org/10.1007/s12088-016-0637-x
2. Lichtenthaler HK, Buschmann C (2001) Chlorophylls and carotenoids: measurement and characterization by UV-VIS spectroscopy. Curr Protocol Food Anal Chem. https://doi.org/10.1002/0471142913.faf0403s01
3. Barredo J, García-Estrada C, Kosalkova K, Barreiro C (2017) Biosynthesis of astaxanthin as a main carotenoid in the heterobasidiomycetous yeast *Xanthophyllomyces dendrorhous*. J Fungi 3:44. https://doi.org/10.3390/jof3030044
4. Bhosale P, Bernstein PS (2005) Microbial xanthophylls. Appl Microbiol Biotechnol 68:445–455. https://doi.org/10.1007/s00253-005-0032-8
5. Britton G, Liaaen-Jensen S, Pfander H (2004) Carotenoids. Springer Basel AG, Basel. https://doi.org/10.1007/978-3-0348-7836-4
6. Feltl L, Pacakova V, Stulik K, Volka K (2005) Reliability of carotenoid analyses: a review. Curr Anal Chem 1:93–102. https://doi.org/10.2174/1573411052948424
7. Galasso C, Corinaldesi C, Sansone C (2017) Carotenoids from marine organisms: biological functions and industrial applications. Antioxidants (Basel, Switzerland) 6(96). https://doi.org/10.3390/antiox6040096
8. Fiedor J, Burda K (2014) Potential role of carotenoids as antioxidants in human health and disease. Nutrients 6:466–488. https://doi.org/10.3390/nu6020466
9. Khachik F (2006) Distribution and metabolism of dietary carotenoids in humans as a criterion for development of nutritional supplements. Pure Appl Chem 78:1551–1557. https://doi.org/10.1351/pac200678081551
10. Parker RS (1989) Carotenoids in human blood and tissues. J Nutr 119:101–104
11. Young AJ (1991) The photoprotective role of carotenoids in higher plants. Physiol Plant 83:702–708. https://doi.org/10.1111/j.1399-3054.1991.tb02490.x
12. Blomhoff R, Blomhoff HK (2006) Overview of retinoid metabolism and function. J Neurobiol 66:606–630. https://doi.org/10.1002/neu.20242
13. Walter MH, Strack D (2011) Carotenoids and their cleavage products: biosynthesis and functions. Nat Prod Rep 28:663–692. https://doi.org/10.1039/c0np00036a
14. Vílchez C, Forján E, Cuaresma M et al (2011) Marine carotenoids: biological functions and commercial applications. Mar Drugs 9:319–333. https://doi.org/10.3390/md9030319
15. Hammond BR, Renzi LM (2013) Carotenoids. Adv Nutr 4:474–476. https://doi.org/10.3945/an.113.004028.part
16. Sathasivam R, Ki J-S (2018) A review of the biological activities of microalgal carotenoids and their potential use in healthcare and cosmetic industries. Mar Drugs 16(1):E26. https://doi.org/10.3390/md16010026
17. Granado F, Olmedilla B, Blanco I (2003) Nutritional and clinical relevance of lutein in human health. Br J Nutr 90:487. https://doi.org/10.1079/BJN2003927
18. Masaki H (2010) Role of antioxidants in the skin: anti-aging effects. J Dermatol Sci 58:85–90. https://doi.org/10.1016/j.jdermsci.2010.03.003
19. Breithaupt DR (2008) Xanthophylls in poultry feeding. In: Britton G, Liaaen-Jensen S, Pfander H (eds) Carotenoids, Vol. 4 Natural functions. Birkhäuser Basel, Basel, pp 255–264
20. Higuera-Ciapara I, Félix-Valenzuela L, Goycoolea FM (2006) Astaxanthin: a review of its chemistry and applications. Crit Rev Food Sci Nutr 46:185–196. https://doi.org/10.1080/10408690590957188
21. Schweiggert RM, Carle R (2016) Carotenoid production by bacteria, microalgae, and fungi. In: Kaczor A, Baranska M (eds) Carotenoids: Nutrition Analysis Technology. John Wiley & Sons, Ltd, Chichester, pp 217–240
22. Wang X, Willén R, Wadström T (2000) Astaxanthin-rich algal meal and vitamin C inhibit *Helicobacter pylori* infection in BALB/

cA mice. Antimicrob Agents Chemother 44:2452–2457

23. Park JS, Chyun JH, Kim YK et al (2010) Astaxanthin decreased oxidative stress and inflammation and enhanced immune response in humans. Nutr Metab (Lond) 7:18. https://doi.org/10.1186/1743-7075-7-18
24. Yasui Y, Hosokawa M, Mikami N et al (2011) Dietary astaxanthin inhibits colitis and colitis-associated colon carcinogenesis in mice via modulation of the inflammatory cytokines. Chem Biol Interact 193:79–87. https://doi.org/10.1016/j.cbi.2011.05.006
25. Fassett RG, Coombes JS (2009) Astaxanthin, oxidative stress, inflammation and cardiovascular disease. Future Cardiol 5:333–342. https://doi.org/10.2217/fca.09.19
26. Li J, Zhu D, Niu J et al (2011) An economic assessment of astaxanthin production by large scale cultivation of *Haematococcus pluvialis*. Biotechnol Adv 29:568–574. https://doi.org/10.1016/j.biotechadv.2011.04.001
27. Comhaire FH, El Garem Y, Mahmoud A et al (2005) Combined conventional/antioxidant "Astaxanthin" treatment for male infertility: a double blind, randomized trial. Asian J Androl 7:257–262. https://doi.org/10.1111/j.1745-7262.2005.00047.x
28. Hussein G, Goto H, Oda S et al (2006) Antihypertensive potential and mechanism of action of astaxanthin: III. Antioxidant and histopathological effects in spontaneously hypertensive rats. Biol Pharm Bull 29:684–688
29. Tominaga K, Hongo N, Karato M, Yamashita E (2012) Cosmetic benefits of astaxanthin on humans subjects. Acta Biochim 59:43–47
30. Naguib YM (2000) Antioxidant activities of astaxanthin and related carotenoids. J Agric Food Chem 48:1150–1154
31. Hussein G, Sankawa U, Goto H et al (2006) Astaxanthin, a carotenoid with potential in human health and nutrition. J Nat Prod 69:443–449. https://doi.org/10.1021/np050354+
32. Alvarez V, Rodríguez-Sáiz M, de la Fuente JL et al (2006) The crtS gene of *Xanthophyllomyces dendrorhous* encodes a novel cytochrome-P450 hydroxylase involved in the conversion of beta-carotene into astaxanthin and other xanthophylls. Fungal Genet Biol 43:261–272. https://doi.org/10.1016/j.fgb.2005.12.004
33. Green AS, Fascetti AJ (2016) Meeting the vitamin A requirement: the efficacy and importance of β-carotene in animal species. Sci World J 2016:7393620. https://doi.org/10.1155/2016/7393620
34. Capa-Robles W, Paniagua-Michel J, Soto JO (2009) The biosynthesis and accumulation of beta-carotene in *Dunaliella salina* proceed via the glyceraldehyde 3-phosphate/pyruvate pathway. Nat Prod Res 23:1021–1028. https://doi.org/10.1080/14786410802689689
35. da Costa Cardoso LA, Kanno KYF, Karp SG (2017) Microbial production of carotenoids: a review. Afr J Biotechnol 16:139–146. https://doi.org/10.5897/AJB2016.15763
36. Verwaal R, Wang J, Meijnen J-P et al (2007) High-level production of beta-carotene in *Saccharomyces cerevisiae* by successive transformation with carotenogenic genes from *Xanthophyllomyces dendrorhous*. Appl Environ Microbiol 73:4342–4350. https://doi.org/10.1128/AEM.02759-06
37. Cardoso LAC, Karp SG, Vendruscolo F et al (2017) Biotechnological production of carotenoids and their applications in food and pharmaceutical products. In: Nikolic G (ed) Carotenoids. InTech, Rijeka, Croatia, pp 126–140
38. Strobel M, Tinz J, Biesalski H-K (2007) The importance of beta-carotene as a source of vitamin A with special regard to pregnant and breastfeeding women. Eur J Nutr 46(Suppl 1):I1–I20. https://doi.org/10.1007/s00394-007-1001-z
39. Schagen SK, Zampeli VA, Makrantonaki E, Zouboulis CC (2012) Discovering the link between nutrition and skin aging. Dermatoendocrinology 4:298–307. https://doi.org/10.4161/derm.22876
40. Olson JH, Erie JC, Bakri SJ (2011) Nutritional supplementation and age-related macular degeneration. Semin Ophthalmol 26:131–136. https://doi.org/10.3109/08820538.2011.577131
41. Kritchevsky SB (1999) beta-Carotene, carotenoids and the prevention of coronary heart disease. J Nutr 129:5–8
42. Garewal HS, Katz RV, Meyskens F et al (1999) Beta-carotene produces sustained remissions in patients with oral leukoplakia: results of a multicenter prospective trial. Arch Otolaryngol Head Neck Surg 125:1305–1310
43. Kelly Y, Sacker A, Marmot M (2003) Nutrition and respiratory health in adults: findings from the health survey for Scotland. Eur Respir J 21:664–671
44. Haxo F (1950) Carotenoids of the Mushroom *Cantharellus cinnabarinus*. Bot Gaz 112:228–232
45. Gharibzahedi SMT, Razavi SH, Mousavi SM (2013) Microbial canthaxanthin: perspectives on biochemistry and biotechnological production. Eng Life Sci 13:408–417. https://doi.org/10.1002/elsc.201200153
46. Esatbeyoglu T, Rimbach G (2017) Canthaxanthin: from molecule to function.

Mol Nutr Food Res 61:1–49. https://doi.org/10.1002/mnfr.201600469
47. Brizio P, Benedetto A, Righetti M et al (2013) Astaxanthin and canthaxanthin (xanthophyll) as supplements in rainbow trout diet: in vivo assessment of residual levels and contributions to human health. J Agric Food Chem 61:10954–10959. https://doi.org/10.1021/jf4012664
48. Zaheer K (2017) Hen egg carotenoids (lutein and zeaxanthin) and nutritional impacts on human health: a review. CyTA—J Food 15:474–487. https://doi.org/10.1080/19476337.2016.1266033
49. Nwachukwu ID, Udenigwe CC, Aluko RE (2016) Lutein and zeaxanthin: production technology, bioavailability, mechanisms of action, visual function, and health claim status. Trends Food Sci Technol 49:74–84. https://doi.org/10.1016/j.tifs.2015.12.005
50. Gong M, Bassi A (2016) Carotenoids from microalgae: a review of recent developments. Biotechnol Adv 34:1396–1412. https://doi.org/10.1016/j.biotechadv.2016.10.005
51. Gong M, Wang Y, Bassi A (2017) Process analysis and modeling of a single-step lutein extraction method for wet microalgae. Appl Microbiol Biotechnol 101:8089–8099. https://doi.org/10.1007/s00253-017-8496-x
52. Šivel M, Kráčmar S, Fišera M et al (2014) Lutein content in marigold flower (*Tagetes erecta* L.) concentrates used for production of food supplements. Czech J Food Sci 32:521–525
53. Manayi A, Abdollahi M, Raman T et al (2016) Lutein and cataract: from bench to bedside. Crit Rev Biotechnol 36:829–839. https://doi.org/10.3109/07388551.2015.1049510
54. Milani A, Basirnejad M, Shahbazi S, Bolhassani A (2017) Carotenoids: biochemistry, pharmacology and treatment. Br J Pharmacol 174:1290–1324. https://doi.org/10.1111/bph.13625
55. Story EN, Kopec RE, Schwartz SJ, Harris GK (2010) An update on the health effects of tomato lycopene. Annu Rev Food Sci Technol 1:189–210. https://doi.org/10.1146/annurev.food.102308.124120
56. Khoo HE, Prasad KN, Kong KW et al (2011) Carotenoids and their isomers: color pigments in fruits and vegetables. Molecules 16:1710–1738. https://doi.org/10.3390/molecules16021710
57. Ma T, Deng Z, Liu T (2016) Microbial production strategies and applications of lycopene and other terpenoids. World J Microbiol Biotechnol 32:15. https://doi.org/10.1007/s11274-015-1975-2
58. Hernández-Almanza A, Montañez J, Martínez G et al (2016) Lycopene: progress in microbial production. Trends Food Sci Technol 56:142–148. https://doi.org/10.1016/j.tifs.2016.08.013
59. Rao AV, Rao LG (2007) Carotenoids and human health. Pharmacol Res 55:207–216. https://doi.org/10.1016/j.phrs.2007.01.012
60. Block G, Patterson B, Subar A (1992) Fruit, vegetables, and cancer prevention: a review of the epidemiological evidence. Nutr Cancer 18:1–29. https://doi.org/10.1080/01635589209514201
61. Schwarz S, Obermüller-Jevic UC, Hellmis E et al (2008) Lycopene inhibits disease progression in patients with benign prostate hyperplasia. J Nutr 138:49–53
62. Sesso HD, Buring JE, Norkus EP, Gaziano JM (2005) Plasma lycopene, other carotenoids, and retinol and the risk of cardiovascular disease in men. Am J Clin Nutr 81:990–997
63. Durairajanayagam D, Agarwal A, Ong C, Prashast P (2014) Lycopene and male infertility. Asian J Androl 16:420–425. https://doi.org/10.4103/1008-682X.126384
64. Zhang Y, Liu Z, Sun J et al (2017) Biotechnological production of zeaxanthin by microorganisms. Trends Food Sci Technol 71:225–234. https://doi.org/10.1016/j.tifs.2017.11.006
65. Sajilata MG, Singhal RS, Kamat MY (2008) The carotenoid pigment zeaxanthin: a review. Compr Rev Food Sci Food Saf 7:29–49. https://doi.org/10.1111/j.1541-4337.2007.00028.x
66. Cataldo VF, López J, Cárcamo M, Agosin E (2016) Chemical vs. biotechnological synthesis of C13-apocarotenoids: current methods, applications and perspectives. Appl Microbiol Biotechnol 100:5703–5718. https://doi.org/10.1007/s00253-016-7583-8
67. Liau B-C, Hong S-E, Chang L-P et al (2011) Separation of sight-protecting zeaxanthin from Nannochloropsis oculata by using supercritical fluids extraction coupled with elution chromatography. Sep Purif Technol 78:1–8. https://doi.org/10.1016/J.SEPPUR.2011.01.008
68. Li X-R, Tian G-Q, Shen H-J, Liu J-Z (2015) Metabolic engineering of *Escherichia coli* to produce zeaxanthin. J Ind Microbiol Biotechnol 42:627–636. https://doi.org/10.1007/s10295-014-1565-6
69. Mares J (2016) Lutein and zeaxanthin isomers in eye health and disease. Annu Rev Nutr 36:571–602. https://doi.org/10.1146/annurev-nutr-071715-051110

Chapter 3

Carotenoids: From Plants to Food and Feed Industries

Panagiota Langi, Sotirios Kiokias, Theodoros Varzakas, and Charalampos Proestos

Abstract

In this review, carotenoids from plants are described, and their natural existence is addressed. Carotenoids are 40-carbon isoprenoid molecules that produce the red, yellow, and orange pigmentation found in nature. Various plants, microalgae, bacteria, and fungi are natural sources of carotenoids and are presented in detail. The chemistry of carotenoids and their classification is also described along with the effect of carotenoids on human health which is explained with focus on lutein—zeaxanthin, astaxanthin, canthaxanthin, capsanthin, and lycopene. Clinical studies suggest that carotenoid consumption is associated with lower risk of cardiovascular disease, cancer, and eye disease. Finally, another issue discussed is the role of carotenoids in animals and their feed with focus on birds, fish and crustaceans, livestock, and poultry.

Key words Carotenoids, Plants, Food, Feed industry

1 Introduction

Carotenoids are natural pigments which are found in plants, algae, fungi, birds, and fish flesh cuticle of crustaceans or insects. They are referred to as pigment, because of their characteristic colors that range in the yellow to red spectrum [1]. For instance, the yellowish-orange color of carrots is due to one of most familiar carotenoids and vitamin A precursor, β-carotene, from which the name of the whole class of these naturally occurring pigments come from [2].

Carotenoids can be classified into two groups according to their function: xanthophylls, including lutein and zeaxanthin, and carotenes, such as α-carotene, β-carotene, and lycopene [3]. They cannot be synthesized in vivo by humans or animals, and they are consumed only through diet.

Carotenoid pigments are a group of bioactive compounds that are of interest to the food scientists, nutritionists, and food industries due to their anti-inflammatory and antioxidant effects. They are recognized as playing an important role in the prevention of

Carlos Barreiro and José-Luis Barredo (eds.), *Microbial Carotenoids: Methods and Protocols*, Methods in Molecular Biology, vol. 1852, https://doi.org/10.1007/978-1-4939-8742-9_3, © Springer Science+Business Media, LLC, part of Springer Nature 2018

human diseases and maintaining good health [1, 2]. Their antioxidant property mediates the harmful effects of free radicals and hence can potentially protect humans from compromised immune response, premature aging, certain cancers, cardiovascular diseases, and/or arthritis. Fruits, juices, and vegetables are the major dietary sources of bioactive compounds such as carotenoids for the human diet, with yellow-orange vegetables and fruits generally rich in β-carotene and α-carotene; orange fruits (such as mandarin) supplying α-cryptoxanthin; and dark green vegetables providing lutein and also tomatoes and products providing lycopene (Table 1). In case of lutein and zeaxanthin, they are also present in high concentrations in egg yolks [2]. More than 600 carotenoids have so far been identified in nature [4]. However, in a typical human diet, only about 40 are present.

Table 1
Examples of major contributors of carotenoids in North American diet [2]

Carotenoid	Food source	Amount (IU, International Units)
β-Carotene	Apricot, dried	17,600
	Carrots, cooked	9771
	Spinach, cooked	5300
	Green collard	5400
	Broccoli, cooked	1300
	Tomato, raw	520
α-Carotene	Carrots, cooked	3723
Lycopene	Tomatoes, raw	3100
	Tomato juice	10,000
	Tomato paste	36,500
	Tomato ketchup	12,390
	Tomato sauce	13,060
β-Cryptoxanthin	Tangerine	1060
	Papaya	470
Lutein	Spinach, cooked	12,475
	Green collard	16,300
	Beet, green	7700
	Broccoli, cooked	1839
	Green peas, cooked	1690

They also play an important role in animal nutrition. All animals need carotenoids in order to have an improved performance, a better product color, and a superior consistency of coloration. For this reason, carotenoids play an important role in animal feed which can be found in nature, or they can be provided to them with supplements [5, 6].

2 Carotenoids

2.1 Natural Existence of Carotenoids

Although they seem to be hidden, carotenoids are one of the most widespread and ubiquitous lipid-soluble pigments in nature. They are present in leaves, fruits, vegetables, flowers, roots, and seeds and are mostly concentrated in fatty tissues such as internal fat and egg yolk in terrestrial animals, whereas in fish they are found either in fat or linked to the flesh. There is a particularly large variety of carotenoid structures in algae and bacteria, which can synthesize a wide range of them. Various plants, microalgae, bacteria, and fungi are natural sources of carotenoids.

2.1.1 Plants

Carotenoids fulfill many processes that are necessary for normal growth and development in plants, but they are also responsible for the breathtaking variety of yellow to red colors we see in leaves and flowers as well as to provide several aromas in plants. In many flowers, the presence of carotenoids is fully expressed as they are found accumulated in chromoplasts in the absence of chlorophyll. Carotenoids found in terrestrial plants mainly correspond to red and yellow xanthophylls such as lutein, zeaxanthin, capsanthin, violaxanthin, or neoxanthin, all of which belong to the xanthophyll cycle and are involved in protection of photosynthesis. The marigold flower is a good example of the natural occurrence of lutein yellow [7].

Content and types of carotenoids in plants depend on several pre- and pro-harvesting factors, genotype, ripening time, cultivation method and climatic conditions, and processing. Different parts of the same plant may also contain different types and amount of carotenoids. For instance, peel of the fruits is generally richer in carotenoids compared to pulp [3].

2.1.2 Algae

On the other hand, microalgae are a source of useful products such as carotenoid pigments, polyunsaturated fatty acids, vitamins, lipids, and proteins. Algae are simple unicellular organisms (microalgae) alone or grouped together in colonies or found relatively complex pluricellular tissues (macroalgae). These organisms can be classified into different phyla according to their dominant photosynthetic pigment and include green algae (Chlorophyta), brown algae (Phaeophyta), and red algae (Rhodophyta) [7]. Some of the specific advantages that microalgae cultivation offers, compared to traditional plant-based sources, include a faster cultivation, processing, and harvesting cycle and the ability to be cultured on waste materials.

The antioxidants astaxanthin, β-carotene, lutein, lycopene, and canthaxanthin, found in microalgae, are the major carotenoids of commercial value. In spite of the observed advantages, the large-scale and cost-effective manufacture of the carotenoids from algae is currently quite challenging both as far as production and downstream extraction and purnication are concerned. An integrated bioprocessing approach using microalgae thus needs to consider both the upstream production of microalgae and the downstream harvesting and extraction of carotenoids. The existence of rigid cell walls in many algal species poses difficulties as this prevents full recovery of bioactive compounds. This can, therefore, slow down the overall bioprocess significantly [8].

2.1.3 Bacteria

Several bacteria have the ability to produce carotenoids which are accumulated in chromatophores and are linked to the cytoplasmic membrane. Practically only *Halobacterium* contain carotenoids, in the form of β-carotene, giving a bright red cell suspension [7]. *Paracoccus carotinifaciens*, isolated from soil, is Gram-negative, aerobic, orange-pigmented, rod-shaped, motile by peritrichous flagella, and astaxanthin-producing. This organism produces carotenoids, mainly astaxanthin, and does not produce bacteriochlorophyll. The ubiquinone system is Q-10. Analysis of the 16S rRNA sequence of *Paracoccus carotinifaciens* showed it to belong to the alpha-3 subclass of the *Proteobacteria*, forming a cluster with the species of the genus *Paracoccus*.

2.1.4 Fungi

Compared to other groups of organisms, several fungal groups are able to produce and accumulate remarkably high levels of carotenoids intracellularly, a property that certain microalgae also have. The mushroom *Cantharellus cinnabarinus* is one of the best examples of occurrence of orange-red oxycarotenoid canthaxanthin in nature [7]. Compared to many other groups of organisms such as bacteria, plants, and animals, several fungal phyla including species within *Zygomycetes*, *Ascomycetes*, *Basidiomycetes*, and the asexual *Deuteromycetes* produce high levels of carotenoids [9].

2.2 Chemistry of Carotenoids and Classification

Carotenoids were named by their discoverers for some special property or for their source, for example, carotene (from carrots), cryptoxanthin (hidden pigment), and zeaxanthin (from maize, *Zea mays*). Carotenes and xanthophylls are the major carotenoid subgroups (oxycarotenoids) [7].

Carotenoids are lipophilic compounds, and therefore they are hydrophobic molecules with very low solubility in water functioning in hydrophobic areas of the cell. Most carotenoids share C_{40} ($C_{40}H_{56}$) tetraterpenoid pigments, and they consist of eight isoprenoid units joined together in a specific way that the organization of isoprenoid units is reserved at the center of the molecule. This results in a parent $C_{40}H_{56}$ hydrocarbon skeleton conjugated by alternating double and single bods. Double bond conjugation is linked with photochemical properties and chemical reactivity toward molecules including free radicals.

Carotenoids may exist in different configurations due to isomerism around the C-C bonds, obtaining trans and cis isomers (Table 2). Isomers differ not only in their melting points, solubility, and stability but also in ultraviolet characteristics [8, 10, 11].

Carotenoids can be found in free form or esterified with organic fatty acid such as palmitic, estearic, linoleic, etc. The natural functions and properties of carotenoids depend on the molecular structure. The conjugated polyene chromophore presented in carotenoid molecule determines the light absorption and light-harvesting properties. Therefore, chromophore is the part of a carotenoid molecule responsible for its color and photoprotective actions. The wavelengths of the colors of the carotenoids fall between 400 and 600 nm [7].

As mentioned above, carotenoids can be classified into two groups: xanthophylls, including lutein and zeaxanthin, and carotenes, such as α-carotene, β-carotene, and lycopene (Fig. 1). As antioxidants, carotenoids are in general sensitive to light, oxygen, and heat, which can lead to difficulties in storage and handling. Xanthophylls, which are hydrocarbon only, are relatively hydrophilic compounds due to the presence of hydroxyl groups and keto groups at the end rings [8, 11].

Despite the diversity of the carotenoid family, less than 30 carotenoids play important roles in photosynthesis, most of which are located in the thylakoid membranes. The carotenoids absorb light and quench excess energy in photosynthetic metabolism. Some primary carotenoids like lutein serve as accessory pigments which can transfer absorbed energy to chlorophylls, therefore expanding the light-absorbing spectrum of algae or plants. Secondary carotenoids like astaxanthin and canthaxanthin play a role in cell-protective mechanisms. Primary carotenoids are tightly

Table 2
Differences between cis and trans carotenoid isomers

Cis isomers	Trans isomers
More readily solubilized	More thermostable
Can be more readily absorbed and transported	More prevalent in the nature
Different molecular geometries	Linear and rigid molecules

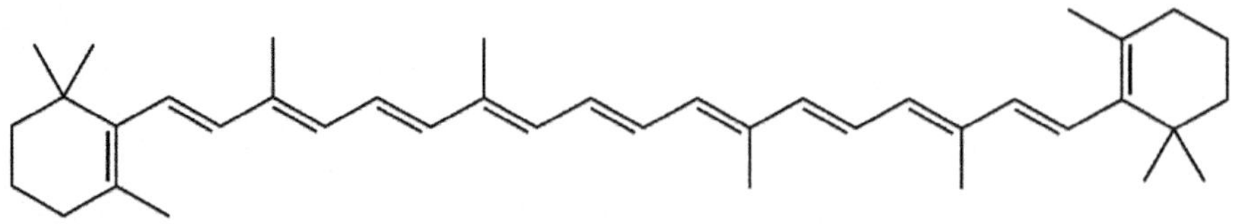

β-carotene

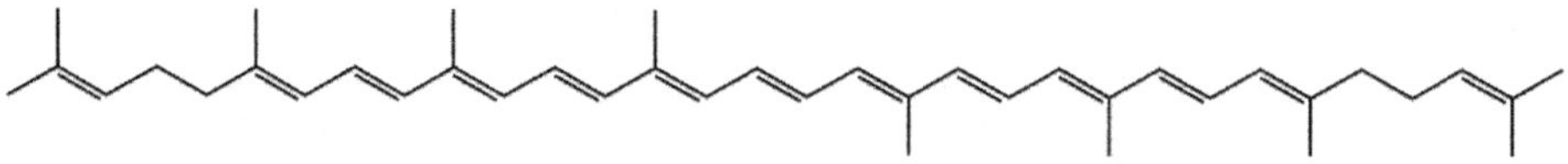

Lycopene

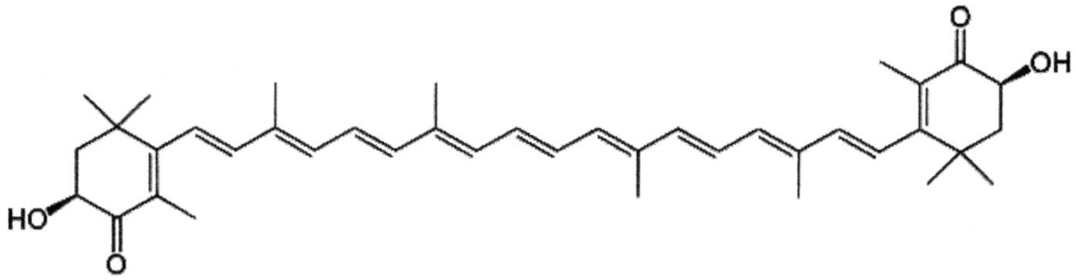

Astaxanthin

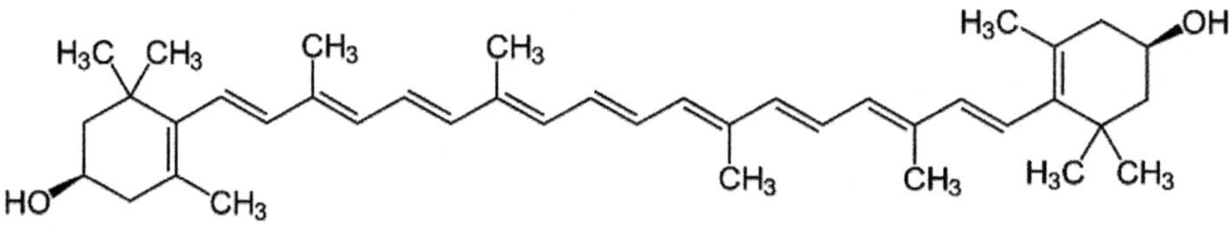

Zeaxanthin

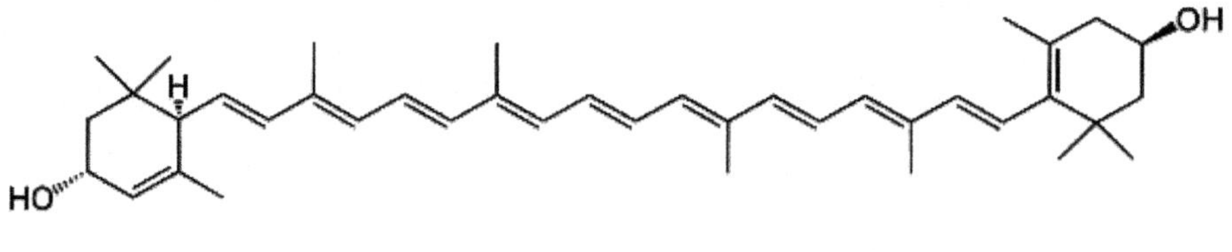

Lutein

Fig. 1 Chemical structure of some common carotenoids [3]

associated with structural and functional components in the cellular photosynthetic apparatus, while the secondary carotenoids are produced to high levels and are dispensed in oily droplets. Most carotenoids are found in ester or diester form; therefore, saponification is needed after the extraction of pigments.

Because of their antioxidant property, carotenoids are able to protect cells from reactive radicals, prevent lipid peroxidation, and promote the stability and functionality of the photosynthetic apparatus. Carotenoids can also promote the integrity of membranes, which is essential for cell survival. In particular, they improve the cell membrane fluidity under high-temperature or high light conditions [8].

3 Carotenoids and Human Health

Apart from allowing improving the aspect of food and making it more appealing according to the consumer preferences, the presence of carotenoids in food offers a variety of health benefits (Table 3).

3.1 Lutein–Zeaxanthin

The marigold flower extract is the main source of lutein and its stereoisomer zeaxanthin. In humans, the lutein and zeaxanthin are believed to function in two ways: firstly, as an antioxidant, thereby protecting from oxidative damage, and secondly as a filter of ultraviolet light. Evidence shows that the consumption of lutein and zeaxanthin is related to a lower incidence of age-related macular degeneration (AMD) and cataracts. Lutein is also able to protect skin from UV-induced damage and may reduce the risk of cardiovascular disease. Furthermore, there is strong epidemiologic evidence that lutein can protect against the development of certain types of cancer. Additionally, lutein has been indicated to improve the immune response. Lutein and zeaxanthin are also becoming increasingly important in the nutraceutical market since they are now understood to play a significant role in eye health, preventing cataract and macular degeneration. Generally, these two xanthophylls are not considered toxic and are relatively safe for human consumption. They are the macular carotenoids creating the macular pigment at the back of the eye. Brain lutein concentrations have been shown to be lower in mild cognitive impairment patients than in those with normal cognitive function [32].

3.2 Astaxanthin

A positive role in many human health problems has been associated to the consumption of astaxanthin, such as UV-light protection and anti-inflammatory properties, mainly linked to its antioxidant power. Moreover, positive human clinical trials supporting eye health have shown that astaxanthin helps diabetic retinopathy, macular degeneration, eye strain and fatigue, and seeing in fine detail. Moreover, it improves the ability of protective white blood cells to surround and destroy infecting organisms. Astaxanthin also protects human lymphocytes and neutrophils against the oxidant stresses imposed by the actions of certain white blood cells without reducing the killing effects of white blood cells

Table 3
Health benefits of six carotenoids confirmed by human studies [8]

Compound	Health benefits	References
Astaxanthin	Strong antioxidant property Anti-inflammatory effects Anticancer Cardiovascular health	Fasano et al. (2014) [12] Chew et al. (1999) [13] Li et al. (2011) [14] Park et al. (2010) [15] Pashkow et al. (2008) [16]
Lutein	Prevents cataract and age-related macular degeneration Antioxidant property Anticancer Prevents cardiovascular diseases	Manayi et al. (2015) [17] Granado et al. (2003) [18] Bone and Landrum (2003) [19] Cha et al. (2008) [20] Vijayapadma et al. (2014) [21] Alves-Rodrigues and Shao (2004) [22]
β-Carotene	Prevent night blindness Antioxidant property Prevents liver fibrosis	Dufossé et al. (2005) [23] Virtamo et al. (2014) [24] Shaish et al. (2006) [25] Virtamo et al. (2014) [24]
Lycopene	Anticancer Prevents cardiovascular diseases Radiation protection	Viuda-Martos et al. (2014) [26] Srinivasan et al. (2009) [27] Devasagayam et al. (2004) [28]
Canthaxanthin	Creates tan color Antioxidant property	Zhang et al. (2014) [29]
Fucoxanthin	Anti-obesity Antioxidant property	Abidov et al. (2010) [30] Nanba and Toyooka (2008) [31]

themselves. Epidemiological studies show that increased intake of carotenoids such as astaxanthin typically lowers risk of many different types of cancer. In this sense, the positive role of astaxanthin in the prevention of different types of cancer, such as colon and breast cancer, is shown. Astaxanthin is best recognized for the pinkish color in aquatic fish and shrimps. It is the strongest antioxidant in carotenoids and thus exhibits stronger antioxidant activity than vitamin E and β-carotene. It has been reported to have the potential to enhance antibody production, antiaging, and sun-proofing, and it also demonstrates anti-inflammatory effects when administered with aspirin [7, 8].

3.3 Canthaxanthin

Canthaxanthin has been reported to act as an antioxidant, to potentiate immune response, and to enhance gap junctional communication between cells directly or through the formation of 4-oxo-retinoic acid, which can also stimulate the retinoic acid receptor. All these functions are considered as possible mechanisms involved in the antitumor effect of this compound. Canthaxanthin may protect people from some blood disorder diseases. However, it was reported to be possibly unsafe in large quantities [7].

3.4 Capsanthin

It has been shown that ingestion of paprika extract obtained from red peppers, mainly rich in the carotenoid capsanthin, upregulates different hepatic genes positively involved in the glucose metabolism and the expression of low-density lipoprotein receptors in rodents. On the other hand, a downregulation of some genes involved in cholesterol catabolism has also been observed. This would lead to a promotion of glucose and fatty acid metabolism and stabilization of blood lipid level disorders. Capsanthin has also been shown to increase plasma HDL cholesterol, which has been related to a lower incidence of cardiovascular diseases [7].

3.5 Lycopene

Lycopene was marketed as an antioxidant and was proposed for treatment of cardiovascular diseases and prostate cancer. Although there were no differences in the levels of β-carotene, lutein, cryptoxanthin, vitamin E, and vitamin A between the cancer patients, the control levels of lycopene were significantly lower in the cancer patients [2].

4 Carotenoids and Animals

4.1 The Role of Carotenoids in Animals and Their Feed

With the development of optimized animal nutrition, the industry created tools to formulate their feed at the best costs to provide animals with required nutrients. Carotenoids are included in the feed as part of the nutrients that support animal health, as well as product quality. The variety of biological effects of carotenoids in animals can be classified into functions, actions, and association [33]. Main carotenoids used in animal nutrition include astaxanthin, β-carotene, β-apo-8′-carotenoic acid ethyl ester, lutein, zeaxanthin, canthaxanthin, and capsanthin. These additives are foreseen for use in feed and/or water for different animal species. Feed needs to provide a balanced diet, specific for each animal type in order to achieve a high performance.

Historically, animal production involved mainly fresh feed, which provided the basic macronutrients but lacked sufficient quantities of carotenoids, vitamins, and minerals. Today, carotenoids are commercially produced for feed and can either be obtained by total synthesis, extracted from natural sources, or biosynthesized by microorganisms [7]. In animals, carotenoid pigments have several important biological activities from nutritional and physiological standpoints, and they can metabolize some of them into vitamin A (retinol) [4].

The lipophilic character of carotenoids makes them more stable to feed processing and therefore can serve to quicken the appetite of visually oriented animals, for example, poultry. The yellow to red coloring effect of carotenoids is of interest in egg yolk, skin, and plumage of ornamental fish and birds [8]. They can also form complexes with proteins and lipoproteins producing an even wider range of colors, including green and blue hues. They have also been shown to have excellent antioxidant properties, often helping to avoid oxidative stress, which poses health risks, especially for immunity and reproduction [7].

Some dietary carotenoids (α-carotene or β-carotene) can be split by mammals into molecules of vitamin A, a crucial health-promoting agent. These carotenoids therefore are referred to as provitamin A. Carotenoid precursors of vitamin A can prevent visual disorders of animals, an approach that has already been followed with the development of "golden rice" in human nutrition [34].

In marine animals, marine sponges, sea urchins, shells of krill, prawn, and some variety of fish, we can find carotenoids. The characteristic pink to orange pigmentation of salmon and trout flesh is also derived from carotenoid deposition, mainly astaxanthin, which is obtained from krill in the diet. The most obvious case in wild animals is the diversity of carotenoids found in bird feathers and skin, as well as in reptile and amphibian skin. These animals owe their color to carotenoids in their diet [7].

4.2 Carotenoids from Animals

4.2.1 Birds

In birds with ornamental-colored plumages, the development of carotenoid-pigmented plumage (yellow, orange, and red) and some of the brighter hues is due to their diet, since they eat vegetables rich in carotenoids [11].

In general, red, orange, and yellow to violet colors are the consequence of feather colorization via carotenoids. Some examples include the flamingo and the songbirds. A flamingo's pink or reddish feather, leg, and facial coloration come from a diet high in alpha and beta carotenoid pigments, including β-carotene and canthaxanthin. Feather pigmentation due to carotenoids is found in different songbirds such as the yellow warbler or the common yellowthroat, which owe their yellow plumage coloration to the lutein in their diets [7].

Birds are also able to accumulate carotenoids from the diet in the egg yolk very efficiently. Lutein, zeaxanthin, β-carotene, and canthaxanthin are important carotenoids deposited in their eggs, which results in a wide range of yellow to orange hues in egg yolk. One important factor affecting pigmentation in birds is animal health. Illness greatly reduces the absorption of carotenoids. Normally, an unhealthy or diseased bird would be pale since it is less efficient in absorbing and depositing carotenoids throughout its body. In addition, unhealthy animals would metabolize the already accumulated carotenoids to activate their immune defenses.

Birds can discriminate a greater variety of colors than humans: as some birds can see into the ultraviolet range, visual stimuli are important. This has to do with the fact that here other stimuli are less well developed. Contrary to mammals, smell and taste stimuli play a minor role. Carotenoids are fat-soluble compounds, which means that their absorption is affected by the level of fat in the diet. Consequently, a diet low in fats may cause feathers to be improperly colored since the animal is not absorbing enough carotenoids [7].

4.2.2 Fish and Crustaceans

Most people consider salmonids as a "red" fish. The major carotenoid in wild salmonids is astaxanthin which originates from the feed they live on, small crustaceans or other fish with small crustaceans in their digestive system. Many marine and freshwater animals, including fish and crustaceans, owe their bright coloration to carotenoids as well. The carotenoids in crustaceans are mainly from algae and in fish from plankton, from other fish with small crustaceans in their digestive system, or from small crustaceans they eat. The richest sources of carotenoids are found in the algae and tiny animals such as shrimp, mollusks, and insect larvae that live in the mud at the bottom of shallow pools that make up the bulk of a flamingo's diet.

Carotenoids are also important on improving immunity in marine species. It has been shown that salmonids having a higher content of astaxanthin are more resistant to bacterial and fungi diseases. In relation to crustaceans, astaxanthin has shown to increase the survival rate. A study in shrimps demonstrated that there is a significant decrease in mortality of adult shrimp fed a carotenoid-enriched diet in comparison with individuals receiving carotenoid-free diets. Carotenoids in crustaceans are frequently linked to protein molecules and may provide blue or green color. Dietary carotenoids, among them astaxanthin, are responsible for the characteristic red color of shrimps and other crustaceans when cooked. This red color originates from the release of astaxanthin from the carotenoproteins when denatured by the heat of cooking.

Salmonids and crustaceans are major contributors to the world production of farmed finfish and shellfish. To meet consumers' demand for pigmented flesh and exoskeleton of fish or crustaceans, intensive aquaculture diets are fortified with carotenoids. The coloration of salmonids flesh or shrimp carapaces is considered a criterion of quality.

Carotenoids are added to fish feed, so that farmed salmon ingest it in the same way as wild salmon. In these cases, astaxanthin and its salts are generally used alone or more rarely in combination with canthaxanthin to enhance the pink-red color of the flesh. The efficacy of pigmentation depends on factors such as form and concentration of carotenoid source, diet composition and dietary fat content and quality, fish size and physiological state, state of sexual maturation, and genetic background.

In the case of salmonids (e.g., salmon and trout), the aim of using red oxycarotenoids in their diets is to enhance the orange-pink color of the flesh. In the case of other fish species, skin pigmentation is the target. For flesh pigmentation, the carotenoid most commonly used is astaxanthin, although in some markets it can be combined with canthaxanthin. These carotenoids are normally supplemented to fish from the age of 6 months onward.

In crustaceans, carotenoids are used to enhance the pigmentation of the exoskeleton. Similarly, to fish, the carotenoid most commonly used is astaxanthin, but also β-carotene is used, although in some markets it is combined with canthaxanthin. In crustaceans, astaxanthin is deposited below the cuticle, over the flesh [7].

4.2.3 Poultry

The origins of pigmentation of egg yolk and poultry are diverse, but they are mainly related to health aspects. They are grounded in the traditional familiar farming systems where animals have access to grass, maize, and high carotenoid diets. Carotenoids from these diets are responsible for the pigmentation of egg yolk, skin (and fat), as well as legs, beak, comb, and feathers. When birds fail to consume these xanthophylls, the hues of their egg yolk or skin become paler.

Therefore, a uniform and good pigmentation generally means good health and good practical hygienic conditions. For marketing of poultry products, in many countries of the world, a bright yellow or yellow-orange color in egg yolk and broiler skin is often associated to freshness and health and has become an indicator of high-quality products. Animals cannot synthesize carotenoids, and the colors demanded by consumer cannot be achieved only with the sources of xanthophylls. For that reason, carotenoids began to be added to feed to achieve the desired color and conform to customer preference.

There are many factors that influence the coloration of poultry tissues and can be classified as (1) external, those that affect the carotenoid before the animal consumes it, or (2) internal, the ability of the animal to metabolize (genetic ability) and deposit subcutaneous fat. Also, health is a determinant.

As pigmentation is affected by several factors, in some cases higher doses of xanthophylls can be necessary to achieve the assigned color values. Since lipid metabolism is slow in broilers during the high growth period and pigments are deposited in nonspecific tissues (skin, shank, etc.), higher dosages are required for pigmentation compared to the laying animals. Red xanthophylls are necessary to give a golden color. Only yellow birds can deposit the carotenoids in fat and skin. Most white birds are unable to store, and thus it is impossible to pigment their tissues. Carotenoids can also be deposited in the shanks of certain poultry. In such a case, colored legs are perceived by the consumers as a sign of bird's good bird health and high quality [7].

4.2.4 Livestock

β-carotene is found in high concentrations in the ovary and plays a direct role in ovarian function. Cows with low β-carotene levels therefore tend to have poor fertility and stand to benefit from supplementation. Companies dealing with livestock products like milk increase their profit if fertility is improved, because (a) shorter calving intervals mean higher milk yields and more calf sales; (b) replacement costs are lower, as more cows get back in calf; (c) semen, labor, and veterinary costs reduced; and (d) cows complete lactations in better condition [35].

5 Conclusions

Carotenoid pigments are a group of bioactive compounds that are of interest to the food scientists, nutritionists, and food industries due to their positive impact on human health and currently are being studied to understand their effect on human chronic diseases. Carotenoids are responsible for the attractive color of most fruit and vegetables, having diverse biological functions and activities. Antioxidant properties of carotenoids are linked with their ability to scavenge free radicals and singlet oxygen.

They color our foods and provide pleasure to our eating experience. They contribute to our vitamin A requirement and the reduction of oxidative stress. They may even contribute to seeing those pigmented foods more clearly, even if it is no longer in the bush but rather in the supermarket.

Dark green vegetables, particularly leafy greens, are nutrient-dense foods that may have a range of health benefits when included in a varied, balanced diet. Scientists worldwide continue to identify and characterize substances in wild and cultivated green vegetables that may be responsible for associations of plant components with decreased risk for chronic disease, including cancer, heart disease, diabetes, and eye diseases. Plants have evolved a vast number of carotenoids as essential compounds for their development and survival. Animals and humans, as plant consumers, have also evolved to take advantage of the unique properties of carotenoids. The mechanisms for the absorption and metabolism of carotenoids are well established.

In fact, there is an economic potential to foster the use of plant pigments as natural antioxidants to enhance the quality and extend the shelf life of animal products. Because of an increased environmental conscience and an increasing market for natural feed/food additives in developed countries and the chance for developing countries to use sound knowledge of effective plant-derived feed additives, available without high costs, the topic of plant pigments in animal nutrition should be given high priority.

References

1. Kiokias S, Proestos C, Varzakas T (2016) A review of the structure, biosynthesis, absorption of carotenoids-analysis and properties of their common natural extracts. Curr Res Nutr Food Sci 4(Special issue):25–37
2. Rao AV, Rao LG (2007) Carotenoids and human health. Pharmacol Res 55:207–216
3. Kumar Saini R, Nile SH, Park SW (2015) Carotenoids from fruits and vegetables: chemistry, analysis, occurrence, bioavailability and biological activities. Food Res Int 76:735–750
4. Fernández-Garcia E et al (2012) Carotenoids bioavailability from foods: from plant pigments to efficient biological activities. Food Res Int 46:438–450
5. DSM in Animal Nutrition & Health, 'Carophyll-because color matters
6. DSM in Animal Nutrition & Health, 'Carotenoids in animal nutrition and health'
7. Amaya E et al (2014) Carotenoids in animal nutrition. Fefana Publication
8. Gong M, Bassi A (2016) Carotenoids from microalgae: a review of recent developments. Biotechnol Adv 34:1396–1412
9. Echavarri-Erasum C, Johnson EA (2002) Fungal carotenoids. Appl Mycol Biotechnol 2:45–85
10. Jomova K, Valko M (2013) Health protective effects of carotenoids and their interactions with other biological antioxidants. Eur J Med Chem 70:102–110
11. Zakynthinos G, Varzakas T (2016) Carotenoids: from plants to food industry. Curr Res Nutr Food Sci 4(Special issue):38–51
12. Fasano E et al (2014) Antioxidant and anti-inflammatory effects of selected two human immortalized keratinocyte lines. Biomed Res Int 2014:1–11
13. Chew BP et al (1999) A comparison of the anticancer activities of dietary β-carotene, canthaxanthin and astaxanthin in mice in vivo. Anticancer Res 19:1849–1853
14. Li J et al (2011) An economic assessment of astaxanthin production by large scale cultivation of *Haematococcus pluvialis*. Biotechnol Adv 29:568–574
15. Park JS et al (2010) Astaxanthin decreased oxidative stress and inflammation and enhanced immune response in humans. Nutr Metab (Lond) 7:18
16. Pashkow FJ, Watumull DG, Campbell CL (2008) Astaxanthin: a novel potential treatment for oxidative stress and inflammation in cardiovascular disease. Am J Cardiol 101:S58–S68
17. Manayi A et al (2015) Lutein and cataract: from bench to bedside. Crit Rev Biotechnol 8551:1–11
18. Granado F, Olmedilla B, Blanco I (2003) Nutritional and clinical relevance of lutein in human health. Br J Nutr 90:487–502
19. Bone R, Landrum J (2003) Lutein and zeaxanthin dietary supplements raise macular pigment density and serum concentrations of these carotenoids in humans. J Nutr 133:992–998
20. Cha K, Koo S, Lee D (2008) Antiproliferative effects of carotenoids extracted from *Chlorella ellipsoidea* and *Chlorella* vulgaris on human colon cancer cells. J Agric Food Chem 56:10,521–10,526
21. Vijayapadma V et al (2014) Protective effect of lutein against benzo(a)pyrene-induced oxidative stress in human erythrocytes. Toxicol Ind Health 30:284–293
22. Alves-Rodrigues A, Shao A (2004) The science behind lutein. Toxicol Lett 150:57–83
23. Dufossé L et al (2005) Microorganisms and microalgae as sources of pigments for food use: a scientific oddity or an industrial reality? Trends Food Sci Technol 16:389–406
24. Virtamo J et al (2014) Effects of α-tocopherol and β-carotene supplementation on cancer incidence and mortality: 18-year postintervention follow-up of the alpha-tocopherol, beta-carotene cancer prevention study. Int J Cancer 135:178–185
25. Shaish A et al (2006) 9-cis-β-carotene-rich powder of the alga *Dunaliella bardawil* increases plasma HDL-cholesterol in fibrate-treated patients. Atherosclerosis 189:215–221
26. Viuda-Martos M et al (2014) Tomato and tomato byproducts. Human health benefits of lycopene and its application to meat products: a review. Crit Rev Food Sci Nutr 54:1032–1049
27. Srinivasan M et al (2009) Lycopene: an antioxidant and radioprotector against γ-radiation-induced cellular damages in cultured human lymphocytes. Toxicology 262:43–49
28. Devasagayam TP et al (2004) Free radicals and antioxidants in human health: current status and future prospects. J Assoc Physicians India 52:794–804
29. Zhang W et al (2014) Attached cultivation of *Haematococcus pluvialis* for astaxanthin production. Bioresour Technol 158:329–335
30. Abidov M et al (2010) The effects of Xanthigen™ in the weight management of

obese premenopausal women with non-alcoholic fatty liver disease and normal liver fat. Diabetes Obes Metab 12:72–81

31. Nanba K, Toyooka S (2006) The allelic distribution of a single nucleotide polymorphism in the PDCD5 gene locus of Japanese non-small cell lung cancer patients. Mol Med Rep 1:667–671
32. Jayne V et al (2015) Carotenoids and health in older people. Maturitas 80:63–68
33. Esteban R et al (2015) Versatility of carotenoids: an integrated view on diversity, evolution, functional roles and environmental interactions. Environ Exp Bot 119:63–75
34. Faehnrich B et al (2016) Phytogenic pigments in animal nutrition: potentials and risks. J Sci Food Agric 96:1420–1430
35. DSM in Animal Nutrition & Health, 'β-carotene—more than a pretty color'

Chapter 4

Express Analysis of Microalgal Secondary Carotenoids by TLC and UV-Vis Spectroscopy

Galina S. Minyuk and Alexei E. Solovchenko

Abstract

We elucidate the peculiarities of express analysis of secondary carotenoids from microalgae and their preliminary identification using TLC and UV-Vis spectroscopy with emphasis on biotechnologically significant producers of the value-added pigment astaxanthin. Advantages and limitations of the method are described to underline the value of TLC as a potential companion method to mainstream separation techniques such as HPLC. Special attention is paid to common errors and pitfalls of the method and possible work-arounds, as well as to overall strategy of the analysis, sample preparation, and material selection.

Key words Secondary carotenoids, Keto-carotenoids, Microalgae, Thin-layer chromatography, TLC, Absorption spectrophotometry, Astaxanthin, Solvent selection

1 Introduction

Carotenoids (Car) are a diverse (ca. 800 members) group of C_{40}-colored lipid-soluble molecules ubiquitous in photoautotrophic organisms [1, 2]. These pigments are divided, according to their substituent composition, into carotenes, the simple hydrocarbon compounds, and xanthophylls containing oxygen atoms within hydroxy, epoxy, or keto groups.

According to their functions, Car could be divided into two major groups. Photosynthetic or primary Car are closely associated with photosynthetic apparatus, that is, with thylakoid membranes [3]. Under stressful conditions certain microalgae accumulate Car which are deposited in oil bodies outside thylakoids; these pigments (called secondary Car) do not participate in photosynthesis [4].

Carlos Barreiro and José-Luis Barredo (eds.), *Microbial Carotenoids: Methods and Protocols*, Methods in Molecular Biology, vol. 1852, https://doi.org/10.1007/978-1-4939-8742-9_4,

Secondary Car of microalgae found extensive use in the food and feed industry [5, 6]. Thus, astaxanthin (3,3′-dihydroxy-ß-carotene-4,4′-dione) is believed to be the most powerful natural antioxidant [7] which does not exert a prooxidant effect typical of other carotenoids or pose the threat of hypervitaminosis [8]. Microalgae such as *Haematococcus pluvialis* represent the riches source of natural astaxanthin [9]. Analysis of carotenoids is essential for elucidating physiology of the microalgae, screening of the biodiversity for new efficient algal strains, and improving productivity of existing cultures with regard to the valuable Car pigments.

Although HPLC-DAD/MS/MS and NMR are currently the mainstream methods of separation and/or identification of Car, it is not always warrantable and often not feasible in small (field) laboratories and production facilities. In these cases, the time-proven simplicity and reliability often make TLC the method of choice for routine (express) assessment of Car in microalgae, preliminary purification, and fractionation of complex Car mixtures for further analysis [10, 11]. Added value of TLC includes low cost of analysis and equipment, streamlined sample preparation, easier preparative separation, and simultaneous analysis of many samples under the same conditions increasing the robustness of the analysis [11, 12].

Below we will consider in detail the selection of the materials, preparation of samples, and TLC separation of secondary Car from the chlorophyte *Bracteacoccus minor* (Schmidle ex Chodat), Petrová (1931), Sphaeropleales, and Chlorophyceae [13].

2 Materials

2.1 Plant Material

Culture of green carotenogenic microalgae *Bracteacoccus minor* (Schmidle ex Chodat), Petrová (1931), strain SAG 221-1 (the University of Göttingen Department Experimental Phycology and Culture Collection of Algae (EPSAG)), or any other astaxanthin synthesizing green microalgae.

2.2 Equipment

1. TLC silica gel 60 aluminum sheets (Table 1) (*see* **Note 1**).
2. Microdespensers or capillary micropipettes with accuracy ± 1% and volume range 20–100 μL (*see* **Note 2**).
3. Rectangular TLC developing tanks with flat bottom and non-slip or latch-style lids (*see* **Note 3**).
4. Centrifuge and compatible glassware.
5. Vortex.

Table 1
Features of several common silica gel 60 TLC plates

Product features	Merck[a] TLC silica gel 60 aluminum sheets	Merck TLC silica gel 60 aluminum sheets with CZ[b]	*Macherey-Nagel* Alugram® Xtra SIL G	Alugram® Xtra SILGUR with CZ
Sorbent	**High purity unmodified silica gel 60**			
Plate size, cm	5 × 10; 10 × 20; 20 × 20		10 × 20; 20 × 20	
Mean pore size, Å	60	60	60	60
Pore volume, mL/g	0.74–0.84	0.74–0.84	0.75	0.75
Specific surface area, m^2/g	480–540	480–540	500	500
Particle size, μm	9.5–11.5	9.5–11.5	5–17	5–17
Layer thickness, μm	175–225	160–220	200	200
Deviation of layer thickness per plate, μm	≤30	≤40	No data	No data
Sorbent in concentrating zone (CZ)	–	C18 silica gel	–	kieselgur
Layer thickness of CZ, μm	–	90–140	–	No data
Fluctuation in the thickness of the CZ, μm	–	≤30	–	No data
Reference	http://www.merckmillipore.com		http://www.mn-net.com	

[a]TLC plates with similar characteristics can be obtained from Sigma-Aldrich, EMD Millipore, Whatman, Analtech, Sorbtech, etc.
[b]With concentrating zone

6. Porcelain mortar and pestle.
7. Spectrophotometer.
8. Standard 1 cm spectrophotometric cells.
9. Porter homogenizer.
10. Pipettors and tips (0.1–1 mL).
11. Scanner or camera for documenting the chromatograms.
12. Scissors for cutting TLC plates.
13. Airtight test tubes and beakers.
14. Tweezers.

15. Fume hood.
16. Lab oven.
17. Desiccator cabinet tank.
18. Rotary evaporator.
19. Filter paper medium speed (sheets).
20. 15 mL conical centrifuge tubes (e.g., Corning™, Falcon™).

2.3 Reagents

1. Hexane ($CH_3 (CH_2)_4CH_3$) for HPLC, ≥95.5(GC).
2. Acetone (CH_3COCH_3) for HPLC, ≥99.9(GC).
3. Benzene (C_6H_6) for HPLC, ≥99.9.
4. Solvent mixture #1: hexane/acetone (9:1, v/v) (prepared immediately before use).
5. Solvent mixture #2: hexane/benzene/acetone (10:7:1.6, v/v) (prepared immediately before use).
6. Ethanol (CH_3CH_2OH) for spectroscopy or absolute for analysis.
7. Calcium chloride ($CaCl_2$) anhydrous, granular, ≥93.0%.
8. Silica gel (SiO_2) granular with moisture indicator.
9. Commercially available carotenoid standards (Table 2).
10. Liquid nitrogen (N_2), high purity grade, ≥99.998%.
11. Nitrogen Gas (N_2), high purity grade, compressed, ≥99.998%.
12. Quartz sand white (SiO_2), ≥99.995% trace metals basis.
13. Sodium borohydride ($NaBH_4$) for analysis.
14. Tetrahydrofuran (C_4H_8O, THF) for spectroscopy Uvasol®.
15. Acetic anhydride ($(CH_3CO)_2O$) ACS reagent, ≥98.0%.
16. Pyridine (C_5H_5N) anhydrous, 99.8%.
17. Butylated hydroxytoluene (BHT, $((CH_3)_3C)_2C_6H_2(CH_3)OH$), >99.0% (GC).
18. Silver nitrate ($AgNO_3$) ACS reagent ≥99.0%.
19. Distilled water.

3 Methods

3.1 Extraction of Car from Microalgal Biomass

To prevent the oxidation and isomerization of Car (which are manifested by the appearance of strong absorption in the UV part of the spectrum, aberrant spots on TLC plates, and distorted absorption spectra), it is important to avoid as much as possible oxygen, light, alkaline media, and their combinations; for details, *see* [10–12]. Specific difficulties of Car extraction from microalgal cells include the need of breaking of the tough cell wall. Possible work-arounds for achieving complete extraction in this case include

Table 2
Commercially available carotenoid standards

Common name of the carotenoid	IUPAC name	Supplier	Grade
all-trans-Astaxanthin	3*S*,3′*S*,*all-trans*-3,3′-dihydroxy-β,β-carotene-4,4′-dione, (3*S*,3′*S*)-3,3′-dihydroxy-β,β-carotene-4,4′-dione	S[a]	Analytical standard ≥97% (HPLC)
(*rac./meso*)-Astaxanthin dipalmitate	(3*RS*,3′*RS*)-3,3′-dihexadecanoyloxy-β,β-carotene-4,4′-dione	C	HPLC 98%, synth., cryst.
(*rac./meso*)-Astaxanthin monopalmitate	(3*RS*,3′*RS*)-3,-hexadecanoyloxy-β,β-caroten-4,4′-dione	C	HPLC 97%, synth., cryst.
Astacene	3,3′-dihydroxy-2,3,2′,3′-tetradehydro-β,β-carotene-4,4′-dione	C	HPLS 99%, synth. cryst.
Canthaxanthin (trans)	β,β-carotene -4,4′-dione	S	Analytical standard
9Z-Canthaxanthin	(9Z)-β,β-carotene-4,4′-dion 96%	C	HPLS 96%, synth. cryst.
13Z-canthaxanthin	(13Z)-β,β-carotene-4,4′-dion	C	HPLS 97%, synth. cryst.
(3R)-Adonirubin	(3R)-3-hydroxy-β,β-carotene-4,4′-dione	C	HPLS 98%, synth. cryst.
(3S,3′R)-Adonixanthin	(3S,3′R)-3,3′-dihydroxy-β,β-carotene-4,4′-dione	C	HPLS 99%, synth. cryst.
β-Cryptoxanthin	(3R)-β,β-carotene-3-ol	C	HPLS 97%, synth. cryst.
β-Cryptoxanthin palmitate	(3R)-3-palmitoxy-β,β-carotene	C	HPLS 95%, synth. cryst.
Echinenone	(β,β-carotene-4-one)	C	HPLS 98%, synth. cryst.
3-Hydroxyechinenone	(rac.)-3-hydroxyechinenone (3RS)-3-hydroxy-β,β-carotene-4-one	C	HPLS 97%, synth. cryst.
α-Carotene	β,ε-carotene	S	Analytical standard ≥ 98.0% (HPLC)
α-Carotene	β,ε-Carotene	C	HPLS 97%, synth. cryst.
β-Carotene	β,β-Carotene	S	Pharmaceutical secondary standard

(continued)

Table 2 (continued)

Common name of the carotenoid	IUPAC name	Supplier	Grade
β-Carotene	β,β-Carotene	C	HPLS 96%, synth. cryst.
Lutein	β,ε-Carotene-3,3′-diol	S	analytical standard ≥97% (HPLC)
Lutein (zeaxanthin-free)	(3R,3R′,6′R)-β,ε-carotene-3,3′-diol	C	HPLS 99%, synth. cryst.
Neoxanthin	(3*S*,3′*S*,5*R*,5′*R*,6*R*,6′*S*,9′-*cis*)-6,7-didehydro-5′,6′-epoxy-5′,6′-dihydro-β,β-carotene-3,3′,5(6*H*)-triol	S, C	Analytical standard ≥97% (HPLC)
Zeaxanthin	β,β-carotene-3,3′-diol	S	Analytical standard ≥97% (HPLC)
Zeaxanthin, (3R,3R′)	(3R,3′R)-β,β-carotene-3,3′-diol	C	HPLS 97%, synth. cryst.
Zeaxanthin palmitate	(3R,3′R) -3,3′-dihaxadecanoiloxy-β,β-carotene	C	HPLS 98%, synth. cryst.
Antheraxanthin	(3S,5R,6S,3R)-5,6-epoxy-5,6-dihydro-β,β-carotene-3,3′-diol	~C	HPLS 95%, isolated. cryst.

[a]S, Sigma-Aldrich Corp.; C, CaroteNature GmbH, Switzerland

grinding cells frozen in liquid nitrogen with quartz sand and/or extraction with solvent mixtures such as methanol/hexane, 3:1 [14], hexane/isopropanol, 6:4 [15], or hexane/acetone/ethanol, 100:70:70 [16].

1. Sample 15–20 mL of microalgal cell suspension (dry weight content 0.5–1.5 g/L).
2. Centrifuge the sample (5 min at 1000 × *g*) and discard the supernatant.
3. Wash the pellet by vortexing in 15–20 mL of distilled water to remove the remaining nutrient media component.
4. Centrifuge the sample (5 min at 1000 × *g*) and discard the supernatant. The following steps should be carried out under dimmed light and preferably in a cold room (+4 °C).
5. Grind the pellet with a precooled porcelain mortar, and pestle in 100% acetone (*see* **Note 4**) with addition of BHT (0.01% w/v) with quartz sand (*see* **Note 5**). This operation can also be done in a Potter homogenizer, a Retsch mill (3 min at 25 Hz), or a similar apparatus. Repeat this step until complete removal of the color from the cell residue.
6. Centrifuge the pooled homogenates (5 min at 3000–4000 × *g*), discard the pellet, and evaporate to dryness on a rotary evaporator at a temperature below 40 °C (*see* **Note 6**).
7. To avoid the interference from water traces, further dry the residue in a stream of N_2 and then redissolve it in 0.5–1.0 mL (V_1 in Eqs. 1–3) of HPLC-grade acetone. The resulting sample is further used for determination of total Car (*see* Subheading 3.2) and TLC analysis (*see* Subheading 3.4).

3.2 Spectrophotometric Assay of Total Car in Microalgal Cells

1. Take a 50–100 μL (V_2 in the Eqs. 1–3) aliquot from the sample obtained as a result of the procedure described above to a sealable test tube.
2. Make up the sample to 5–10 mL (V_3 in the Eqs. 1–3) with 100% acetone.
3. Record absorbance spectrum in the range 400–750 nm in a 1 cm cell.
4. Calculate the chlorophyll a and b as well as total carotenoid concentrations using equations by Lichtenthaler [17]:

$$[\mathrm{Chl}0]=\frac{\left(12.24\left[D_{661.6}-D_{750}\right]-2.04\left[D_{644.8}-D_{750}\right]\right)\times V_3\times V_1}{V_2\times V_4} \quad (1)$$

$$[\mathrm{Chl}b]=\frac{\left(20.13\left[D_{644.8}-D_{750}\right]-4.19\left[D_{661.6}-D_{750}\right]\right)\times V_3\times V_1}{V_2\times V_4} \quad (2)$$

$$[Car]=\frac{\left(1000\left[D_{470}-D_{750}\right]-1.9[\mathrm{Chl}a]-63.14[\mathrm{Chl}b]\right)\times V_3\times V_1}{214\times V_2\times V_4} \quad (3)$$

where:

($Chla$), ($Chlb$), and (Car)—concentrations of chlorophyll a, chlorophyll b, and total Car in the diluted extract, μg mL

V_1—volume of the redissolved extract, mL

V_2—volume of the aliquot taken from the redissolved sample (V_1), mL

V_3—volume of the made-up aliquot (V_2), mL

V_4—volume of the cell suspension sample, mL

$D\lambda$—optical density of the extract at wavelength λ

3.3 TLC Plate Preparation, Sample Application, and Analysis Setup

Carry out all procedures in a thermally controlled room (20–22 °C) at dim light, under fume hood with proper ventilation. Open fire is not allowed during manipulations with organic solvents. All labware in contact with the organic solvents should be made of glass, Teflon, or metal; avoid plastic tips, beakers, etc.

1. Select and check TLC plates (*see* **Note 7**) carefully (they should be free from scratches and other damages to the sorbent layer, especially on the edges).
2. Carefully mark each plate with a needle or a blunt black graphite pencil: mark the start and the solvent front lines (approx. 1.5–2 cm from the lower and upper edges of the plate), and put the sample names above the corresponding lanes (if several samples will be run on a single plate).
3. Activate the plates in an oven for 30 min at 110 °C and keep the activated TLC plates in a desiccator cabinet tank above freshly baked $CaCl_2$ or silica gel granules (4 h at 180 °C). The activated plates should be used during the same day.
4. Mix the solvents (*see* **Notes 8** and **9**) constituting the mobile phase(s) for the TLC plate development using glass pipettes (or automatic with glass tips) in graduated glass cylinder(s) with airtight (ground glass) stoppers in a glass beaker with an airtight cap. Solvent mixture #1 and solvent mixture #2 should be always freshly prepared.
5. Saturate the TLC tanks with the solvent vapors for 20 min (*see* **Note 10**). To do this, fill the TLC tank 1 with the solvent mixture #1 (25–30 mL for TLC plate 20 × 20 cm) about 20 min before the start of the development. The TLC tank 2 should be filled with the solvent mixture #2 about 10 min after start of the development in the tank 1 (i.e., approximately 20 min before the start of the development in the tank 1).
6. Seal the filled TLC tanks airtightly, and store in a dimly lit place, and cover them from sunlight with an opaque material (black paper, metal foil, etc.).
7. Take the 50–150 μL aliquot of the sample extract depending on the plate width and the concentration of the sample (this is

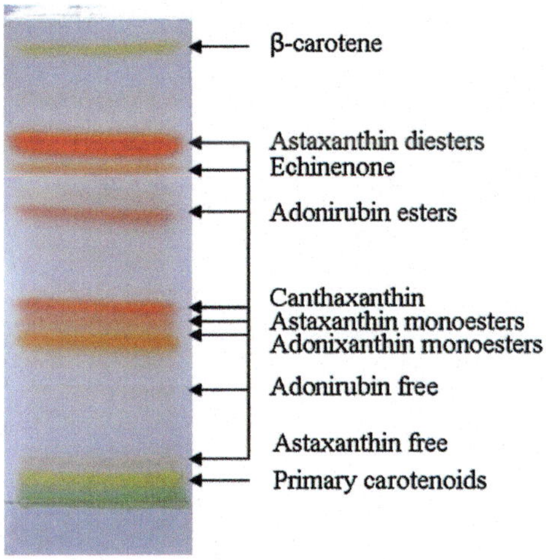

Fig. 1 TLC of carotenoid extract from of *Bracteacoccus minor* sequentially developed in two solvent systems: (1) hexane-acetone (9:1); (2) benzene-hexane-acetone (10:7:1.6)

V_2 in the Eq. 4) and apply it to the TLC plate using one of the capillary/applicators described in the Subheading 2.2 above. Recommended drop volume is approximately 5 μL; anyway, the diameter of the spot after application should be 2–2.5 mm. The sample extract should be applied dropwise, placing the drops side by side to achieve a neat line of drops (*see* **Notes 11** and **12**).

8. Immediately after application, place the TLC plate into the TLC tank #1.
9. When the solvent front reaches the front line marked on **step 1**, take the plate out of the TLC tank 1, dry in a stream of an inert gas for 3–5 s, and place it in the TLC tank 2.
10. When the solvent front reaches the front line marked on the **step 1**, take the plate out of the TLC tank 2 and take a picture of it using a scanner or a camera (Fig. 1).
11. Cut the TLC plate into stripes, each of them containing a single spot. Immediately place the stripes into 2 mL aliquots of acetone (or other suitable solvent, depending on the Car) in the airtight test tubes.
12. Stir the test tubes vigorously on a shaker for 30–40 s to elute the pigments from the plate.
13. From each test tube, take out the plate stripe with tweezers, wash the eluate remaining on the stripe back to the test tube with the respective solvent, and discard the plate stripe.
14. Centrifuge the test tubes (5 min at 3000–4000 × *g*) to remove the particles of the silica gel and transfer the supernatant to

Table 3
Absorption maxima of the Car from Chlorophyceae in different solvents (nm)

Carotenoid	Solvent: Acetone	Hexane (H) or light petroleum (LP)	Benzene	Ethanol	Chloroform
		Secondary Car			
Adonirubin (Phoenicoxanthin)	474[b] 474[d]	471[g] 462 (H)[d] 466 (H)[a] 469[e] 460[f]	482[d] 483[e] 478[f]	478[a] 464[f] 477 (99,5%)[e]	479 [a] 486[e] 480[f]
Adonixanthin (β-doradexanthin)	479[d]	460 (H)[i] 462[g] 465[h] 460 (LP)[a]	468[d] 472[f]	465[a] 460[j]	481[f]
Astacene	482[e]	473(LP)[e]		478[e]	494[e]
Astaxanthin	475[l] 477[d] 477[b]; 478[k] 480[a]	466–467 (H)[e] 468(H)[m] 472 (H)[d] 468 (LP)[a]	484[d] 485[a]	478[a] 479[k] 480[j]	485[a]
Canthaxanthin	468[c] 472[d] 460[k]	467[d] 466 (PL)[a] 468	484[a]	474[a] 476[n] 476 (95%)[j]	482[a]
β-cryptoxanthin		425, 449, 476 (PL)[a] 428, 444, 473	435, 463, 489[a]	428, 450, 478[a]	435, 459, 485[a]
Echinenone	456[k] 462[d] 460[a] 460[b]	458 (482)[a] 460[d]	470[a] 472[d]	461[a] 453–461[p] 466[h] 462 (95%)[j]	471[a] 473[k]

3′-hydroxyechinenone	459 (471)[l] 458 (475)[o] 463[d] 464[b]	452[d] 456[p] 465[e] 457[f]	465[d] 469[r] 472[p] 473[f]	470[e] 466[f]	482[e] 472[f]
Primary Car					
Antheraxanthin	(422), 448, 475[c]	422, 455, 472 (H, LP)[a] 420, 444, 472 (H)[c]		422, 444, 472[a] 422, 444, 473 (95%)[j] 422, 444, 472[c]	430, 456 484[a]
β-carotene (β,β-carotene)	429, 452, 478[a] (428), 453, 482[b] 452, 479[d]	(425), 450, 477 (H, LP)[a] 451, 489 (H)[d]	435, 462, 487[a] 463, 489[d]	(428), 451, 480[s] 450, 476[a]	435, 461, 485[a]
α-carotene (β,ε-carotene)	424, 448, 476 [a] 424, 448, 476[s]	423, 445, 473 (H, LP)[a] 420, 442, 472 (H)[s] 421, 445, 474 (PL)[s] 420, 444, 473[t]	432, 456, 485[a]	423, 444, 474 (95%)[j] 423, 444, 473[a]	433, 457, 484[a] 432, 457, 485[s]
Lutein	425, 448, 476[c]	421, 445, 474 (LP)[a] 420, 445, 475 (H)[s] 420, 444, 474 (LP)[s] 421, 445, 474 (H)[c]	432, 458, 487[a]	422, 444, 475 (95%)[j] 422, 445, 474[a] 422, 445, 474[c]	435, 458, 485[a]
Neoxanthin	416, 440, 470[a] 413, 438, 466[c]	416, 438, 467 (LP)[a] 411, 435, 463[c]	426, 453, 483[a]	414, 439, 466 (95%)[j] 415, 439, 467[a] 413, 437, 466[c]	423, 448, 476[a]
Violaxanthin	415, 438, 467[c]	416, 440, 465 (LP)[a] 417, 440, 470[c]	427, 456, 483[a]	418, 440, 470 (95%)[j] 417, 440, 469[c] 419, 440, 470[a]	426, 449, 478[a]
Zeaxanthin	430, 452. 479[a]; (428), 454, 481[c]	424, 449, 476 (LP)[a] (424), 450, 478 (H)[c]	440, 463, 491[a]	450, 479 (95%)[j] 428, 450, 478[a]; (428), 450, 478[c]	433, 462, 493[a]

Compiled from Britton [20][a], Andrewes, Phaff [37][b], Egeland, Garrido [19][c], Qin, Liu [38][d], Cooper, Davis [39][e], Bernhard, Englert [40][f], Matsuno, Maoka [41][g], Fraser, Shimada [42][h], Castillo, Negre-Sadargues [43][i], Weber [44][j], Kopczynski, Lenzer [45][k], Grung, Metzger [46][l], Laatsch and Schwab [47][m], Lorquin, Molouba [48][n], Holt and Krogmann [49][o], Rowan [50][p], Yokoyama and Miki [51][r], Foppen [52][s], Bjørnland and Aguilar-Martinez [53][t]

clean test tubes. Record the volume of the supernatant (V_5 in the Eq. 4) and take its absorbance spectrum or just read the absorbance at the corresponding wavelength (Table 3).

15. Calculate the carotenoid content of each fraction (per unit volume of the algal cell suspension) according to the following equation:

$$\left[\mathrm{Car}^*\right] = \frac{D_{\lambda\,\mathrm{max}} \times V_5 \times V_1 \times 10000}{\mathrm{E}^{1\%}_{1\mathrm{A}m} \times V_{2*} \times V_4} \tag{4}$$

where:

(Car*)—the carotenoid content (μg/mL of the algal cell suspension)

$D_{\lambda\mathrm{max}}$—absorbance of the eluate in the characteristic maximum in corresponding solvent (abs. units)

$^{1\%}_{1\mathrm{cm}}$—specific extinction coefficient (Table 4)

V_1—sample extract volume, mL

V_{2*}—volume of the aliquot applied to the TLC plate, mL

V_5—volume of the corresponding carotenoid fraction eluate, mL

V_4—volume of the cell suspension sample taken for extraction, mL

3.4 Identification of the Carotenoid Fractions

Ideally, comprehensive identification of the carotenoid pigments requires, in addition to UV-Vis spectroscopy, a complex of advanced methods including mass spectrometry, nuclear magnetic resonance, circular dichroism, and Raman spectroscopy [1, 10]. Although it is not always realistic and not necessary either to apply this complex of methods routinely. Furthermore, the composition and physical and chemical properties of the secondary Car of (green) microalgae are well known [18]. These circumstances make the separation and identification of the Car by such simple method as TLC much more straightforward. However, the primary focus of this chapter is separation of the complex Car mixtures rather than identification, so we will just briefly recapitulate the simple tests handy for identification of carotenoid fractions obtained by means of TLC.

The first idea about carotenoid composition of the sample can be obtained by visual assessment of the developed TLC plate. All primary (photosynthetic) Car possess yellow coloration on the white background of silica gel. By contrast, most of the secondary Car (excepting β-cryptoxanthin) exhibit different shades of orange and red. The keto-Car possessing two keto-groups at the 4,4′-position in their molecules (astaxanthin, adonirubin, and canthaxanthin) have magenta color, whereas the keto-Car with a single keto group (adonixanthin, echinenone, 3′-hydroxyechine-

Table 4
Specific extinction coefficients ($E^{1\%}_{1cm}$) of Car common for carotenogenic green algae

Carotenoid (abbreviation)	Solvent	λ_{max}, nm	$E^{1\%}_{1cm}$
Neoxanthin (Neo)	Ethanol	442	2380[c]
	Ethanol	438	2470[a]
	Ethanol	438	2270[b]
Violaxanthin (Viola)	Acetone	442	2400[a]
	Ethanol	443	2550[c]
	Ethanol	440	2550[a]
Antheraxanthin (Anth)	Ethanol	446	2350[a]
Lutein (Lut)	Ethanol	445	2550[a]
	Diethyl ether	445	2480[a]
	Diethyl ether	445	2600[a]
	Dioxane	453	2520[c]
Zeaxanthin (Zea)	Acetone	452	2340[a]
	Petroleum ether	449	2348[a]
	Ethanol	450	2480[a]
	Ethanol	453	2450[c]
β-Carotene (ββ-Car)	Acetone	454	2500[b]
	Petroleum ether	450	2592[a]
	Ethanol	450	2620[a]
	Chloroform	465	2396[a]
α-Carotene (βε-Car)	Acetone	448	2700[b]
	Petroleum ether	444	2800[a]
	Hexane	445	2710[a]
β-Cryptoxanthin (Cryp)	Hexane	452	2460[c]
	Hexane	450	2460[a]
	Petroleum ether	449	2386[a]
	Ethanol	452	2500[c]
Echinenone (Echin)	Petroleum ether/cyclohexane	458	2158[a]
		461	2110[d]
Canthaxanthin (Can)	Cyclohexane	469	2200[d]
	Petroleum ether	466	2200[a]
Astaxanthin (Asta)	Acetone	477	2177[e]
	Acetone	480	2100[f]
	Chloroform	486	1692[e]
	Methanol	477	2307[e]
	Dimethyl sulfoxide	492	1981[e]
	Dimethyl sulfoxide	492	2220[f]
	Dimethylformamide	486	2065[e]
	Hexane	470	2100[f]

Compiled from Britton [20][a], Jeffrey, Mantoura [23][b], Egeland, Garrido [19][c], Surmatis, Walser [54][d], Aquasearch Inc [55][e], Li, Miao [56][f]

Table 5
Absorption maxima of the keto-Car (before and after reduction with $NaBH_4$) in ethanol

Carotenoid	Intact λ_{max}, nm	Reduced λ_{max}, nm
Adonixanthin	460[a]	449, 478[a]
Adonirubin	464[b]	(428), 451, 478[b]
Astaxanthin	480[a]	450, 474[a]
Canthaxanthin	476[a]	450, 474[a]
Echinenone	462[a]	450, 474[a]
3′-Hydroxyechinenone	470[c]	(430), 451, 478[c]

Adapted from Weber [44][a], Bernhard, Englert [40][b]; Cooper, Davis [39][c]

none) possess red-orange color [10, 19] (*see* **Note 13**). Hydroxyl groups in the 3,3′ position affect the spectral properties of the keto-Car only slightly but vastly increase their affinity to silica gel. Therefore, free keto-Car possessing 3,3′ hydroxyls and 4,4′ keto groups display, after developing with hexane-based solvent systems, low Rf values close to those of the primary Car. On the contrary, mono- and di-acyl esters of the keto-Car are more mobile so they are better separated from each other as well as from the free keto-Car.

The Car of the most of microalgal species are represented by carotenes and xanthophylls with characteristic (three-headed) absorption maxima in the blue part of the spectrum, 400–490 nm [20–22]. In most cases, the secondary Car typical of Chlorophyceae can be identified by combination of their UV-Vis absorbance spectra, TLC mobility (Rf), and color chemical tests revealing the type, number, and/or positions of the functional groups in their molecules. The UV-Vis spectra of the substance in question are recorded in several solvents. The positions and the amplitude ratios of the maxima are then compared with reference values (Table 3). Since reference values from different sources may vary, the comparison of the spectra of the compound in question and its pure standard recorded under the same conditions gives more reliable results (*see* Table 2 and [23]).

3.4.1 The Chromatographic Test

1. Carry out cochromatography of the total Car or Car fractions with pure Car standards using the method described above.
2. Check coincidence of the TLC mobility (Rf) of the compound in question and those of the corresponding standard. The Rf match suggests the same polarity and double bond number in

both compounds [10]. Cochromatographing of the carotenoid extracts of unknown composition with those from algal species with carotenoid composition is well established (e.g., *Haematococcus pluvialis* [24] and *Chromochloris* (Chlorella) *zofingiensis* (Dönz) [25]).

3.4.2 The Test for the Presence of 4,4′-Keto Groups Conjugated with the Main Polyene Chain

In this reaction, the keto groups are reduced to hydroxyls resulting in dramatic change of the absorption spectrum shape and a shift of the maxima to shorter wavelengths (by 10–30 nm and the number of keto groups in the carotenoid molecule (Table 5).

1. To reduce the carotenoid molecules with $NaBH_4$ [26], dissolve the studied fraction of Car in 1–2 mL of ethanol or tetrahydrofuran, and record the absorption spectrum of the solution in the range 400–750 nm.
2. Add several crystals of $NaBH_4$ to the solution in the cuvette and put it to a refrigerator for 3 h.
3. Repeat the measurement of the absorption spectrum. Transformation of the weakly resolved spectra of keto-Car to typical three-headed spectra indicates the positive test. A hypsochromic shift of 10–16 nm suggests the presence of a single keto group in the molecule of tested keto-Car. A more profound hypsochromic shift (up to 30 nm) is typical of keto-Car possessing keto groups in both of their β-rings.

3.4.3 The Test for the Presence of Hydroxyl Groups

1. Take the studied Car sample to dryness on the rotary evaporator, and dissolve the dry sample in 2 mL of pyridine; add 0.2 mL of acetic anhydride.
2. Let the mixture stand in darkness at room temperature for 21 h.
3. Transfer the mixture to petroleum ether by placing it to a separation funnel and adding water.
4. Discard the lower (water) phase. Collect the upper (hexane) layer, and wash it with water.
5. Collect the washed upper phase, dry over anhydrous $NaSO_4$, and evaporate on a rotary evaporator to a final volume of 0.5 mL.
6. Apply the sample to a TLC plate (silica gel G) along with the untreated initial Car sample; develop the plate with toluene/methanol mixture (19:1, v/v).
7. Positive test (formation of esters) is evidenced by a considerable increase in Rf of the treated Car sample in comparison of the untreated sample [26].

3.4.4 The Test for the Presence of Ester Bonds

The test for the presence of ester groups is based on alkaline hydrolysis of the tested Car dissolved in hexane or light petroleum-ether (40–60 °C) resulting in a considerable decrease of the tested Car.

1. To prevent oxidative degradation of the Car, the reaction is carried out in darkness, in the atmosphere of nitrogen at 4 °C with addition of 0.1% BHT to the reaction mixture [26].
2. To the solution of the tested Car in hexane containing 0.1% BHT, add an equal portion of 10% KOH in methanol.
3. Sparge the headspace of the reaction vessel with nitrogen for 1 min to remove oxygen, and let the mixture stand overnight at 4 °C in darkness (up to 16 h).
4. Transfer the mixture to the separation funnel, add half of the volume of distilled water, shake well, and incubate till complete phase separation. Discard the lower (water) phase.
5. Repeat this procedure until complete removal of KOH (as indicated by a pH strip test).
6. Dry the Car fraction over anhydrous $NaSO_4$ (1 g per 10 mL of the Car solution) for 1 h, and evaporate on a rotary evaporator (or in a stream of an inert gas) to 0.5 mL.
7. Apply the resulting aliquot to a TLC plate along with the initial untreated sample. Develop with mixture of hexane and acetone (7.5:2.5, v/v).
8. Positive test is evidenced by a dramatic decrease of Rf of the resulting product in comparison with the initial unsaponified Car.

3.4.5 The Test for the Presence of β- and ε-Rings

The presence of β- or ε-cycles in molecules of Car is revealed with $AgNO_3$.

1. Carry out the TLC of the tested sample as described in Subheading 3.3.
2. Dry the developed TLC plate in a stream of nitrogen.
3. Spray the TLC plates with methanolic solution of $AgNO_3$ (2.8 g $AgNO_3$ in 100 mL MeOH).
4. Check the color. The Car containing only β-rings (e.g., zeaxanthin) develop red coloration, whereas the Car containing ε-rings (e.g., lutein) remain yellow [27].

4 Notes

1. Currently, silica gel is the most widespread adsorbent for TLC of Car. Aluminum oxide, the second most commonly used sorbent, is not suitable for separation of Car containing 3-hydroxyl and 4-keto groups (e.g., astaxanthin, adonixanthin, and adonirubin (since these Car are readily oxidized on Al_2O_3 yielding α-diketones, which are very difficult to elute from the TLC plates [10]). Combinations of normal-phase unmodified silica

gel stationary phase and a polar solvent-based mobile phase are currently the most common combinations in TLC. On the contrary, reverse-phase silica gel (C8, C18) together with polar solvent systems are widespread in HPLC. Therefore, these chromatographic techniques complement each other very well. Currently there are many brands of commercially available TLC plates. To apply highly diluted samples, use TLC plates with concentrating zone (Table 1).

2. It is also possible to apply samples to TLC plates using conventional pipettes in combination with 100–200 μL long tips used in electrophoresis for application of samples to the gel, for example, Eppendorf GELoader®.
3. Flat-bottom TLC tanks with well-ground lid will keep the solvent front line from distortions [11]. Box tanks with snap fixation of the lid, for example, Latch-lid™ (Sigma-Aldrich) or KONTES (Kimble Chase, USA), proved to be especially handy.
4. Avoid alkaline hydrolysis as much as possible since it brings about irreversible oxidation of astaxanthin into astacene.
5. If quartz sand is unavailable, it can be substituted by glass sand obtained by grinding clean microscopy coverslips, glass Pasteur pipettes, or a similar small glassware.
6. Do not store Car extracts in acetone since the pigments degrade rapidly in this solvent. The same is true for dichloromethane and diethyl ether. For long-term storage, the extracts should be evaporated to dryness and redissolved in hexane or petroleum ether and sealed under atmosphere of nitrogen. Store the extracts at the temperature −20 °C or below.
7. The TLC plates stored at a humidity >60% and/or taken from a previously open packages should be washed (developed) in a TLC tank using a polar solvent, for example, acetonitrile or the actual solvent mix used for the analysis.
8. Individual pure solvents are seldom used for separation of complex Car mixtures excepting the coarse fractionation or sample purification [10]. Binary mixtures of hexane and polar solvents are commonly used in silica gel TLC. One of more simple solvent systems (hexane/acetone; 7.5:2.5[1]) was used by Grung and D'Souza [28] for analysis of Car in mature aplanospores of *Haematococcus pluvialis* (Volvocales); corresponding Rf were β-carotene, 1.0; echinenone, 0.87; astaxanthin diesters, 0.75–0.85; astaxanthin monoesters, 0.6; canthaxanthin, 0.44; astaxanthin, 0.33; and lutein, 0.28. More polar systems containing acetone were also sug-

[1] Here and below, the proportions of mixture component are specified by volume.

gested: hexane/acetone, 7:3 [14] including ternary systems like hexane/acetone/chloroform, 2:2:1 [29] and acetone/ petroleum ether/chloroform, 2:2:1 [30]. The following binary solvent systems were used for the separation of Car extracted with 90% acetone and saponified in (a) hexane/ petroleum ether/acetone, 8:2; (b) hexane/acetone, 7:3; and (c) benzene/ethyl acetate, 7:3 v/v [31]. A higher separation efficiency was achieved using a "semi-gradient" TLC [10] involving sequential development in two solvent systems of increasing polarity. This approach tested in *Bracteacoccus minor* (Sphaeropleales) [13] allowed to resolve all primary and secondary Car including free and esterified species.

9. Composing of an efficient mobile phase for TLC is seldom trivial. Although comprehensive theories and manuals were developed for TLC [11, 12, 32], the mobile phase is selected and then optimized from scratch for each particular combination of analytes, sorbents, and available solvents. New software for TLC and HPLC mobile phase composition can be of some help [33, 34]. Generally, mixtures of Car of unknown composition are separated several times on silica gel TLC using solvents of increasing polarity to reveal the key groups of Car and select an algorithm of their further analysis. Preliminary open-column separation of complex Car mixtures into fractions dramatically differing in their polarity often can streamline the subsequent TLC analysis [10, 26].

   Solvents for TLC mobile phases should:

   (a) Be free of water and other impurities, especially those capable of oxidation of Car. HPLC-grade solvents are normally the best choice.
   (b) Be able to dissolve both polar and less polar Car.
   (c) Have a sufficient eluotropic capability.
   (d) Be volatile enough to evaporate quickly from the developed TLC plate.

   Optimization of the mobile phase for TLC should be carried out taking into account the physical-chemical properties of solvents (Table 6). Thus, the higher the permeability coefficient (k (mm^2/s), defined as the rate of eluent propagation in a solid stationary phase), the longer the development time and high TLC spot swelling.

   Another important parameter is the eluotropic strengths (sorption energy) of the solvent, ε, defined relatively to that of pentane, ε^0 (Table 6). According to the ε values, solvents are

Table 6
Physical-chemical properties of organic solvents (at 20 °C) commonly used for separation of Carotenoids on silica gel TLC plates [11, 32, 35, 36]

Solvent	Boiling point (°C)	Viscosity η, (mPa·s)	Surface tension, γ, dyne per cm	Dielectric constant (ε^1)	Steam pressure, mbar (ρ)	Solvent strength (ε^0)
Pentane	36.1	0.24	16.0	1.84	430	0
Hexane	68.9	0.31	18.4	1.88	120	0
Heptane	98.4	0.40	20.4	1.92	36	0
Cyclopentane	80.7	1.84	25.5	2.02	78	0.03
Carbon tetrachloride	76.5	0.94	27.0	2.24	90	0.14
Benzene	80.1	0.63	28.9	2.25	75	0.25
Toluene	110.6	0.57	—	2.4	22	0.22
Chloroform	61.7	0.56	27.1	4.81	158	0.31
Ethyl acetate	77.7	0.45	23.9	6.00	73	0.48
Dichloromethane	40.0	0.42	26.5	9.08	340	0.3
2-Butanol	117.3	2.8	24.6	17.8	—	—
Methyl ethyl ketone	79.6	0.43	24.6	18.5	79	—
1-Propanol	94.4	2.15	23.8	20.1	14	0.63
Acetone	56.2	0.32	12.7	20.7	175	0.5
Tetrahydrofuran	67.0	0.47	—	—	150	0.53
Ethanol	78.5	1.16	22.8.	24.3	44	0.68
Methanol	65	0.58	22.6	32.63	96	0.7
Acetonitrile	81.6	0.36	29.3	37.5	73	0.6

As could be seen from Table 2, TLC allows using of a wider array of solvents. Thus, it is possible to use eluents with a high vapor pressure such as dichloromethane or diethyl ether, which cannot be used in HPLC

arranged into eluotropic series; the exact order of solvents in an eluotropic series depends on the specific sorbent used [10–12, 35]. Affinity of common secondary carotenoids to silica gel is characterized in Table 7.

10. It is critical to achieve a level line of the solvent front. Common causes of the solvent line distortion that should be avoided include (i) tilted dipping of the TLC plate into the solvent mix, (ii) excessive filling of the TLC tank (the solvent mix level

Table 7
The adsorption affinity of the main carotenoid groups under TLC on silica gel G (in ascending order, adapted from [10])

Carotenoid groups	Secondary Car of green algae
Carotenes (hydrocarbons)	β-Carotene
Monohydroxycarotenoid acyl esters	β-Cryptoxanthin acyl esters
Dihydroxycarotenoid diacyl esters	Zeaxanthin diacyl esters
Dihydroxyketocarotenoid diacyl esters	Astaxanthin, adonixanthin diacyl esters
Monoketocarotenoids	Echinenone
Monohydroxycarotenoids	β-Cryptocanthin
Dihydroxycarotenoid monoacyl esters	Zeaxanthin monoacyl esters
Diketocarotenoids	Canthaxanthin
Dihydroxyketocarotenoid monoacyl esters	Astaxanthin, monoacyl esters
Monohydroxy/monoketocarotenoids	3′-Hydroxyechinenone
Dihydroxycarotenoids	Zeaxanthin
Dihydroxyketocarotenoids	Astaxanthin, adonixanthin

should be below the start line), (iii) contact of the plate with the filter paper covering the walls of the tank, (iv) traces of water in the solvent, and (v) irregular edge(s) of the TLC plate.

11. To accelerate evaporation and decrease the oxidation of the sample with oxygen from the air, the TLC plate in the place of the sample application should be blowed by the stream of an inert gas (e.g., nitrogen or argon) using a hose with a glass tip. The upper part of the plate can be covered with a clean glass slide.
12. The amount of the sample applied as well as the quality of its application is critical for the quality and efficiency of separation. Overloading of the plate or insufficient application of the sample leads to swelling of the spots already at the start and to distortion of the solvent front and hence to a poor separation.
13. The Car adsorbed on the silica gel TLC plate are readily oxidized in the air and hence change their color. Therefore, the separations should be documented by taking their photos through the glass wall of the TLC tank or immediately upon removal of the developed plate from the tank. To improve the vapor saturation of the TLC tank, inner surfaces of the walls of

tall TLC tanks (e.g., those for 20 × 20 cm plates) should be covered with filter paper. Avoid contacts of the paper with the TLC plate.

Acknowledgment

Financial support of Russian Science Foundation (grant # 14-50-00029) is gratefully appreciated.

References

1. Britton G (1985) General carotenoid methods. Methods Enzymol 111:113–149
2. Takaichi S (2013) Tetraterpenes: carotenoids. In: Ramawat KG, Mérillon J-M (eds) Natural Products. Springer, Berlin, Heidelberg
3. Green B, Durnford D (1996) The chlorophyll-carotenoid proteins of oxygenic photosynthesis. Annu Rev Plant Biol 47:685–714
4. Solovchenko A (2013) Physiology and adaptive significance of secondary carotenogenesis in green microalgae. Russ J Plant Physiol 60:1–13
5. Lorenz RT, Cysewski GR (2000) Commercial potential for *Haematococcus* microalgae as a natural source of astaxanthin. Trends Biotechnol 18:160–167
6. Boer L (2013) Biotechnological production of colorants. In: Biotechnology of food and feed additives. Springer, Berlin, Heidelberg
7. Guerin M, Huntley M, Olaizola M (2003) *Haematococcus* astaxanthin: applications for human health and nutrition. Trends Biotechnol 21:210–216
8. Otani H (2013) Site-specific antioxidative therapy for prevention of atherosclerosis and cardiovascular disease. Oxid Med Cell Longev 2013:79689. https://doi.org/10.1155/2013/796891
9. Sussela M, Toppo K (2006) *Haematococcus pluvialis*—a green alga, richest natural source of astaxanthin. Curr Sci 90:1602–1603
10. Britton G (2008) TLC of carotenoids. In: Waksmundzka-Hajnos M, Sherma J, Kowalska T (eds) Thin layer chromatography in phytochemistry. CRC Press, Boca Raton, FL
11. Hahn-Deinstrop E (2007) Applied thin layer chromatography: best practice and avoidance of mistakes, 2nd edn. Wiley, New York
12. Spangenberg B, Poole CF, Weins C (2011) Quantitative thin-layer chromatography: a practical survey. Springer, Heidelberg
13. Minyuk G, Chelebieva E, Chubchikova I (2014) Secondary carotenogenesis of the green microalga *Bracteacoccus minor* (Chodat) Petrova (Chlorophyta) in a two-stage culture. Inter J Algae 16:354–368
14. Sarada R et al (2006) An efficient method for extraction of astaxanthin from green alga *Haematococcus pluvialis*. J Agric Food Chem 54:7585–7588
15. Dong S et al (2014) Four different methods comparison for extraction of astaxanthin from green alga *Haematococcus pluvialis*. Sci World J 2014:694305. https://doi.org/10.1155/2014/694305
16. Domınguez-Bocanegra A et al (2004) Influence of environmental and nutritional factors in the production of astaxanthin from *Haematococcus pluvialis*. Bioresour Technol 92:209–214
17. Lichtenthaler H (1987) Chlorophyll and carotenoids: pigments of photosynthetic biomembranes. Methods Enzymol 148:350–382
18. Lemoine Y, Schoefs B (2010) Secondary ketocarotenoid astaxanthin biosynthesis in algae: a multifunctional response to stress. Photosynth Res 106:155–177
19. Egeland E et al (2011) Data sheets aiding identification of phytoplankton carotenoids and chlorophylls. In: Roy S et al (eds) Phytoplankton pigments: characterization, chemotaxonomy and applications in oceanography. Cambridge University Press
20. Britton G (1995) UV/visible spectroscopy. In: Britton G, Liaaen-Jensen S, Pfander H (eds) Carotenoids. Birkhauser, Basel
21. Young A (1993) Occurrence and distribution of carotenoids in photosynthetic systems. In: Young A, , Britton G (eds) Carotenoids in photosynthesis. Chapman and Hall: London
22. Goodwin TW (1961) Biosynthesis and function of carotenoids. Annu Rev Plant Physiol 12:219–244

23. Jeffrey S, Mantoura R, Bjørnland T (1997) Data for the identification of 47 key phytoplankton pigments. In: Jeffrey SW, RFC M, Wright SW (eds) Phytoplankton pigments in oceanography: guidelines to modern methods. UNESCO Publishing, Paris
24. Boussiba S (2000) Carotenogenesis in the green alga *Haematococcus pluvialis*: cellular physiology and stress response. Physiol Plant 108:111–117
25. Liu J et al (2014) *Chlorella zofingiensis* as an alternative microalgal producer of astaxanthin: biology and industrial potential. Mar Drugs 12:3487–3515
26. Rodriguez-Amaya DB (2001) A guide to carotenoid analysis in foods. ILSI Press, Washington
27. Isaksen M, Francis GW (1990) Silver ion spray reagent for the discrimination of the β- and ε-end groups in carotenoids on thin-ayer chromatograms. Chromatographia 29:363–365
28. Grung M et al (1992) Algal carotenoids 51 secondary carotenoids 2 *Haematococcus pluvialis* aplanospores as a source of (3S, 3′ S)-astaxanthin esters. J Appl Phycol 4:165–171
29. Chekanov K et al (2014) Accumulation of astaxanthin by a new *Haematococcus pluvialis* strain BM1 from the White Sea coastal rocks (Russia). Mar Drugs 12:4504–4520
30. Elumalai S, Santhose BI, Kanna GR (2014) Extraction of carotenoid and thin layer chromatography (TLC), GC–MS, FT–IR and HPLC analysis of pharmaceutically important pigment astaxanthin from a new strain of *Haematococcus pluvialis*. Week Sci Res J 2:2321–7871
31. Liu B, , Lee Y(1999) Composition and biosynthesis pathways of carotenoids in the astaxanthin-produsing green alga *Chlorococcum* sp. Biotechnol Lett 21:1007–1010
32. Snyder LR (1968) Principles of adsorption chromatography; the separation of nonionic organic compounds. Dekker, New York
33. Härmälä P et al (1992) Retention behaviour of closely related coumarins in thin-layer chromatographic preassays for high-performance liquid chromatography according to the "PRISMA" model. J Chromatogr A 593:329–337
34. Nyiredy S (2002) Planar chromatographic method development using the PRISMA optimization system and flow charts. J Chromatogr Sci 40:553–563
35. Barwick VJ (1997) Strategies for solvent selection—a literature review. TrAC 16:293–309
36. Weins C (2010) Quantitative thin-layer chromatography: a practical survey. Springer, Heidelberg
37. Andrewes AG, Phaff HJ, Starr MP (1976) Carotenoids of Phaffia rhodozyma, a red-pigmented fermenting yeast. Phytochemistry 15:1003–1007
38. Qin S, Liu G-X, Hu Z-Y (2008) The accumulation and metabolism of astaxanthin in *Scenedesmus obliquus* (Chlorophyceae). Process Biochem 43:795–802
39. Cooper RD et al (1975) Carotenoids and related compounds. Part XXXII. Synthesis of astaxanthin, phoenicoxanthin, hydroxyechinenone, and the corresponding diosphenols. J Chem Soc Perkin Trans 1:2195–2204
40. Bernhard K et al (1981) Synthese von optisch aktiven, natürlichen Carotinoiden und strukturell verwandten Naturprodukten. IX. Synthese von (3R)-Hydroxyechinenon, (3R,3′ R)-und (3R,3′S)-Adonixanthin, (3R)-Adonirubin, deren optischen Antipoden und verwandten Verbindungen. Helv Chim Acta 64:2469–2484
41. Matsuno T, Maoka T, Ikuno Y (1986) Comparative biochemical studies of carotenoids in fish–XXVII. Carotenoidsin the eggs of three species of cyprinidae. Comp Biochem Physiol B:Comp Biochem 83:335–337
42. Fraser PD, Shimada H, Misawa N (1998) Enzymic confirmation of reactions involved in routes to astaxanthin formation, elucidated using a direct substrate in vitro assay. Eur J Biochem 252:229–236
43. Castillo R, Negre-Sadargues G, Lenel R (1982) General survey of the carotenoids in crustacea. In: Britton G, Goodwin TW (eds) Carotenoid chemistry and biochemistry. Proceedingof the 6th International Symposium on Carotenoids, Liverpul July 1981. Pergamon Press, Oxford, p 211
44. Weber A (1975) Chlorophylle und Carotinoide der Chaetophorineae (Chlorophyceae; Ulotrichales). Arch Mikrobiol 102:45–52
45. Kopczynski M et al (2005) Ultrafast transient lens spectroscopy of various C 40 carotenoids: lycopene, β-carotene, (3R, 3′R)-zeaxanthin, (3R, 3′R, 6′R)-lutein, echinenone, canthaxanthin, and astaxanthin. Phys Chem Chem Phys 7:2793–2803
46. Grung M, Metzger P, Liaaen-Jensen S (1989) Primary and secondary carotenoids in two races of the green alga *Botryococcus braunii*. Biochem Syst Ecol 17:263–269
47. Laatsch H, Schwab D (1984) Identification and quantification of carotenoids and carot-

enoid esters from the foraminiferan *Allogromia laticollaris* Arnold. J Foramin Res 14:36–49
48. Lorquin J, Molouba F, Dreyfus BL (1997) Identification of the carotenoid pigment canthaxanthin from photosynthetic bradyrhizobium strains. Appl Environ Microbiol 63:1151–1154
49. Holt TK, Krogmann DW (1981) A carotenoid-protein from cyanobacteria. Biochim Biophys Acta (BBA)—Bioenerg 637:408–414
50. Rowan KS (1989) Photosynthetic pigments of algae. Cambridge University Press, Cambridge
51. Yokoyama A, Miki W (1995) Composition and presumed biosynthetic pathway of carotenoids in the astaxanthin-producing bacterium *Agrobacterium aurantiacum*. FEMS Microbiol Lett 128:139–144
52. Foppen FH (1971) Tables for the identification of carotenoid pigments. Chromatogr Rev 14(3):133–298
53. Bjørnland T, Aguilar-Martinez M (1976) Carotenoids in red algae. Phytochemistry 15:291–296
54. Surmatis J et al (1970) New methods for the synthesis of oxygenated carotenoids. Helv Chim Acta 53:974–990
55. Aquasearch Inc (1999) Derivation of astaxanthin light absorption coefficients in different solvents. TR1004011 Aquasearch Inc, Kailua-Kona
56. Li Y et al (2012) Accurate quantification of astaxanthin from *Haematococcus* crude extract spectrophotometrically. Chinese J Oceanol Limnol 30:627–637

Chapter 5

Batch Cultivation for Astaxanthin Analysis Using the Green Microalga *Chlorella zofingiensis* Under Multitrophic Growth Conditions

Jin Liu

Abstract

Astaxanthin represents a high-value ketocarotenoid that has been gaining great attention from both the science and public communities in recent years. Microalgae are the primary source of natural astaxanthin. *Chlorella zofingiensis*, a freshwater and oleaginous green microalga, is capable of growing well photoautotrophically, heterotrophically, and mixotrophically for high-density biomass and astaxanthin production. Here we describe the astaxanthin production pipeline using *C. zofingiensis* as cell factories under different trophic conditions, including strain preparation, inoculation, and cultivation, biomass harvest and dewatering, and astaxanthin extraction, determination, and quantification.

Key words *Chlorella zofingiensis*, Microalga, Astaxanthin, Autophototrophic, Heterotrophic, Thin-layer chromatography, TLC, High-performance liquid chromatography, HPLC

1 Introduction

Astaxanthin (3,3′-dihydroxy-β,β-carotene-4,4′-dione) is a high-value red ketocarotenoid with strong antioxidant activity and possesses a broad range of applications in food, feed, nutraceutical, and pharmaceutical industries [1, 2]. Astaxanthin has a chemical formula of $C_{40}H_{52}O_4$ and is present in nature predominantly in the form of 3S, 3'S (Fig. 1), primarily sourced from green microalgae [3]. *Haematococcus pluvialis* and *Chlorella zofingiensis* represent the most promising algal producers of natural astaxanthin [2, 4]. The ability of *H. pluvialis* to accumulate astaxanthin up to 4% of its dry biomass [5], the highest known content in nature, has led to research focus on *H. pluvialis* for commercial astaxanthin production. However, it grows relatively slowly with a low biomass yield and is susceptible to contamination by other fast-growing organisms [3, 6, 7]. In addition, extremely high light illumination is required for astaxanthin induction in this alga thus hindering its

Carlos Barreiro and José-Luis Barredo (eds.), *Microbial Carotenoids: Methods and Protocols*, Methods in Molecular Biology, vol. 1852, https://doi.org/10.1007/978-1-4939-8742-9_5,

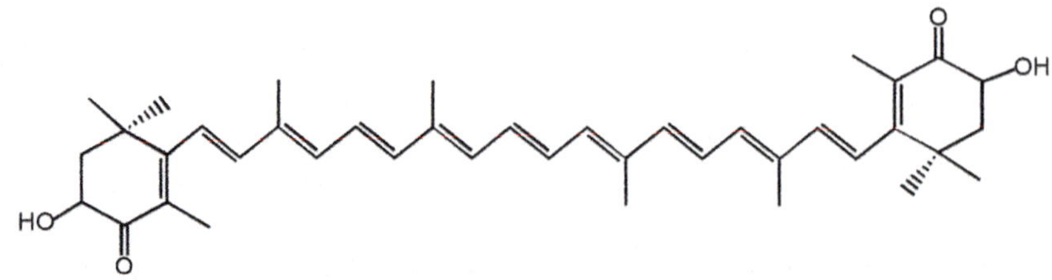

Fig. 1 The chemical structure of astaxanthin

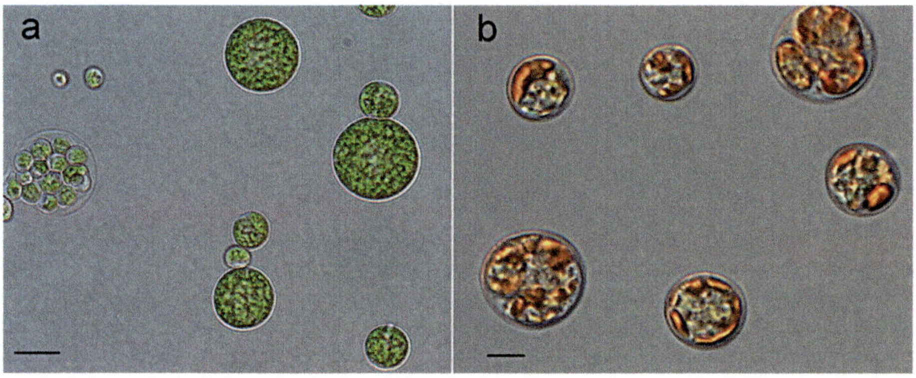

Fig. 2 Light micrographs of ***Chlorella zofingiensis*** cells under favorable (**a**) and stress (**b**) conditions. Bars, 5 μm

commercial application [8, 9]. By contrast, *C. zofingiensis* grows fast photoautotrophically, heterotrophically, and mixotrophically, is easy to be cultured and scaled up both indoors and outdoors, and can achieve high cell densities [2, 10–12]. These robust biotechnological traits enable *C. zofingiensis* to be a promising alternative producer of astaxanthin.

C. zofingiensis is a fresh water green microalga; its cell is nonmotile and in unicellular and spherical form, with the cell size ranging from 2 to 15 μm in diameter (Fig. 2). The astaxanthin production by *C. zofingiensis* under different trophic conditions has been reviewed [2]. Generally, heterotrophy gives rise to high cell density (up to 53 g/L) but with low astaxanthin content (less than 2 mg/g); by contrast, photoautotrophy with high light benefits astaxanthin accumulation, which can reach up to 6 mg/g. To combine the advantages of both heterotrophy and photoautotrophy, a two-stage heterotrophy-to-photoautotrophy (H2P) strategy has been developed: *C. zofingiensis* cells are first grown heterotrophically to allow for ultrahigh biomass (biomass accumulation stage), and the biomass are then subject to photoautotrophic growth for astaxanthin biosynthesis (astaxanthin induction stage). Here we present the procedures to grow *C. zofingiensis* for astaxanthin accumulation under different trophic conditions and the downstream procedures for biomass harvest and dewatering and astaxanthin extraction and determination.

2 Materials

Prepare solutions using deionized or ultrapure water (prepared by purifying, to attain a sensitivity of 18 MΩ cm at 25 °C) and analytical grade reagents. Prepare and store all reagents at room temperature (unless indicated otherwise). Diligently follow all waste disposal regulations when disposing waste materials.

2.1 Algal Cultivation

1. *Chlorella zofingiensis* (ATCC 30412) purchased from the American Type Culture Collection (ATCC, Rockville, MD, USA).
2. Erlenmeyer flasks: Glass, 100 and 250 mL.
3. Glass columns (250 mL), artificial lights, air pump, compressed CO_2, gas flow meter and proportioner.
4. Orbital shaker.
5. Biosafety cabinet, chemical hood.
6. Autoclave.
7. Modified BG-11 medium (MBG): MBG medium is used for the maintenance and photoautotrophic cultivation of *C. zofingiensis*. Prepare the following stock solutions using deionized water: stock solution 1 ($NaNO_3$, 90 g/L), stock solution 2 (K_2HPO_4, 40 g/L), stock solution 3 ($MgSO_4{\cdot}7H_2O$, 7.5 g/L), stock solution 4 ($CaCl_2{\cdot}2H_2O$, 3.6 g/L), stock solution 5 (citric acid, 0.6 g/L; and ferric ammonium citrate, 0.6 g/L), stock solution 6 (Na_2CO_3, 2 g/L), and stock solution 7 (H_3BO_3, 2.86 g/L; $MnCl_2{\cdot}4H_2O$, 1.81 g/L; $ZnSO_4{\cdot}7H_2O$, 0.22 g/L; $NaMoO_4{\cdot}2H_2O$, 0.39 g/L; $CuSO_4{\cdot}5H_2O$, 0.08 g/L; $Co(NO_3)_2{\cdot}6H_2O$, 0.05 g/L; and Na_2EDTA, 1 g/L). Transfer 10 mL solutions 1–6 and 1 mL solution 7 into a cylinder. Add water to a volume of 900 mL. Mix and adjust pH to 7.0 with 1.0 M HCl or NaOH. Make up to 1 L with water and autoclave. For the nitrogen-depleted MBG medium (MBG-N), omit solution 1. For solid medium, add 15 g/L agar prior to autoclave.
8. CZ-M1 medium: CZ-M1 medium is used for the heterotrophic cultivation of *C. zofingiensis*. Prepare the following stock solutions using deionized water: stock solution 1 (KNO_3, 101 g/L), solution 2 ($Na_2HPO_4{\cdot}H_2O$, 62 g/L; and $NaH_2PO_4{\cdot}2H_2O$, 8.9 g/L), stock solution 3 ($MgSO_4{\cdot}7H_2O$, 24.7 g/L), stock solution 4 ($CaCl_2{\cdot}2H_2O$, 14.7 g/L), and stock solution 5 ($FeSO_4{\cdot}7H_2O$, 6.95 g/L; H_3BO_3, 0.061 g/L; $MnSO_4{\cdot}H_2O$, 0.16 g/L; $ZnSO_4{\cdot}7H_2O$, 0.287 g/L; $CuSO_4{\cdot}5H_2O$, 0.0025 g/L; and $Co(NO_3)_2{\cdot}6H_2O$, 0.01235 g/L). Transfer 10 mL solutions 1–4, 1 mL solution 5, and 40 g glucose into a cylinder. Add water to a volume of 900 mL. Mix and adjust pH to 6.8 with 1.0 M HCl or NaOH. Make up to 1 L with water and autoclave.

2.2 Algal Biomass Harvest and Drying

1. Centrifuge tubes: Polypropylene, conical bottom, 50 mL.
2. Centrifuge: Equipped with a 6 × 50 mL conical fixed-angle rotor.
3. Freeze-dryer.

2.3 Astaxanthin Extraction and Analysis

1. Pestle and mortar.
2. Liquid nitrogen.
3. Nitrogen gas.
4. Glass vials: Amber, with PTFE-Lined caps, 12 mL.
5. Solvent A: Acetone, high-performance liquid chromatography (HPLC) grade. Solvent A is used for astaxanthin extraction.
6. Solvent B: Hexane/tert-butylmethyl ether/acetic acid (80:20:2, by volume). All organic solvents are HPLC grade. Solvent B is used as the development reagent for the thin-layer chromatography (TLC) analysis of astaxanthin.
7. Solvent C: acetonitrile/methanol/0.1 M Tris–HCl (pH 8.0) (84:2:14, by volume); solvent D, methanol/ethyl acetate (68:32, v/v). All organic solvents are HPLC grade. Filter Tris–HCl using a Millipore filter (0.22 μm) prior to preparing solvent C. Solvent C and solvent D are used as the mobile phase for the HPLC analysis of astaxanthin.
8. TLC plates: silica gel 60, 10 × 10 cm, 0.25 mm thickness.
9. HPLC: Waters HPLC system equipped with a Waters Spherisorb® ODS2 analytical column (5 μm, 4.6 × 250 mm).

3 Methods

Conduct all procedures at room temperature unless otherwise specified. All materials used for algal cultivation are sterilized by autoclave. All procedures with respect to algal cultivation are conducted in a biosafety cabinet.

3.1 *Chlorella zofingiensis* Maintenance

1. Maintain this alga at 16 °C on an agar plate of MBG medium with dim light (less than 5 μE/m^2/s), and subculture every 3 months (*see* **Note 1**).

3.2 Photoautotrophic Cultivation of *Chlorella zofingiensis*

1. Inoculate a single colony of *C. zofingiensis* from the agar plate in Subheading 3.1, **step 1**, into a 100 mL flask containing 10 mL MBG medium and grow under illumination of 30 μE/m^2/s at 25 °C with orbital shaking of 125 rpm for 6 days to reach the late stationary growth phase (*see* **Note 2**).
2. Transfer the 10 mL culture from Subheading 3.2, **step 1**, into a 250 mL glass column containing 90 mL MBG medium, and allow to grow for 6 days to reach the late stationary growth

Fig. 3 Algal cultures in the 250 mL columns aerated with 1.5% CO_2

phase, with continuous illumination of 70 μE/m^2/s and aeration of 1.5% CO_2 (*see* **Note 3** and Fig. 3).

3. Transfer the 100 mL cultures into two 50 mL centrifuge tubes, centrifuge at 3000 × *g* for 5 min to pellet the cells, and discard the supernatant.
4. Rinse the cell pellet with 50 mL MBG-N medium, centrifuge at 3000 × *g* for 5 min to pellet the cells, and discard the supernatant.
5. Repeat **step 4** in Subheading 3.2 twice.
6. Resuspend the cell pellet with 1000 mL MBG or MBG-N medium, and transfer 100 mL to each 250 mL glass column (1/10 dilution of the cultures from Subheading 3.3, **step 2**).
7. Place the cultures suspended with MBG medium under the illumination of 350 μE/m^2/s or the cultures suspended with MBG-N medium under the illumination of 70 μE/m^2/s for 6 days to allow alga growth and astaxanthin accumulation (*see* **Note 4**).

3.3 Heterotrophic Cultivation of *Chlorella zofingiensis*

1. Inoculate a single colony of *C. zofingiensis* from the agar plate in Subheading 3.1, **step 1**, into a 100 mL flask containing 10 mL CZM1 medium, and grow without light at 25 °C with orbital shaking of 125 rpm for 6 days.
2. Transfer the 10 mL culture from Subheading 3.3, **step 1**, into a 250 mL flask containing 90 mL CZM1 medium, and allow to grow for 4 days.

3. Inoculate 10 mL of the seed culture from Subheading 3.3, **step 2**, into 250 mL flasks containing 90 mL CZM1 medium, and grow in the dark at 25 °C with orbital shaking of 125 rpm for 8 days to allow algal growth and astaxanthin accumulation (*see* **Note 5**).

3.4 Heterotrophy-to-Photoautotrophy (H2P) Cultivation of *Chlorella zofingiensis*

1. Transfer the cultures from Subheading 3.3, **step 3**, into 250 mL glass columns, and grow under the illumination of 350 μE/m^2/s for 4 days to allow astaxanthin accumulation (*see* **Note 6**).

3.5 Algal Biomass Harvest, Drying, and Astaxanthin Extraction

1. Transfer 50 mL cultures from Subheading 3.2, **step 7**, Subheading 3.3, **step 3**, or Subheading 3.4, **step 1**, into 50 mL centrifuge tubes; centrifuge at 3000 × *g* for 5 min.
2. Discard the supernatant, rinse the cell pellet with ultrapure water, and centrifuge at 3000 × *g* for 5 min.
3. Repeat **step 2** in Subheading 3.5 twice.
4. Put the harvested algal samples in a freeze-dryer (−50 °C), and dry for 2 days (*see* **Note 7**).
5. Weight 20 mg of the lyophilized algal cells and transfer to a mortar. Grinding vigorously in the presence of liquid nitrogen to break cell wall (*see* **Note 8**).
6. Add solvent A (2 mL) into the mortar, and grind again for 1 min.
7. Wash off the ground cells, transfer into a 12 mL glass vial, and bring to a final volume of 6 mL with solvent A. Vortex with the full speed for 1 min (*see* **Note 9**).
8. Centrifuge at 5000 × *g* for 5 min to separate the astaxanthin-containing solvent phase from the algal residue (*see* **Note 10**).
9. Transfer the upper solvent phase into a new glass vial, dry under nitrogen gas flow, and resuspend in 0.2 mL solvent A for immediate analysis or stored at 20 °C (*see* **Note 11**).

3.6 TLC Analysis of Astaxanthin

1. Pipette 10 μL sample from Subheading 3.5, **step 9**, at the bottom of a TLC plate (1 cm from the edge), and allow to dry for 5 min (*see* **Note 12**).
2. Pour solvent B in a glass container with a depth of up to 0.8 cm.
3. Place the TLC plate in the container, seal with a glass lid, and allow to develop until the solvent reach 1 cm below the top of the TLC plate (*see* **Note 13** and Fig. 4).
4. Allow to dry for 5 min in the hood.

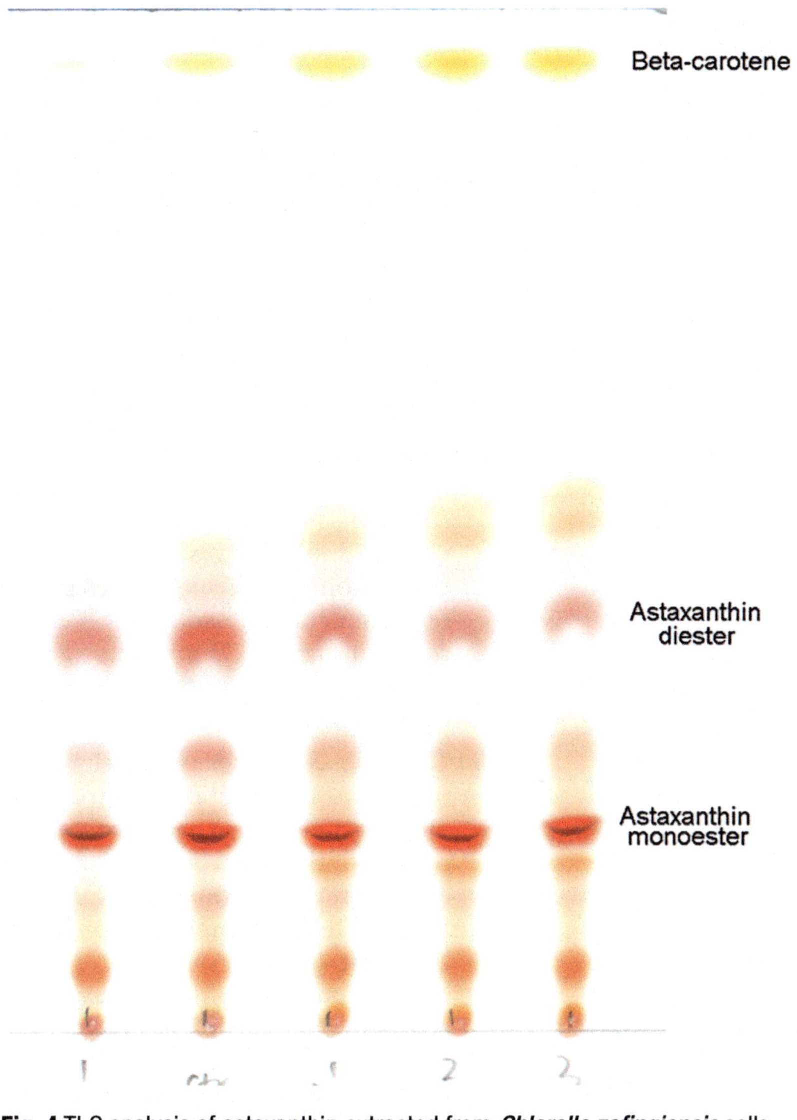

Fig. 4 TLC analysis of astaxanthin extracted from ***Chlorella zofingiensis*** cells

5. Scan TLC plate and save as a TIFF file. The spots circled are esterified astaxanthin.

3.7 HPLC Analysis of Astaxanthin

1. Load 10 μL sample from Subheading 3.5, **step 9**, onto the HPLC system and elute at a flow rate of 1.2 mL/min with a linear gradient from 100% solvent C to 100% solvent D over a 15 min period, followed by 10 min of solvent D [12].
2. Identify individual carotenoids by their absorption spectra and typical retention times compared to carotenoid standards.
3. Quantify astaxanthin using the standard curve.

4 Notes

1. Seal the petri dish tightly to avoid the drying of the agar and the possible contamination of *C. zofingiensis* by other organisms especially bacteria. In some cases, antibiotics (e.g., ampicillin and kanamycin, 100 μg/mL) can be supplemented to avoid bacteria growth. *C. zofingiensis* is not sensitive to those antibiotics.
2. The colony of *C. zofingiensis* used for inoculation should be green and big enough. Otherwise, the cells are not in good condition and may take longer to reach the late stationary growth phase.
3. Compressed air and CO_2 are mixed at a ratio of 100:1.5 through a gas proportioner to make 1.5% CO_2, which passes through a hollow glass rod right above the bottom of the glass column and produces bubbles to aerate the algal cultures. The generation of 2–4 air bubbles per second is enough. To protect from contamination, the 1.5% CO_2-enriched air is subject to an air filter (0.22 μm) prior getting into the cultures. Keep an eye on the gas proportioner to avoid the fluctuation of ratio of air and CO_2.
4. *C. zofingiensis* cells synthesize and accumulate astaxanthin under stress conditions. Nitrogen depletion and high light illumination are the commonly used stresses. High light stress refers to that the light intensity provided to the cultures exceeds the saturation value leading to the growth photoinhibition. The photoinhibition of *C. zofingiensis* is observed when the light intensity is above 250 $\mu E/m^2/s$. In the current experiment, the light intensity of 350 $\mu E/m^2/s$ is used, as it enables *C. zofingiensis* to produce more astaxanthin. The light is provided from one side of the glass columns by using artificial fluorescence lamp, and the light intensity on the surface of the columns is determined by using a light meter.
5. The heterotrophic cultivation of *C. zofingiensis* is in fact an aerobic fermentation process, in which glucose serves as the only carbon and energy sources. To sustain enough oxygen for algal growth, the volume of the cultures is generally kept no more than 40% of the flask volume. A high initial carbon/nitrogen ratio benefits astaxanthin accumulation in heterotrophic *C. zofingiensis* cultures. In the current experiment, glucose is supplemented in excess at a concentration of 40 g/L. The algal cultures utilize glucose and nitrate rapidly for biomass production. Nitrate is consumed within 4 days, leading to a nitrogen-depleted but glucose-rich situation, which induces astaxanthin accumulation. Therefore, the green algal cultures turn orange.

6. As there is still residual glucose in the culture medium, try to be aseptic when manipulating the transferring process. The cultures can be aerated by air or 1.5% CO_2-enriched air. High light is used to further promote intracellular astaxanthin accumulation. The light intensity is dependent on the cell concentration of the cultures.
7. The algal cells should be dried completely. Acetone is used for astaxanthin extraction. If water is present in the cells, it will go into the acetone extracts and interfere the drying of the extracts under nitrogen gas and downstream HPLC analysis. Generally, drying for 2 days is enough. Prior to loading onto the freeze-dryer, the algal cells (in centrifuge tube) should be frozen at −20 °C.
8. Take care and avoid the loss of algal cells when weighting and transferring algal cells. The algal cells can be easily spilled out of the mortar when adding liquid nitrogen. To address it, wet the cells with a trace amount of acetone right before adding liquid nitrogen. Add liquid nitrogen up to two-thirds of the mortar volume, and grind the cells till the liquid nitrogen is completely evaporated; repeat twice. As the algal cells have a thick cell wall, fine sands can be added to facilitate cell wall breaking. Astaxanthin is sensitive to light, so perform these procedures under dim light (e.g., less than 5 $\mu E/m^2/s$).
9. Try to transfer the algal cells completely. Wash with 1–2 mL solvent A each time, pipette and transfer. Repeat multiple times to facilitate the transferring.
10. After the centrifugation, the algal residue at the bottom of the vials should be colorless. Otherwise, the extraction of astaxanthin is not complete and should be repeated.
11. Try to avoid bringing algal residue when transferring the upper solvent phase to the new glass vial. The drying under nitrogen gas may introduce moisture. In this case, drying in freeze-dryer is necessary before resuspension in solvent A.
12. Try to load the 10 μL sample onto the same point, and make spot as small as possible, which is beneficial to the separation of astaxanthin on the TLC plate. To achieve this, spot one drop each time with a 10 μL pipette, air dry, spot another drop, and repeat until done.
13. Take care and make sure the spotted sample is above the solvent level when placing the TLC plate into the container. Generally, the development takes 15–20 min for a 10 cm high TLC plate. Beta-carotene (yellow) runs fastest and appears on the top. Keep an eye on it, and take the TLC plate out when beta-carotene reaches ca. 1 cm away from the top of the TLC plate.

Acknowledgments

This study was partially supported by a grant from the National Natural Science Foundation of China (Project No.: 31571807) and a start-up grant from the National Youth Thousand Talents Program.

References

1. Ambati R, Phang S-M, Ravi S, Aswathanarayana R (2014) Astaxanthin: sources, extraction, stability, biological activities and its commercial applications—a review. Mar Drugs 12:128–152
2. Liu J, Sun Z, Gerken H, Liu Z, Jiang Y, Chen F (2014) *Chlorella zofingiensis* as an alternative microalgal producer of astaxanthin: biology and industrial potential. Mar Drugs 12:3487–3515
3. Lorenz RT, Cysewski GR (2000) Commercial potential for *Haematococcus* microalgae as a natural source of astaxanthin. Trends Biotechnol 18:160–167
4. Guerin M, Huntley ME, Olaizola M (2003) *Haematococcus* astaxanthin: applications for human health and nutrition. Trends Biotechnol 21:210–216
5. Boussiba S (2000) Carotenogenesis in the green alga *Haematococcus pluvialis*: Cellular physiology and stress response. Physiol Plant 108:111–117
6. Olaizola M (2000) Commercial production of astaxanthin from *Haematococcus pluvialis* using 25,000-liter outdoor photobioreactors. J Appl Phycol 12:499–506
7. Hata N, Ogbonna JC, Hasegawa Y, Taroda H, Tanaka H (2001) Production of astaxanthin by *Haematococcus pluvialis* in a sequential heterotrophic-photoautotrophic culture. J Appl Phycol 13:395–402
8. Fabregas J, Otero A, Maseda A, Domínguez A (2001) Two-stage cultures for the production of astaxanthin from *Haematococcus pluvialis*. J Biotechnol 89:65–71
9. Imamoglu E, Dalay MC, Sukan FV (2009) Influences of different stress media and high light intensities on accumulation of astaxanthin in the green alga *Haematococcus pluvialis*. N Biotechnol 26:199–204
10. Liu J, Sun Z, Zhong Y, Gerken H, Huang J, Chen F (2013) Utilization of cane molasses towards cost-saving astaxanthin production by a *Chlorella zofingiensis* mutant. J Appl Phycol 25:1447–1456
11. Liu J, Sun Z, Gerken H, Huang J, Jiang Y, Chen F (2014) Genetic engineering of the green alga *Chlorella zofingiensis*: a modified norflurazon-resistant phytoene desaturase gene as a dominant selectable marker. Appl Microbiol Biotechnol 98:5069–5079
12. Liu J, Mao X, Zhou W, Guarnieri MT (2016) Simultaneous production of triacylglycerol and high-value carotenoids by the astaxanthin-producing oleaginous green microalga *Chlorella zofingiensis*. Bioresour Technol 214:319–327

Chapter 6

Preparative Recovery of Carotenoids from Microalgal Biomass

María del Carmen Cerón García, Cynthia Victoria González López, José María Fernández Sevilla, and Emilio Molina Grima

Abstract

Carotenoids are widespread substances with important physiological roles, and some of them, such as lutein, astaxanthin, or vaucherioxanthin, are high-value products that can be used as high-quality food color and antioxidants, and some have an alleged role in the prevention of disorders such as AMD. Carotenoid extracts are currently obtained from plant sources, but microalgae have been demonstrated to be a competitive source likely to become an alternative. The extraction of carotenoids from microalgae possesses specific problems that arise from the different structure and composition of the source biomass. Here is presented a method for the recovery of carotenoid extracts from microalgal biomass in the kilogram scale.

Key words Carotenoid, Microalgae, Extraction, Preparative scale

1 Introduction

Microalga have attained particular attention in recent years since they are an enormous biological resource, representing one of the most promising sources for new products and application for commercialization [1]. Some of the especial advantages microalgae cultivation offers, compared to traditional plant-based sources, include a faster cultivation, processing, and harvesting cycle and the ability to be cultured on waste materials.

Carotenoids have increasing attention during the last decade due to their antioxidant activity and potential function preventing adverse health conditions in humans. At present, the major carotenoids of market interest are β-carotene, astaxanthin, lutein (with zeaxanthin), lycopene, and canthaxanthin. Constituting astaxanthin (24%), lutein (16%) and β-carotene (14%) as the most recognized and make-up almost half of the carotenoid market [2]. For example, astaxanthin is known to have health benefits such as strong antioxidant property [3], anti-inflammatory effects [4],

Carlos Barreiro and José-Luis Barredo (eds.), *Microbial Carotenoids: Methods and Protocols*, Methods in Molecular Biology, vol. 1852, https://doi.org/10.1007/978-1-4939-8742-9_6,

anticancer [5], cardiovascular health [6], or even as antiaging beauty supplement and skin protector [7]. Lutein prevents cataract and age-related macular generation [8] and protective effect on other diseases, such as cardiovascular and cancer [9]. On the one hand, β-carotene is reported to prevent night blindness [10], has antioxidant property, or even prevents liver fibrosis [11]. On the other hand, vaucherioxanthin has antioxidant property too [12]. Many recent reviews have previously discussed microalgae and their product applications; however, there has been less focus on the downstream processing aspects. It has been described a process to obtain carotenoids from microalgae [13]. Microalgal carotenoid production has excellent potential with major challenges to be overcome [14]. There is still for innovating to achieve better engineering to make the processes more cost competitive compared with others processes [15]. This chapter describes a procedure for the recovery of the carotenoid fraction contained in microalgal biomass that has been developed for different species with a hard cell wall, such as *Scenedesmus almeriensis*, *Chlorella* sp., *Haematococcus pluvialis*, or *Nannochloropsis gaditana*. The procedure proposed also allows recovering of the free fatty acids as an additional final step.

2 Materials

The chemicals used are analytical grade. Solutions have to be prepared using deionized water. All reagents used in the analytical determinations are also of analytical grade. Dry nitrogen and argon are used as sweeping gas or inert atmosphere. Solutions have to be prepared fresh as short in advance as possible. The use of inert atmosphere by sweeping with nitrogen and the use of airtight flasks are advised throughout. Storage under argon atmosphere is recommended. Material size and reactive amounts are given, considering a 50 g batch of dry biomass or using wet biomass containing 20% w/w of dry biomass.

2.1 Microalgal Biomass and Chemicals

1. Dry microalgal biomass from different microalgae, including *Scenedesmus almeriensis*, *Chlorella* sp., *Haematococcus pluvialis*, and *Nannochloropsis gaditana*, as carotenoids oil-rich substrates.
2. Hexane (95% purity).
3. Ethanol (96% v/v).
4. HCl (37%).
5. Methanol (99.9% purity).
6. Distilled water.

2.2 Cell Breakage

1. Benchtop ball mill of 5 L working volume (18 cm inner diameter, 15 cm depth approx.) filled with 50 ceramic balls 27 mm in diameter (2700 kg/m^3 density).
2. 50 g alumina, aluminum oxide, type A-5 (Sigma-Aldrich Inc., St. Louis, MO).

2.3 Alkaline Treatment

1. KOH solution in ethanol-water: mix 23.53 g of KOH (85% purity) with 1600 mL of aqueous ethanol (ethanol 96%). Once prepared, store at 60 °C until used.
2. Screw-topped bottle.
3. Cylindrical jacketed glass reaction tank 1 L working volume (340 mm height, 100 mm diameter, ref. Pobel 802420) provided with an airtight 3-neck lid (ref Pobel 806058) and an outlet at the bottom.
4. Agitator Heidolph RZR 2020.
5. Teflon, Rushton-type stirrer (six blades, 50 mm diameter for a base tank working volume of 1 L).
6. Thermo Scientific Neslab RTE7 recirculating thermostatic bath.

2.4 Extraction

1. Hexane reagent grade.
2. Ethanol reagent grade 96% v/v.
3. Separating funnel 2 L.
4. Buchner funnel 90 mm diameter and flash.
5. Filter paper (FILTER-LAB 1510) cut to adequate size for the Buchner funnel.
6. Nylon net filter 10 μm (Millipore NY1009000) for solvent filtration.
7. Vacuum pump Buchi V-700.

2.5 Solvent Removal

1. Rotary evaporator Buchi R-210.
2. Vacuum pump Buchi V-700.

2.6 Purification

1. Rotary evaporator Buchi R-210.
2. Acetone reagent grade.
3. Ethanol absolute.

2.7 Resuspension in Oil

1. Rotary evaporator Buchi R-210.
2. Vegetable oil. Olive oil is suggested.

2.8 Single-Step Alkaline Extraction

1. Water-ethanol-hexane (WEH) extracting solution: mix distilled water, ethanol (96% v/v reagent grade), and hexane (reagent grade) in a proportion of 10:84:6 v/v/v. Add 20 g of high-grade KOH alkaline agent and dissolve.

2. Jacketed tank, agitator, and recirculator as described in Subheading 2.2.
3. Buchner funnel and flask, filtering media, and vacuum pump as described in Subheading 2.4.

3 Methods

Biomass were obtained from outdoor conditions and closed photobioreactors. Wet paste should be processed immediately or can be stored under −20 °C for up to 6 months as longer delay induces significant degradation of the carotenoid fraction. Solutions are to be prepared fresh as short in advance as possible. Timing is critical in the extraction procedures. The use of inert atmosphere (nitrogen or preferably argon) is advised throughout and requested in specific steps. All the procedures described here are to be carried out at 60 °C unless otherwise stated. Here it is described a procedure for 50 g of biomass (dry weight), which can be scaled up or down by keeping the proportions described. A scheme of the overall process is shown in Fig. 1a.

3.1 Cell Breakage

1. Dry microalgal biomass, typically freeze-dried, to be used for the extraction procedure. After the dying process, the biomass is likely to form lumps of irregular size that can hamper the procedure. In this case, break the lumps mechanically and use a sieve, if needed, to ensure a particle size under 500 μm.
2. Weigh the same alumina as the sample weight.
3. Add the alumina and mix thoroughly by hand or in a tumbler, until both solids are undistinguishable to the eye.
4. Place in a ball mill and process for 30 min. The device used in this work is a horizontal ceramic cylinder 20 cm inner diameter rotating at 120 rpm. The drum is filled to half its working volume with 4 cm diameter ceramic balls (*see* **Note 1**).
5. Transfer the milled biomass and alumina to a 1 L working volume jacketed glass tank thermostated at 60 °C. Process without delay.

3.2 Alkaline Treatment

1. With the milled biomass in the tank, add enough 40% KOH (w/w biomass) aqueous ethanol solution to complete 1.6 L. The KOH solution should be at 60 °C. Close the tank with the stirrer in place. Flush with argon.
2. Stir at 220 rpm with a Rushton-type turbine in order to produce a vigorous agitation. The use of internal baffles is advised to avoid the formation of a vortex. Process for 60 min.
3. Transfer to a 4 L working volume separating funnel (*see* **Note 2**).

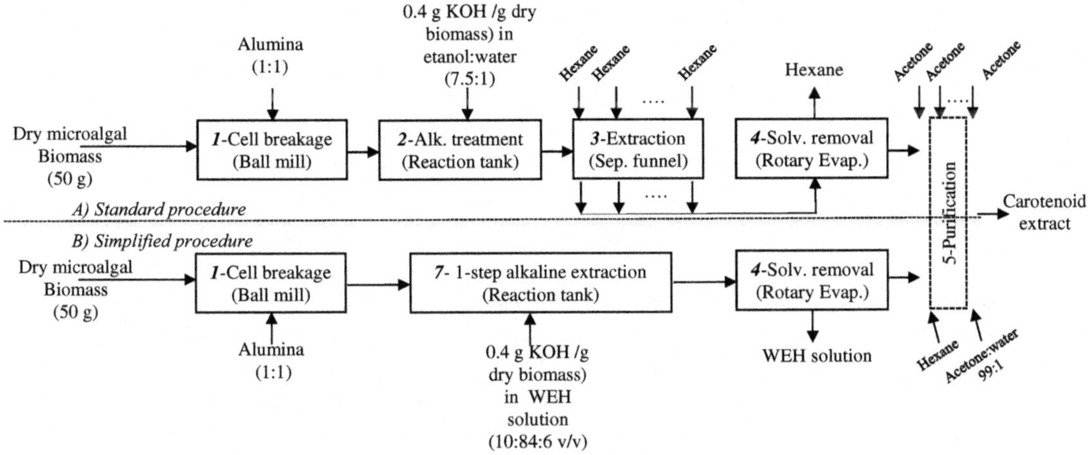

Fig. 1 Processing scheme for the recovery of carotenoid oil extracts from microalgal biomass at preparative scale. Two alternative extractions are proposed. The first one (**a**), depicted in the upper part, combines an alkaline treatment and a multiple extraction with hexane (a modification of method proposed by Hita et al. 2015 [16]), while the second alternative (**b**), shown in the lower part, is based in the use of an especially formulated extraction mixture which greatly reduces the requirements of solvent and the size of the equipment

3.3 Extraction

1. Add hexane in a 1:1 proportion to the hydroalcoholic phase. Special care needs to be taken with overpressure (*see* **Note 3**).
2. Shake vigorously for 10 min. Let rest and allow the two phases to separate. If the separation is impeded by the formation of emulsions, *see* **Note 4**.
3. Recover the hexane phase. Transfer back the aqueous phase to the separating funnel and repeat the extraction with hexane. Two extractions will achieve a recovery yield over 50%, and a 75% can be expected with four extractions and a nearly full recovery with eight extraction steps.
4. Filter through a 60 μm filter using a Buchner funnel and flask attached to a vacuum pump (*see* **Note 5**).
5. The hexane phases recovered can be pooled or kept separately and processed in batches. It is suggested to pool them and the carotenoid content evaluated at this point as described by Cerón-García et al. 2007 [3].

3.4 Solvent Removal

1. Place a suitable volume of the hexane phase in a rotary evaporator, approximately 25% of the evaporator flask capacity. The hexane phase can be added in portions as the solvent is eliminated.
2. Set the bath temperature to 35 °C and the rotation speed to 30 rpm (for a 2 L flask, *see* **Note 6**). Start the vacuum until 125 mbar. The condenser temperature should be as low as possible to ensure a complete recovery of the solvent.

3. Add a new portion of hexane phase before the solvent is completely removed.
4. After the last addition of hexane phase, keep on eliminating the solvent until the remaining volume is 50 mL.
5. Move to a suitable flask and finish the solvent removal under N_2 current at room temperature. The result is a resin extract containing up to 55% carotenoids in free form.

3.5 Purification of the Carotenoid Extract

The resin extract obtained after **step 4** of the former subheading may contain a significant part of non-carotenoid matter such as waxes, sterols, or hydrocarbons, depending on the strain of microalgae used and its culture stage. At this point, the carotenoid extract can be purified by the following procedure.

1. During **step 4** of Subheading 3.4, keep on eliminating solvent until the extract becomes a slurry.
2. Stop the vacuum and add 500 mL hexane (e.g., to extract the apolar carotenoids such as β-carotene). Put back the flask in place and stir at 40 °C for 1 h at 150 rpm.
3. Stop and allow settling. The hexane extract should separate neatly from the slurry (*see* **Note 7**). Transfer the hexane extract to a suitable flask.
4. Repeat the extraction with 500 mL acetone/water (99:1) (v/v). Put back the flask in place and stir at 40 °C for 7 h (4 h achieves an 80% of total carotenoids; 7 h treatment yields close to 100%) at 150 rpm (*see* **Note 8**). Transfer the acetone extract to a suitable flask.
5. Pool the acetone and hexane extracts in a clean rotary-vacuum flask and proceed to eliminate the solvent at 57 °C and 80 mbar vacuum pressure.
6. Remove the solvent until dry or dissolve in olive oil as described for hexane.

To separate the free fatty acids from waxes, we use a modification of the procedure proposed by Hita et al. [16]:

1. Add 182 mL of ethanol (96%) and 68 mL of water and mix.
2. Adjust to pH 5 by adding concentrated (HCl) instead of the sulfuric acid proposed by Hita et al. [16].
3. Suspend with hexane (1:1) and keep under slow agitation for 5 min.
4. Separate the hexane phase as described in Subheading 3.4.

3.6 Resuspension in Oil

Obtaining the carotenoid extract as a final product is not advised because it is difficult to handle and the carotenoids present are unstable [10]. Instead, a resuspension in olive oil is recommended as follows.

1. After **step 4** of Subheading 3.4, stop the vacuum and add of olive oil (*see* **Note 9**) to render a solution of carotenoids in olive oil maximum of approximately 2 g/L (more than this value can remain undissolved).
2. Stir at 40 °C for 10 min.
3. Start the vacuum and eliminate the rest of the solvent.
4. Place in an adequate flask. Strip with 0.2 L/L N_2 for 30 min to reduce the solvent levels to food grade requests.

This final product is easy to handle and has been proven to last for months [12] (*see* **Note 10**).

3.7 Alternative Single-Step Alkaline Extraction

This alternative procedure combines the alkaline treatment and the extraction in a single step and greatly reduces the amount of solvents used and increment to 104% the recuperation of carotenoids in saponification step. On the downside, the extracts obtained tend to contain more impurities. To carry out this procedure, take the mixture of microalgal biomass and alumina (50 + 50 g) as obtained after Subheading 3.1 and process as follows.

1. Prepare 2 L of water-ethanol-hexane (WEH) extracting solution.
2. Place the milled biomass and alumina in a 5 L working volume jacketed glass tank thermostated at 60 °C.
3. Add WEH solution to complete 2 L. The solution should be at 60 °C. Close the tank with the stirrer in place. Flush with argon.
4. Stir at 220 rpm with a Rushton-type turbine in order to produce a vigorous agitation. Process for 5 min.
5. Filter using a Buchner funnel and flask. Use 60 μm filtering medium (*see* **Note 11**) o even centrifuging at 8616 × *g* for 10 min.
6. Once the filtration or centrifugation is completed, wash carefully the cake with the remaining 800 mL WEH solution.
7. Transfer the solution to a rotary evaporator flask. Proceed to remove the solvent in the same conditions as described in **step 2** of Subheading 3.4.
8. Continue with this procedure until the extract becomes completely dry (*see* **Note 12**).
9. Stop the vacuum and add first hexane 500 mL and then 500 mL acetone-water as described in Subheading 3.5.

The result is a resin extract containing up to 80–100% carotenoids in free form that can be further purified as indicated in Subheading 3.5.

4 Notes

1. The ball mill reduces both the microalgae and the alumina to a very fine dust bound to become airborne during manipulation that can be easily inhaled. Bear this in mind and take the adequate precautions, including wearing the appropriate protection.
2. The separation funnel nominal volume must be over 2 L in order to hold 1 L of the KOH treated biomass aqueous solution and the same volume of hexane plus enough headspace to allow proper mixing. The total load can be processed using smaller separating funnels in several batches as long as they are processed simultaneously because lutein degrades if left in contact with the alkaline solution.
3. The hexane is added cold, at 4 °C. This helps to avoid overpressure in the separating funnel.
4. Cell debris can help forming very stable emulsion that will impede phase separation. This can be avoided or ameliorated by filtering in a Buchner funnel and flask or even adding a pinch of salt.
5. If filtering has been performed in **step 2** of the extraction procedure, this filtering step can be skipped.
6. The rotation speed of the rotary evaporator is given as an indication and should be adjusted accordingly to the size (diameter) of the flask used to promote a proper agitation.
7. The hexane and acetone extract should be a transparent yellow solution easily separable from the solid debris. Otherwise, the procedure has failed. This may be caused by the presence of excess water that must be further eliminated.
8. The acetone extract could be preserved at 4 °C without light during a night instead of remain 7 h, without agitation in darkness obtaining the same data.
9. Olive oil has been proven to keep lutein unaltered for months even at room temperature [10].
10. Carotenoids are very stable suspended in olive oil. No significant degradation is observed after months of storage.
11. Place a disk of regular filter paper over the recommended filter to retain the coarser particles. This facilitates and speeds up the procedure.
12. Water should be eliminated thoroughly in order to prevent unwanted substances from dissolving in the hexane or acetone. A color green/brown color instead of yellow in the acetone solution indicates the presence of excess water. At this point, the acetone must be eliminated in the rotary evaporator and the extract dried carefully.

References

1. Pulz O, Gross W (2004) Valuable products from biotechnology of microalgae. Appl Microbiol Biotechnol 65(6):635–648
2. Business Communications Company, (2015) The Global Market for Carotenoids—FOD025E. https://www.bccresearch.com/market-research/food-and-beverage/carotenoids-global-market-fod025d.html
3. Cerón MC, García-Malea MC, Rivas J, Acién FG, Fernández JM, Del Río E (2007) Antioxidant activity of *Haematococcus pluvialis* cells grown in continuous culture as a function of their carotenoid and fatty acid content. Appl Microbiol Biotechnol 74:1112e9
4. Chew BP, Park JS, Wong MW, Wong TS (1999) A comparison of the anticancer activities of dietary β-carotene, canthaxanthin and astaxanthin in mice in vivo. Anticancer Res 19:1849–1853
5. Li J, Zhu D, Niu J, Shen S, Wang G (2011) An economic assessment of astaxanthin production by large scale cultivation of *Haematococcus pluvialis*. Biotechnol Adv 29:568–574
6. Park JS, Chyun JH, Kim YK, Line LL, Chew BP (2010) Astaxanthin decreased oxidative stress and inflammation and enhanced immune response in humans. Nutr Metab (Lond) 7:18
7. Capelli B, Cysewski G (2012) Internal Beauty Pill? Sunscreen in a pill? The world's best kept health secret: natural astaxanthin. Cyanotech Corporation
8. Granado F, Olmedilla B, Blanco I (2003) Nutritional and clinical relevance of lutein in human health. Br J Nutr *90*:487–502
9. Demmig-Adams B, Adams WW III (2002) Food and photosynthesis: antioxidants in photosynthesis and human nutrition. Science *298*:2149–2153
10. Dufossé L, Pintea A (2005) Third International Congress on pigments in food (Conference Paper). Food Res Int 38:381–382
11. Virtamo J, Taylor PR, Kontto J, Männistö S, Utriainen M, Weinstein SJ, Huttunen J, Albanes D (2014) Effects of α-tocopherol and β-carotene supplementation on cancer incidence and mortality: 18-year postintervention follow-up of the alpha-tocopherol, beta-carotene cancer prevention study. Int J Cancer 135:178–185
12. Camacho-Rodríguez J, Cerón-García MC, Macías-Sánchez MD (2016) Long-term preservation of concentrated *Nannochloropsis gaditana* cultures for use in aquaculture. J Appl Phycol 28:299
13. Fernández-Sevilla JM, Acién Fernández FG, Molina Grima E (2012) Obtaining lutein-rich extract from microalgal biomass at preparative scale. Microbial carotenoids from bacteria and microalgae: methods and protocols. Methods Mol Biol 892:307–312 Springer Science+Business Media, LLC
14. Gong M, Bassi A (2016) Carotenoids from microalgae: a review of recent developments. Biotechnol Adv 34:1396–1412
15. Fernández-Sevilla JM, Acién Fernández FG, Molina Grima E (2010) Biotechnological production of lutein and its applications. Appl Microbiol Biotechnol 86:27–40
16. Hita Peña E, Robles Medina A, Jiménez Callejón MJ, Macías Sánchez MD, Esteban Cerdán L, González Moreno PA, Molina Grima E (2015) Extraction of free fatty acids from wet *Nannochloropsis gaditana* biomass for biodiesel production. Renew Energy 75:366–373

Chapter 7

Adaptive Laboratory Evolution for Enhanced Carotenoid Production in Microalgae

Yixi Su, Zhiqian Yi, Snædís Huld Björnsdóttir, Sigurdur Brynjolfsson, and Weiqi Fu

Abstract

In order to produce natural pigments with competitive prices, algal strains employed in industrial production need to be improved for increasing the productivity of valuable metabolites, thereby reducing the overall production cost. Adaptive laboratory evolution (ALE) is a traditional method for strain improvement, which has been effectively utilized in bacteria and fungi. With the growing interest in algal biotechnology, attempts have recently been put forward to improve microalgal strains with ALE approach. This chapter describes a stepwise adaptive evolution strategy that enhances carotenoid yield from microalgae.

Key words Carotenoids, Adaptive laboratory evolution (ALE), Abiotic stress, Green algae, Response surface methodology (RSM), *Dunaliella salina*

1 Introduction

At present, chemically synthesized pigments largely dominate the pigment market due to their low cost, but natural products are more attractive to consumers as they exhibit higher bioactivity, whereas the synthetic pigments expose potential safety issues for direct human consumption. With increasing demand, natural beta-carotene, astaxanthin, and lutein from green algae such as *Dunaliella*, *Haematococcus*, and *Chlorella* have gained growing commercial interest for applications in food, cosmetics, and pharmaceutical industries. Nevertheless, production costs of algae-derived carotenoids need to be reduced to be competitive with their synthetic substituents, which requires advances in strain improvement for higher productivity [1–3]. Current carotenoid production usually has to be compromised with algal growth. Many valuable pigments produced by microalgae, for example, astaxanthin and canthaxanthin, are secondary metabolites whose productions to high cellular levels are only stimulated by environmental stresses

Carlos Barreiro and José-Luis Barredo (eds.), *Microbial Carotenoids: Methods and Protocols*, Methods in Molecular Biology, vol. 1852, https://doi.org/10.1007/978-1-4939-8742-9_7, © Springer Science+Business Media, LLC, part of Springer Nature 2018

such as high irradiation, nutrient deficiency, osmotic pressure, etc. [4]. Even though high contents could be achieved, such extreme conditions are adverse to algal growth and consequently limit the overall carotenoid production.

Adaptive laboratory evolution (ALE) is a biotechnological tool for developing desirable traits of industrial microorganisms in terms of robustness to unfavorable conditions such as unusual carbon source, adverse environment, and chemical inhibitors [5]. In general, ALE process involves prolonged iterative cultivation under artificially specified pressure. Based on intrinsic spontaneous mutations in cells, the defined selective pressure preserves mutants with improved adaptability to the given environment, which directs the microbial population to evolve desirable phenotypes over time [6]. Since environmental stimuli are inevitable for carotenoids accumulation, ALE suits the purpose for developing algal stains that are resistant to the growth stresses and thus exhibit improved production performance. Furthermore, ALE in comparison with other strain improvement approaches, that is, genetic engineering, does not require a priori genetic knowledge of the studied organism and therefore has its edge over genetic manipulation in microalgae. Additionally, it could increase the understanding of molecular mechanisms in microalgae by identifying the causative mutations underlying the improved phenotype [5, 7]. In the last 25 years, the ALE approach has been extensively employed with *Escherichia coli* and *Saccharomyces cerevisiae* for strain improvement, whereas its application with non-conventional hosts, e.g., microalgae, is emerging in recent years [8–11].

Batch cultivation in parallel serial cultures and continuous (chemostat) cultures in bioreactor vessels are the two alternative methods of ALE. For batch cultivation, microorganisms undergo successive propagation in shake flasks, with an aliquot of culture being periodically transferred to fresh medium at regular intervals. This method is advantageous for its cheap and easy setup, but batch variations in nutrient supply, environmental conditions, population density, and growth rate are the shortcomings. By contrast, tight control of nutrient supply and environmental conditions provided by chemostat cultivation is expected to avoid fluctuated growth rates and population densities, but the operating cost is much higher than the serial batch cultures [12].

The stress factor that drives the evolution is a key element of ALE processes. In practice, the types of stress vary on the purpose of experiment or the desired feature of the microorganism. Meanwhile, levels of the designed stress need to be suitable for selection, that is, repressing the growth of undesired variants while avoiding death of the entire population. When the microbial population approaches to adaptation to the designed stress (i.e., mutants with high fitness predominate the population), further improvement in the growth rate can barely be observed. ALE can proceed by changing or enhancing the stress.

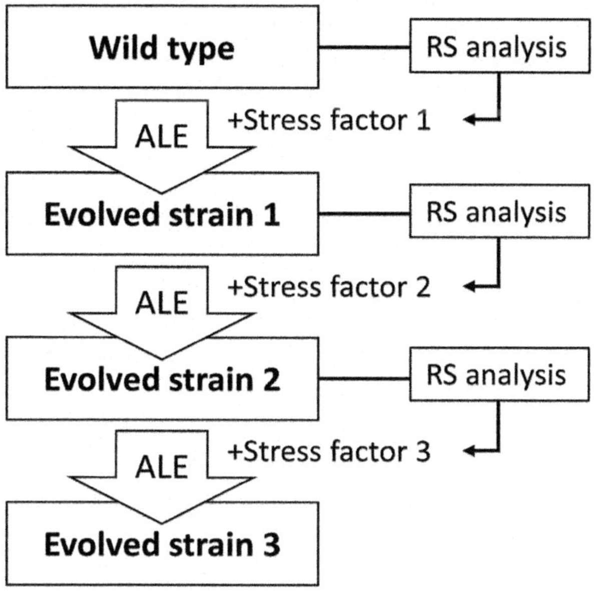

Fig. 1 A workflow of a rational adaptive evolution process for strain improvement: response surface analysis is used to guide the progress of ALE by determining the stress factor that has the largest impact on the desired phenotype

For rational design of experiments, a response surface analysis was proposed to facilitate the selection of abiotic stress factor with appropriate levels [13]. Response surface methodology (RSM) is a mathematical technique for the modeling and analysis of a response of interest influenced by several variables, aiming at optimization of the response [14]. This approach that studies multivariable effect on the objective can help determining the type and level of stress factor for further ALE [13]. Here we demonstrate a stepwise adaptive evolution strategy (Fig. 1) combined with RSM for progressive improvement of carotenoid production in microalgae.

2 Materials

The same basic medium and bubble column photobioreactors are used throughout the main experimental procedures for seeding culture preparation, adaptive laboratory evolution, and response surface analysis experiments, unless stated specifically. Parameter settings and any modification are specified for each step.

2.1 Seed Culture

1. *Dunaliella salina* (UTEX #200) (University of Texas, USA).
2. Gg-8 medium: NaCl 1.5 M; KNO_3 31.2 mM; $MgSO_4{\cdot}7H_2O$ 5.2 mM; H_3BO_3 4.0 mM; $Na_2HPO_4{\cdot}2H_2O$ 0.75 mM; $CaCl_2{\cdot}2H_2O$ 0.085 mM; FeNaEDTA 27.2 μM; $ZnSO_4{\cdot}7H_2O$ 16.0 μM; $MnCl_2{\cdot}4H_2O$ 3.5 μM; $CuSO_4{\cdot}5H_2O$ 3.0 μM; $Na_2MoO_4{\cdot}2H_2O$ 1.5 μM; $CoCl_2$ 1.0 μM; and vitamin B_{12} 0.1 μM.

3. LED lighting: red, 85 μE/m^2/s.
4. Quantum sensor (Model LI-1400, LI-COR biosciences, Lincoln, NE, USA).
5. Bubble column photobioreactor (PBR): cylindrical (H = 30 cm, D = 4 cm), working volume = 300 ± 5 mL.
6. Gas input: 90 mL/min air, 2.5% CO_2.
7. Spectrophotometer (BIOMATE 3, Thermo Scientific, Waltham, USA).

2.2 Adaptive Laboratory Evolution

1. Red LED: centered at 660 nm, 20 nm bandwidth at half peak height. Blue LED: centered at 470 nm, 20 nm bandwidth at half peak height (LUMEX Inc., Taiwan, China) (*see* **Note 1**).
2. Gg-8 medium.
3. Bubble column PBR.
4. Gas input.

2.3 Response Surface Analysis

1. Seed culture: the evolved *D. salina* strain (HI 001) after ALE treatment by light stress, maintained under the same conditions as the previous ALE treatment.
2. Mixed red/blue LED with different proportions of blue light (%): total photon flux of 170 μE/m^2/s; 0, 25, 50.
3. Modified Gg-8 medium with different levels of KNO_3 (mM): 0.2, 31.2, 62.2.
4. Modified Gg-8 medium with different levels of NaCl (M): 0.5, 1.5, 2.5.
5. Red and blue LED.
6. Bubble column PBR.
7. Gas input.

2.4 Biomass Determination

1. Cellulose membrane (pore size, 0.45 μm).
2. Deionized water.
3. Desiccator.
4. Balance.

2.5 UPLC Analysis of Carotenoids

1. Centrifuge.
2. Lyophilizer.
3. Extraction solvent: ethanol/hexane 2:1 (v/v) containing 0.1% (w/v) butylated hydroxytoluene.
4. Hexane.
5. Water.
6. Nitrogen gas.
7. Evaporator (miVac, GENEVAC LTD, Suffolk, England).
8. Methyl tertiary butyl ether: acetonitrile (MTBE/CAN; 50:50).
9. Syringe filters.

10. UPLC apparatus equipped with a TUV detector and mass spectrometer (Q-TOF).
11. ACQUITY UPLC HSS T3 1.8 μm column (2.1 × 150 mm; Waters, Manchester, UK).
12. Separation solvent phase A: ACN/methanol/MTBE (70:20:10, v/v/v), HPLC grade.
13. Separation solvent phase B: 10 mM ammonium acetate, HPLC grade.
14. Lock mass: leucine enkephalin (2 ng/μL) (*m/z* 556.2771) (Sigma-Aldrich, Co. LLC, St. Louis, MO, USA).
15. Standard pigments (Sigma-Aldrich, Co. LLC, St. Louis, MO, USA).

3 Methods

For adaptive laboratory evolution, biomass production by each cultivation cycle is determined to monitor the improvement throughout the process, while measurements for pigment contents by UPLC-UV-MS could be optional depending on the purpose of particular experiments (*see* **Note 2**). For the response surface analysis, data of biomass productivity and carotenoid contents are combined to obtain overall productivity of the targeted carotenoid (i.e., lutein).

3.1 Seed Culture

1. Inoculate at relative low density to keep stains in exponential growth state, for example, OD_{600} = 0.05–0.1.
2. Regulate culture pH between 6.5 and 7.5.
3. Cultivate *D. salina* under low light intensity (85 μE/m^2/s) of red LED lighting till late exponential phase at room temperature.

3.2 Adaptive Laboratory Evolution

1. Stress setup: supply the PBR with combined blue and red LED light (microarray, 1:3) at a total photon flux of 170 μE/m^2/s at a frequency of 10 kHz of flashing light (*see* **Note 3**).
2. ALE: conduct a semi-continuous culture with repeated 5-day cycle under the stressful light conditions. Initiate a new cycle before the stationary phase is reached; thus cells are predominantly grown in exponential phase during each cycle. For starting each new cycle, dilute culture to the same biomass (ca. 0.5 gDCW/L) by removing part of the culture and refilling same volume of fresh medium (*see* **Notes 4–6**).
3. Monitoring growth: measure cell dry weight at the beginning and end of each cycle (*see* Subheading 3.4) (*see* **Note 7**).

3.3 Response Surface Analysis

Box–Behnken design (BBD) (Table 1) is one type of experiment design that requires three-level factors for fitting the second-order model. BBD is efficient in terms of the number of required runs.

In addition, it does not contain any points at the vertices of the experiment region (Fig. 2) and thus avoids unsatisfactory results caused by extreme conditions [15] (*see* **Note 8**).

1. Three abiotic stresses: light quality (percentage of the blue LEDs of the total LEDs), nitrogen availability (KNO_3 concentrations in the medium), and osmotic stress (NaCl concentrations in the medium) are selected as experimental variables in a Box–Behnken-type experimental design (Table 2).
2. Run experiments in batch: cultivate *D. salina* (HI 001) in triplicates in PBRs for 5 days at different combination of levels of the factors.
3. Measurements: for each experiment, determine biomass production (*see* Subheading 3.4) and pigment contents (*see* Subheading 3.5).

   Simulation and prediction: BBD generates 15 measurements at corresponding combinations of stress factor conditions, which are used for modeling.
4. Create a quadratic model (Eq. 1) with three variables: X_1 is the percentage of blue LED (% of total), X_2 is the KNO_3 concentration (mM), and X_3 is the NaCl concentration (M) in the medium. Y is the daily lutein productivity (mg/L/day).

$$Y = \beta_0 + \sum_{i=1}^{k}\beta_i X_i + \sum_{i=1}^{k}\beta_{ii} X_i^2 + \sum_{i<j}\beta_{ij} X_i X_j + \epsilon \qquad (1)$$

5. Use the 15 data points to calculate all coefficients giving (*see* **Note 9**) [13].

$$\begin{aligned} Y = -2.9112 &+ 0.0639X_1 + 0.0676X_2 \\ &+ 4.5330X_3 + 0.000087X_1X_2 \\ &- 0.0048X_1X_3 - 0.0065X_2X_3 \\ &- 0.0012X_1^2 - 0.0011X_2^2 \\ &- 1.3682X_3^2 \end{aligned} \qquad (2)$$

6. Plot the contours that correspond to particular heights of the response surface.

3.4 Biomass Determination

1. Filter 5 mL of cell suspension samples on a cellulose membrane.
2. Wash twice with deionized water.
3. Dry overnight at 60 °C.
4. Weight with balance.

3.5 UPLC Analysis of Carotenoids

1. Centrifuge 0.5 mL cell suspension at 1000 × *g* for 10 min at 4 °C.
2. Lyophilization: freeze dry samples at −80 °C.

Table 1
Coded factor levels for a Box–Behnken design of a three-variable system [15]

Experiments	Variables		
	X_1	X_2	X_3
1	−1	−1	0
2	1	−1	0
3	−1	1	0
4	1	1	0
5	−1	0	−1
6	1	0	−1
7	−1	0	1
8	1	0	1
9	0	−1	−1
10	0	1	−1
11	0	−1	1
12	0	1	1
Central point	0	0	0
Central point	0	0	0
Central point	0	0	0

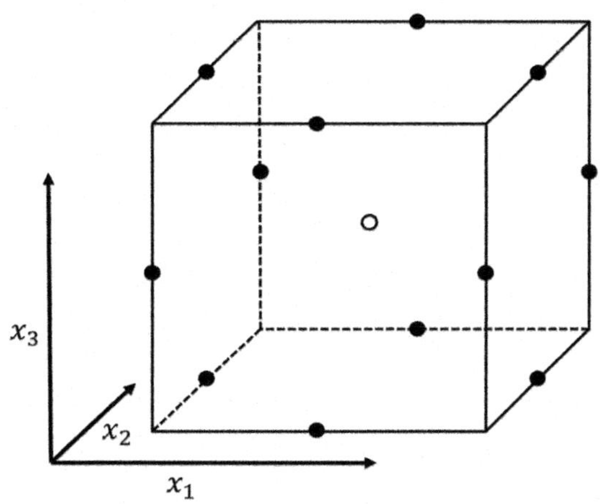

Fig. 2 A Box–Behnken design for three variables [15]

Table 2
Three abiotic-stress factors are subjected to response surface analysis by a Box–Behnken design

	Coded levels of abiotic stresses		
Variables	**Lower limit (−1)**	**Central point (0)**	**Upper limit (1)**
X_1: Blue LED percentage (% of total LEDs)	0	25	50
X_2: Nitrogen concentration (mM)	0.2	31.2	62.2
X_3: NaCl concentration (M)	0.5	1.5	2.5

3. Soak cell pellet in 3 mL extraction solvent (ethanol: hexane 2:1, v/v) containing 0.1% (w/v) butylated hydroxytoluene till colorless by shaking (*see* **Note 10**).
4. Add 2 mL of water and 4 mL of hexane and mix vigorously.
5. Centrifuge at 1000 × *g* for 5 min.
6. Evaporate the upper hexane layer under N_2 at 25 ± 2 °C.
7. Reconstitute in MTBE/CAN (50:50) solution.
8. Perform UPLC separation by reversed phase chromatography using an ACQUITY UPLC 1.8 μm column. The elution flow rate is 0.45 mL/min with a gradient of 60% phase A at 0 min, 75% at 5 min, 100% at 17.5 min, 98% at 24 min, and 60% between 25 and 30 min. A TUV detector is used for UV detection at 450 nm.
9. Operate the mass spectrometer in V mode for high sensitivity using a capillary voltage of 3 kV and a cone voltage of 30 V. Cone and desolvation gas flow are 20 and 800 L/h, respectively, while source and desolvation gas temperature are 100 and 500 °C, respectively.
10. Identify metabolites using retention time, UV detection and exact mass measurement.

4 Notes

1. Because adaptation to the stress is the criteria for selection, the choice of environmental stress for ALE experiment should be designed to affect traits that are relevant to the microorganism growth characteristics.
2. Cell growth is the principal indicator for the progress of ALE. Cellular contents including proteins, carbohydrates,

lipids, and pigments can also be analyzed to evaluate performance of ALE in terms of improvement in metabolite production. Pigments as the products of interest are measured by LC–MS in the present protocol. Other analytical techniques can be used for specific cellular contents.

3. An appropriate level of the designed stress is vital for efficient selection of robust mutants in ALE, especially for the studies on inhibitory chemicals to microalgae, for example, phenol [10]. Therefore, a preliminary assessment on the effect of serial stress degrees on cell growth may be necessary.

4. The conditions for each cycle including initial cell density, nutrient supply, and abiotic environmental factors (e.g., temperature, pH) should be kept as identical as possible. By batch cultivation fluctuated growth rates as a consequence of variation between cycles are usually observed. Furthermore, cultivation conditions other than the selected stress factor should not be introduced with other stresses at the same time, which otherwise may lead to misleading results.

5. Evolution based on spontaneous mutations could be a slow process. To accelerate the process, ALE can be carried out along with other approaches such as genetic modifications, mutagenesis, and sexual reproduction [16].

6. Bacterial contamination may occur during the long-term cultivation, which may cause bioflocculation. Although mixing is usually applied, aggregation of microalgal cells on the bottom of the reactor may take place. In this case, the culture should be fully resuspended when collecting samples and initiating a new cycle.

7. As described in the present protocol, measurement of dry biomass weight (DW) is the basic method for monitoring changes in cell growth through ALE. Alternatively, optical density (OD) that is calibrated to estimate the DW of specific microalgae species can be used for this purpose. Measuring ODs at the beginning and end of each cycle is more efficient for maintaining identical initial cell density and for monitoring cell growth. However, the bioflocculation problem as mentioned in **Note 6** and contaminants may change the calibration, which must be taken into account when using indirect biomass determination. In the case that bacterial contamination occurs, the relationship between OD and DW may need recalibration.

8. Response surface (RS) analysis facilitates the progress of ALE by identifying factors that greatly impact the growth of microalgae, which can lead to the most efficient enhancement.

9. Coefficients of variables can indicate each stress factor's impact on the dependent variable, Y. In this specific case, the coefficient of X_3 (salinity) is much higher than those of X_2 and X_3, suggesting that osmotic stress among the tested factors is

more likely to have the higher influence on growth performance of the evolved strain from the initial ALE process.

10. Extracted pigment samples are quite sensitive to light, oxygen, and heat. Therefore, pigment containers should be kept away from these factors to prevent degradation and stored under −80 °C conditions.

Acknowledgment

This work was supported by Icelandic Technology Development Fund with a grant number of 163922-0611.

References

1. Henríquez V, Escobar C, Galarza J, Gimpel J (2016) Carotenoids in microalgae. In: Stange C (ed) Carotenoids in nature. Springer, Switzerland, pp 219–237
2. Sandmann G (2014) Carotenoids of biotechnological importance. In: Schrader J, Bohlmann J (eds) Biotechnology of isoprenoids. Springer, Switzerland, pp 449–467
3. Kuczynska P, Jemiola-Rzeminska M, Strzalka K (2015) Photosynthetic pigments in diatoms. Mar Drugs 13(9):5847–5881
4. Guedes AC, Amaro HM, Malcata FX (2011) Microalgae as sources of carotenoids. Mar Drugs 9(4):625–644
5. Winkler JD, Kao KC (2014) Recent advances in the evolutionary engineering of industrial biocatalysts. Genomics 104 ((6):406–411
6. Hong K-K (2012) Advancing metabolic engineering through combination of systems biology and adaptive evolution. Chalmers University of Technology, Sweden
7. Banerjee C, Singh PK, Shukla P (2016) Microalgal bioengineering for sustainable energy development: recent transgenesis and metabolic engineering strategies. Biotechnol J 11(3):303–314
8. Fu W, Gudmundsson O, Feist AM, Herjolfsson G, Brynjolfsson S, Palsson BØ (2012) Maximizing biomass productivity and cell density of Chlorella vulgaris by using light-emitting diode-based photobioreactor. J Biotechnol 161(3):242–249
9. Fu W, Guðmundsson Ó, Paglia G, Herjólfsson G, Andrésson ÓS, Palsson BØ, Brynjólfsson S (2013) Enhancement of carotenoid biosynthesis in the green microalga Dunaliella salina with light-emitting diodes and adaptive laboratory evolution. Appl Microbiol Biotechnol 97(6):2395–2403
10. Wang L, Xue C, Wang L, Zhao Q, Wei W, Sun Y (2016) Strain improvement of Chlorella sp. for phenol biodegradation by adaptive laboratory evolution. Bioresour Technol 205:264–268
11. Li D, Wang L, Zhao Q, Wei W, Sun Y (2015) Improving high carbon dioxide tolerance and carbon dioxide fixation capability of Chlorella sp. by adaptive laboratory evolution. Bioresour Technol 185:269–275
12. Dragosits M, Mattanovich D (2013) Adaptive laboratory evolution—principles and applications for biotechnology. Microb Cell Fact 12(1):1
13. Fu W, Paglia G, Magnúsdóttir M, Steinarsdóttir EA, Gudmundsson S, Palsson BØ, Andrésson ÓS, Brynjólfsson S (2014) Effects of abiotic stressors on lutein production in the green microalga Dunaliella salina. Microb Cell Fact 13(1):3
14. Montgomery DC (ed) (2008) Design and analysis of experiments. Response surface methods and designs. John Wiley & Sons, Inc., New Jersey, p 478
15. Ferreira SC, Bruns R, Ferreira H, Matos G, David J, Brandao G, da Silva EP, Portugal L, Dos Reis P, Souza A (2007) Box-Behnken design: an alternative for the optimization of analytical methods. Anal Chim Acta 597(2):179–186
16. Takouridis SJ, Tribe DE, Gras SL, Martin GJ (2015) The selective breeding of the freshwater microalga Chlamydomonas reinhardtii for growth in salinity. Bioresour Technol 184:18–22

Chapter 8

Carotenoid Production by Recombinant *Corynebacterium glutamicum*: Strain Construction, Cultivation, Extraction, and Quantification of Carotenoids and Terpenes

Nadja A. Henke, Jonas Frohwitter, Petra Peters-Wendisch, and Volker F. Wendisch

Abstract

Corynebacterium glutamicum is a workhorse of industrial amino acid production employed for more than five decades for the million-ton-scale production of L-glutamate and L-lysine. This bacterium is pigmented due to the biosynthesis of the carotenoid decaprenoxanthin. Decaprenoxanthin is a carotenoid with 50 carbon atoms, and, thus, *C. glutamicum* belongs to the rare group of bacteria that produce long-chain C50 carotenoids. C50 carotenoids have been mainly isolated from extremely halophilic archaea (Kelly and Jensen, Acta Chem Scand 21:2578, 1967; Pfander, Pure Appl Chem 66:2369–2374, 1994) and from Gram-positive bacteria of the order Actinomycetales (Netzer et al., J Bacteriol 192:5688–5699, 2010). The characteristic yellow phenotype of *C. glutamicum* is due to the cyclic C50 carotenoid decaprenoxanthin and its glycosides. Decaprenoxanthin production has been improved by plasmid-borne overexpression of endogenous genes of carotenogenesis. Gene deletion resulted in the production of the C40 carotenoid lycopene, an intermediate of decaprenoxanthin biosynthesis. Heterologous gene expression was required to develop strains overproducing nonnative carotenoids and terpenes, such as astaxanthin (Henke et al., Mar Drugs 14:E124, 2016) and (+)-valencene (Frohwitter et al., J Biotechnol 191:205–213, 2014). Integration of additional copies of endogenous genes expressed from strong promoters improved isoprenoid biosynthesis. Here, we describe *C. glutamicum* strains, plasmids, and methods for overexpression of endogenous and heterologous genes, gene deletion, replacement, and genomic integration. Moreover, strain cultivation as well as extraction, identification, and quantitative determination of terpenes and carotenoids produced by *C. glutamicum* is detailed.

Key words Carotenoids, Terpenoids, Decaprenoxanthin, Astaxanthin, Valencene

1 Introduction

C50 carotenoids occur in extremely halophilic archaea [1–2] and Gram-positive bacteria of the order Actinomycetales such as *C. glutamicum* [3]. *C. glutamicum* belongs to the order of *Actinomycetales* and possesses the typical outer membrane-like envelope of these Gram-positive bacteria and has been engineered

Carlos Barreiro and José-Luis Barredo (eds.), *Microbial Carotenoids: Methods and Protocols*, Methods in Molecular Biology, vol. 1852, https://doi.org/10.1007/978-1-4939-8742-9_8,

to synthesize nonnative terpenoids [4–5]. The nonpathogenic *C. glutamicum* is related to the pathogenic *C. diphtheriae*, *Mycobacterium leprae*, and *M. tuberculosis*. The GRAS-status *C. glutamicum* has a long history in biotechnology: after its isolation as glutamate-secreting bacterium in the 1950s [6], it has been used for amino acid production. Meanwhile, different proteinogenic amino acids, in particular glutamate and lysine, are produced for the feed and food industry at the million-ton scale. Metabolic engineering of *C. glutamicum* enabled production of various value-added compounds [7], such as non-proteinogenic ω-amino acids [8–11], diamines [12], alcohols [13], or polyhydroxyalkanoates [14, 15]. This bacterium has the ability to grow aerobically on a variety of carbon sources like glucose, fructose, sucrose, mannitol, arabitol, propionate, and acetate [16, 17]. In addition, it has been engineered to grow with alternative carbon sources such as glycerol [18], pentoses [19], amino sugars [20, 21], β-glucans [22], levoglucosan [23], and starch [24].

C. glutamicum is pigmented due to synthesis of the C50 carotenoid decaprenoxanthin and its glucosides. Its potential to produce carotenoids and sesquiterpenes has been explored over recent years [4, 5, 25–28]. The carotenogenic pathway of *C. glutamicum* was identified [29], and several metabolic engineering strategies were applied to convert this biotechnologically established bacterium into a carotenoid producer [30, 31]. *C. glutamicum* strains for the production of the C50 carotenoids decaprenoxanthin, sarcinaxanthin, and bisanhydrobacterioruberin; the C40 carotenoids lycopene, β-carotene, zeaxanthin, canthaxanthin, and astaxanthin; and the sesquiterpene valencene have been described along with methods to extract carotenoids and to capture terpenes by dodecane overlay cultivation.

2 Material

2.1 Genetic Engineering of Carotenoid Production

1. Expression vector pEKEx3 [32].
2. Expression vector pVWEx1 [33].
3. Expression vector pSH1 [4].
4. Expression vector pECXT99A [34].
5. LB medium (*see* Table 1).
6. BHIS-Solution A (*see* Table 1).
7. BHIS-Solution B (*see* Table 1).
8. CgXII minimal medium (*see* Table 1).
9. Trace element solution (for CgXII) (*see* Table 1).
10. Kanamycin stock solution 50 mg/mL.
11. Spectinomycin stock solution 100 mg/mL.

Table 1
Complex and minimal media

Component	Used quantity
LB Medium	
Bacto tryptone	10 g
Yeast extract	5 g
Sodium chloride	10 g
(Agar for plates)	16 g
H_2O	Ad 1000 mL
BHIS-Solution A	
Brain hear infusion	37 g
(Agar for plates)	16 g
H_2O	Ad 500 mL
BHIS-Solution B	
Sorbit	90 g
(Sucrose)	100 g
H_2O	Ad 500 mL
CgXII minimal medium	
$(NH_4)_2SO_4$	20 g
Urea	5 g
KH_2PO_4	1 g
K_2HPO_4	1 g
MOPS buffer	42 g
$CaCl_2$ (water-free) (1 g/100 mL)	1 mL
$MgSO_4$ heptahydrate (25 g/100 mL)	1 mL
H_2O	Add to 800 mL
→ Adjust pH to 7.0 with KOH	
H_2O	Add to 900 mL
Trace element solution (for CgXII)	
$FeSO_4 \times 7H_2O$	1 g
$MnSO_4 \times 7H_2O$	1 g
$ZnSO_4 \times 7H_2O$	0.1 g
$CuSO_4$	0.02 g

(continued)

Table 1 (continued)

Component	Used quantity
$NiCl_2 \times 6H_20$	0.002 g
H_2O	add to 90 mL
→ Solve with addition of HCl to an final pH of ≈1	
H_2O	Add to 100 mL

Table 2 Antibiotics

	Stock solution conc. [mg/mL]	End. conc. E. coli [μg/mL]	End. conc. C. glutamicum [μg/mL]
Kanamycin	50	25	25
Spectinomycin	100	100	100
Tetracycline	5	5	5
Nalidixic	50	50	50

12. Tetracycline stock solution 5 mg/mL.
13. Nalidixic stock solution 50 mg/mL.
14. Shake flasks with two baffles 500 mL.
15. Cellulose nitrate filter (0.45 μm pore size).
16. Tweezers.
17. Rotary shaker.
18. Reaction tubes.
19. Centrifuge.
20. Electroporator (e.g., Gene Pulser).
21. Electroporation cuvette (2 mm).
22. Water bath or thermo block.

2.2 Production and Extraction of Carotenoids

1. Cryovials 2 mL.
2. Glycerol 86%.
3. CgXII minimal medium.
4. Glucose stock solution (20% or 40%).
5. Trace element solution (for CgXII) (*see* Table 2).
6. IPTG solution (1 M).
7. Baffled shake flasks 500 mL.

8. Methanol/acetone (7:3) 0.05% BHT (butylhydroxytoluol).
9. Safe-lock reaction tubes 2 mL.
10. Water bath or thermo block.
11. Dodecane.
12. Centrifugation tubes 15 mL.
13. Rotary shaker.
14. Centrifuge.

2.3 Analysis of Carotenoids

1. UV-cuvettes.
2. Vials and caps.
3. Spectrophotometer.
4. Methanol/water (9:1).
5. Methanol/methyl tert-butyl ether/ethyl acetate (5:4:1).
6. Precolumn (LiChrospher 100 RP18 EC-5, 40 × 4 mm, CS-Chromatographie, Langerwehe, Germany) and main column (LiChrospher 100 RP18 EC-5, 125 × 4 mm, CS-Chromatographie).
7. Precolumn (MultoHigh 100 RP18-5, 10 × 4 mm, CS-Chromatographie, Langerwehe, Germany) and a main column (ProntoSIL 200-5 C30, 250 × 4 mm, CS-Chromatographie).
8. Thermo Scientific TRACE GC ULTRA connected to a Thermo Scientific ISQ single quadrupole mass spectrometer using a TG-5MS column (length, 30 m; I.D., 0.25 mm; film thickness, 0.25 m) (Thermo Scientific, Waltham, MA, Massachusetts, USA).

3 Methods

3.1 Genetic Engineering of Carotenoid Production

3.1.1 Preparation of Electro-Competent Cells

The transformation of *C. glutamicum* with expression plasmids is usually carried out by electroporation. For this purpose, the respective strain of *C. glutamicum* that is targeted for the transformation needs to be made electro-competent carrying out the following protocol:

1. Prepare a culture of the *C. glutamicum* strain in 10 mL LB (with antibiotics if required) in a 50 mL non-baffled shake flask, and incubate overnight at 30 °C on a rotary shaker at 180 rpm.
2. Inoculate 2 × 50 mL BHIS (with antibiotics if required) in 500 mL baffled shake flasks each with 1 mL of the overnight culture, and incubate at 30 °C on a rotary shaker at 180 rpm till an OD_{600} of 0.8–1.2 (usually 4–6 h) (*see* **Note 1**).
3. Add 1.5 mg/L ampicillin (end concentration) to the cultures, and continue shaking for 1–1.5 h.

Table 3
Buffers for preparation of electro-competent cells

Component	EPB1	EPB2
0.5 M HEPES buffer (pH 7.2)	20 mL	2 mL
Glycerol (86%)	35.4 g	42.4 g
Add H_2O to	500 mL	200 mL
→ Autoclave and store at 4 °C		

4. Harvest cells at 4 °C and 5950 × *g* for 7 min using precooled 50 mL centrifugation tubes (*see* **Note 2**).
5. Wash three times with 30 mL EPB1 (*see* Table 3) (ice-cold), and discard supernatant.
6. Resuspend pellets in 1.5 mL of EPB2 (*see* Table 3) (ice-cold), and aliquot them in 150 μL in 1.5 reaction tubes (precooled).
7. Freeze aliquoted cells at −80 °C until further use (*see* **Note 3**).

3.1.2 Transformation

1. Thaw the frozen, electro-competent cells on ice, precool electroporation cuvettes on ice, and preheat 900 μL BHIS in 1.5 mL reaction tubes to 46 °C.
2. Mix 150 μL electro-competent cells and plasmid DNA (expression plasmids, 100–500 ng; deletion plasmids, 500 ng–20 μg) in electroporation cuvettes on ice; incubate for up to 5 min (*see* **Note 4**).
3. Transform cells via electroporation (at 2.5 kV, 200 Ω and 25 μF) using a Gene Pulser Xcell™ (Biorad).
4. Directly after the electroporation, add the preheated BHIS to the cuvette, mix, and transfer the cells to the reaction tube.
5. Inoculate at 46 °C for 6 min.
6. Regenerate the cells for 1–1.5 h at 30 °C with shaking at 450 rpm.
7. Plate the cells (100 μL and rest) on LB plates containing the respective antibiotic(s) (*see* **Note 5**).
8. Incubate the plates at 30 °C; colonies should be visible after 2 days.

3.1.3 Gene Deletion, Replacement, and Genomic Integrations

Targeted deletions, replacements, or integrations in the chromosome of *C. glutamicum* can be performed via double homologous recombination events. For this purpose the suicide vector pK19*mobsacB* is used [35].

Fused up- and downstream regions of the gene to be deleted were cloned into the suicide vector pK19*mobsacB* as described below.

Vector Construction

1. Genomic regions flanking the locus of interest were PCR amplified with a high-fidelity polymerase. For deletion of a gene, 400–500 bp upstream including the first 18 bp of the CDS (flanking region 1) as well as 400–500 bp downstream including the last 36 bp of the CDS (flanking region 2) were amplified.
2. Those two fragments were inserted into the linearized pK19*mobsacB* with a linker sequence of 21 bp in between the two fragments.
3. For integration of a gene of interest, 400–500 bp of the genomic region upstream (FR1)/and downstream (FR2) of the desired integration position were PCR amplified. In addition, the gene of interest was PCR amplified with its native promoter sequence, or the CDS was fused to a promoter sequence of interest. All 3–4 fragments were cloned into the vector as mentioned above.
4. The resulting suicide plasmid can be transferred into the desired *C. glutamicum* strain (recipient) via (i) electroporation or (ii) transconjugation.
5. Electroporation can be performed as described above (*see* Subheading 3.1.1).
6. For transconjugation the created suicide vector is transformed into the donor strain *E. coli* S17-1 (*see* **Note 6**).

The suicide vector pK19*mobsacB* cannot replicate in *C. glutamicum* but possesses two marker genes that allow selection of the two homologous recombination events. First, in the presence of kanamycin, conjugants with chromosomally integrated pK19*mobsacB* (1. homologous recombination) can be selected positively by kanamycin resistance. It has to be checked whether the transformants are still sucrose sensitive. Second, in the presence of sucrose, the second homologous recombination event involving flanking regions 1 or 2, that is, the loss of the pK19*mobsacB*, can be positively selected since cells carrying the *sacB* gene of the suicide vector are killed in the presence of sucrose. It has to be checked whether the cells are sensitive for kanamycin after the second homologous recombination event.

Transconjugation

1. (Day 1) Preparation of fresh cells: Prepare a fresh LB plate for the S17-1 strain carrying the pK19*mobsacB*-derived plasmid (donor) (+Km25) and the *C. glutamicum* strain of interest (recipient) (+appropriate antibiotics).
2. (Day 2) Pre-preculture of donor: Inoculate 50 mL LB (+Km25) with a loop of *E. coli* strain S17-1 in a 500 mL baffled shake flask, and incubate overnight at 37 °C on a rotary shaker at 180 rpm (25 mm stroke).

3. (Day 3) Precultivation of donor and recipient:
   (a) Donor: Inoculate 50 mL LB (+Km25) with 1 mL of overnight culture of *E. coli* strain S17-1 in a 500 mL baffled shake flask, and incubate again overnight at 37 °C on a rotary shaker at 180 rpm(25 mm stroke).
   (b) Recipient: Inoculate 50 mL LB/BHIS (+appropriate antibiotics) (*see* **Note 7**) with a loop of the *C. glutamicum* recipient in a 500 mL baffled shake flask, and incubate overnight at 30 °C on a rotary shaker at 120 rpm (25 mm stroke).
4. (Day 4) Cultivation and transconjugation
5. Cultivation:
   (a) Recipient: Inoculate 50 mL LB/BHIS (+appropriate antibiotics) with 1–5 mL preculture of recipient, and incubate at 30 °C and 120 rpm (25 mm stroke) till OD_{600nm} 1–1.5.
   (b) Donor: Inoculate 50 mL LB (+Km25) with 500–1000 μL preculture of donor, and incubate at 37 °C and 180 rpm (25 mm stroke) till OD_{600nm} 1–1.5.
   (c) Measure OD_{600} in regular intervals (every 1–2 h).
6. If cells are ready, do the following:
   (a) Recipient: Harvest 50 mL *C. glutamicum* culture in sterile falcon at 5950 × *g* and 5 min. Resuspend cell pellet in 5 mL LB, and make aliquots of 800 μL. Do a heat shock for 9 min at 50 °C. Recipients can be used for 2 days now (storage 22–30 °C).
   (b) Donor: Harvest 10 mL S17-1 culture in sterile falcon at 5950 × *g* and 5 min. Gently resuspend the cells in 1 mL LB (*see* **Note 8**).
7. Transconjugation:
   (a) Mix 800 μL recipient with 200 μL donor in 1.5 mL reaction tube (do triplicates), and centrifuge 5 min at 4450 × *g*.
   (b) Remove the supernatant, and gently resuspend the cells by stirring softly with a big pipette.
   (c) Place three cellulose nitrate filters (0.45 μm pore size and autoclaved) on one LB/BHIS plate, and pipette each cell mixture on one cellulose nitrate filter. The suspension has to be applied in portions to allow complete absorption by the filter.
   (d) An "egg-like" drop forms; let it dry for approx. 10 min under the sterile bench. Make sure not to disturb the conjugation by vibrations.
   (e) Incubate the plate (bottom down) for 20 h at 30 °C.

8. (Day 5) Selection of transconjugants (1. homologous recombination: suicide vector integration): Transfer each filter into a 1.5 mL reaction tube, and resuspend the cells in 500 μL LB. Plate 100 μL (and rest) on LB Km15 Nal50. Incubate for 1–2 days to see the conjugants.
9. (Day 7) Purification of conjugants: Pick single colonies again on LB Km15 Nal50 to get rid of contaminating *E. coli* cells.
10. (Day 8) Check recombinants for sucrose sensitivity: Transfer the colonies parallel to (1) LB Km25 and (2) LB Km25 + 10% sucrose to check for functionality of SacB (1. homologous recombination: kick-in).
11. (Day 9) Selection for sucrose-resistant recombinants (2. homologous recombination: suicide vector excision): Strike out the colonies, which are Km resistant but sucrose sensitive on LB 10% sucrose (do duplicates).
12. (Day 10) Check for sucrose resistant and kanamycin sensitive: Pick the cells parallel on (1) LB Km25 and (2) LB 10% sucrose.
13. (Day 11) Categorize homologous recombinants (expect 50% wild type and 50% replacement mutants): Do colony PCR (50 μL) on the ones, which are Km sensitive but sucrose resistant.

Colony PCR and Sequencing

Since homologous recombination statistically results in 50% WT alleles and 50% deletion alleles, genomic constitution of the targeted locus has to be assessed. Therefore, a colony PCR can be performed (make sure primers used for verification differ from those used for suicide vector construction by being more distant from the gene locus, e.g., 100 bp further upstream and 100 bp further downstream). After preparation and filling of the reaction mixtures (usually 20–50 μL), few cells of a colony were transferred into the mixture with a sterile tip. For efficient cell lysis, the initial denaturation step of the PCR is set to 10 min. PCR products can be analyzed in a 1% agarose gel. Afterward PCR products of colonies with the expected size, that is, indicating the expected results for the chosen genetic manipulation (deletion, integration, or replacement), are cleaned up and sequenced.

3.2 Production and Extraction of Carotenoids

3.2.1 Strain Handling

1. Three independently obtained but genetically identical *C. glutamicum* strains ("genetic replicates") are used to inoculate test tubes with 5 mL LB (with respective antibiotics) and incubated overnight at 30 °C in a rotary incubator.
2. On the next day, 650 μL of the overnight culture is mixed with 350 μL sterile glycerol (86%) in cryovials which then can be stored at −80 °C for strain preservation.
3. In parallel, 50 μL of the overnight culture should be plated on LB plates (with respective antibiotics); after overnight incubation at 30 °C, the plates are stored in the fridge at 4 °C.

4. To set up a production culture, an overnight preculture of 10 mL LB (with respective antibiotics) (*see* **Note 9**) in a 50 mL shake flask is inoculated with the respective strain from the LB plate using either a loop or a sterile pipette tip.

3.2.2 Cultivation Conditions

1. For setting up the main culture, the OD_{600} of the preculture is determined, and it is then harvested by centrifugation in 50 mL centrifugation tubes at 5950 × *g* at room temperature for 7 min.
2. Afterward, the cells are washed once in 20 mL CgXII medium and resuspended in CgXII medium using an appropriate volume to obtain an OD_{600} of 100.
3. The main culture (50 mL CgXII in a 500 mL baffled shake flask) is inoculated with 500 μL of the overnight culture for an initial OD_{600} of 1 (*see* **Note 10**).

3.2.3 Extraction of Carotenoids After Cultivation

1. For extraction of carotenoids from *C. glutamicum*, 1 mL of the culture was harvested by centrifugation for 7 min at 19,700 × *g* in a safe-lock reaction tube. Supernatant was discarded, and it is important that cells are well resuspended without remaining clumps before addition of the extraction solvent.
2. The carotenoids can be extracted with 800 μL methanol/acetone (7:3) containing 0.05 vol-% BHT (butylhydroxytoluol) at 60 °C for 15 min with careful vortexing every 5 min.
3. After an additional centrifugation step to spin down cell debris, supernatant was transferred to a new reaction tube. If cells were not fully decolorized, extraction procedure can be repeated by addition of fresh 800 μL solvent (*see* **Note 11**).
4. For analysis solvent phase was again centrifuged for 15 min and 19,700 × *g*. 600–1000 μL of clear solvent phase can be directly used for HPLC analysis or stored at 4 °C for approximately 1 day.

For analysis of extracts by spectrophotometry, carotenoids were extracted with acetone for 1 h at 60 °C with shaking of 700 rpm (stroke 3 mm). The extraction with acetone is less toxic but also less efficient. Thus, with this method only relative comparisons of carotenoid contents are possible.

3.2.4 Capture of Volatile Terpenoids During Cultivation

For the production of volatile short-chain terpenes and terpenoids, an in situ extraction might be required to prevent potential toxic effects of the product as well as evaporation from the culture medium. For terpene production in *C. glutamicum*, a two-phase cultivation setup with an overlay of the production medium with 10% v/v dodecane was established.

1. The dodecane was added aseptically to the production cultures after induction with IPTG.
2. After cultivation, the dodecane phase was separated from the medium by centrifugation in 15 mL centrifugation tubes at 4 °C at 5950 × *g* for 10 min (*see* **Note 12**).
3. After separation, 50–100 μL of the dodecane phase was transferred to a GC glass vial, and the vial was filled to 500 μL with dodecane. Afterward the samples were analyzed via GC–MS.

3.3 Analysis of Terpenoids and Carotenoids

3.3.1 Photometer Quantification of Samples Containing One Major Carotenoid

Analysis of carotenoids with a photometer is fast and easy. However, these measurements only make sense for strains that produce one specific carotenoid rather than a mixture of carotenoids.

For qualitative and quantitative analysis of a specific carotenoid, the absorption spectrum from 450 to 650 nm is recorded, and the extraction solvent, usually acetone, is used as a blank. It is important that UV-cuvettes are used and that all samples were filled in the cuvettes just before measurement. Acetone evaporates quickly from the cuvettes leading to wrong (i.e., too high) carotenoid concentrations over longer incubation times (>5 min).

A first hint on the identity of carotenoids can be obtained with the help of the absorption spectrum; however different carotenoids can possess very similar absorption curves (*see* Table 4).

Carotenoids can be quantified via their absorption maxima. Standardization with commercially available carotenoid standards of high purity is recommended.

3.3.2 HPLC Analysis of Samples Containing Multiple Carotenoids

HPLC analysis is recommended for carotenoids mixtures but of course is also suitable for pure samples. There are many HPLC protocols available. We describe two methods which may have to be optimized depending on carotenoids of interest. In both cases cell extracts have to be debris free in order to safeguard the HPLC

Table 4
Absorption maxima of some carotenoids

Carotenoid	Properties	Absorption maxima [nm]
Astaxanthin	Red, cyclic C40	476
β-Carotene	Orange, cyclic C40	424, 454, 478
Canthaxanthin	Red, cyclic C40	474
Decaprenoxanthin	Yellow, cyclic C50	415, 440, 470
Lycopene	Red, linear C40	440, 474, 502
Zeaxanthin	Orange, cyclic C40	424, 450, 478

column and system. Centrifugation at 19,700 × *g* for 10 min and analysis of the clear supernatant are sufficient, whereas filtration is also possible:

1. As a column system, a precolumn (LiChrospher 100 RP18 EC-5, 40 × 4 mm, CS-Chromatographie, Langerwehe, Germany) and a main column (LiChrospher 100 RP18 EC-5, 125 × 4 mm, CS-Chromatographie) were used. The HPLC protocol ensured a gradient elution for 10 min and a mobile phase composition of (A) methanol and (B) methanol/water (9:1) starting from 0% to 100% of eluent B, followed by 20 min of isocratic elution with 100% B. Afterward, the eluent composition was set back to 0% B for 2.5 min. The injection volume was 50 μL, and the flow rate was kept constant at 1.5 mL/min.
2. As a column system, a precolumn (MultoHigh 100 RP18-5, 10 × 4 mm, CS-Chromatographie, Langerwehe, Germany) and a main column (ProntoSIL 200-5 C30, 250 × 4 mm, CS-Chromatographie) were used. The HPLC protocol ensured a gradient elution for 10 min and a mobile phase composition of (A) methanol and (B) methanol/methyl tert-butyl ether/ethyl acetate (5:4:1) starting from 10% to 100% of eluent B, followed by 20 min of isocratic elution with 100% B. Afterward, the eluent composition was set back to 10% B for 3 min. The injection volume was 50 μL, and the flow rate was kept constant at 1.4 mL/min.

3.3.3 GC–MS Analysis for Terpenoids

GC–MS is a suitable method to identify and quantify volatile terpenoids as, for example, present in dodecane culture overlays. The choice of the column and the program used depends on the characteristics of the product and the extraction solvent; however, for most volatile short-chain terpenes and terpenoids, a nonpolar column like the TG5-MS (Thermo Scientific) is a good choice. As an example, protocols for the measurement of the sesquiterpene (+)-valencene and the triterpene squalene will be presented.

1. Dodecane culture overlays containing (+)-valencene obtained by in situ extraction (*see* above) were analyzed by splitless injection of 1 μL (injector temp 230 °C) with an initial oven temperature of 40 °C that was increased by 10 °C/min to 160 °C followed by an increase of 15 °C/min to 250 °C with a final ramp for 2 min 250 °C. The helium gas flow was set to be constant at 1 mL/min. The MS operating parameters were ionization voltage, 70 eV (electron impact ionization); ion source and interface temperature were 230 °C (*see* **Note 13**).
2. For extraction of squalene, an end point extraction with methanol/acetone (7:3) was used; the resulting methanol/acetone sample was analyzed by splitless injection of 1 μL (injector temp 230 °C) with an initial oven temperature of 40 °C that

was increased by 10 °C/min to 160 °C followed by an increase of 15 °C/min to 330 °C with a final ramp for 2 min at 330 °C. The helium gas flow was set to be constant at 1 mL/min. The MS operating parameters were ionization voltage, 70 eV (electron impact ionization); ion source and interface temperature were 230 °C.

For quantification of the products, calibration curves with commercially available standards were applied (*see* **Note 14**).

4 Notes

1. For transformation with expression plasmids, OD_{600} 0.8 is sufficient; for the transformation with deletion plasmids, OD_{600} 1.2 is preferable.
2. As from here, always work on ice.
3. For deletion best use fresh cells without freezing them.
4. Avoid dirty or too old cuvettes to prevent a short circuit!
5. It is possible to work with a three-vector system: pVWEx1 or pSH1 and pECXT99A and pEKEx3. Since cells suffer from the three antibiotic resistances, it is recommended to lower the concentrations of antibiotics (Km15, Tet5, Spec 100).
6. Make sure that S17-1 is no longer than 3 days on plate; *sacB* picks up mutations fast!!!
7. Depending on the metabolic fitness of the recipient strain, LB or BHIS medium is preferable for cultivation!!!
8. The fitness of the donor strain is usually limiting; handle S17-1 gently!!!
9. Depending on the amount of cells needed for the main culture, precultures might be set up in larger volumes, or multiple precultures are prepared.
10. Application of antibiotics in the main culture is not required, since, for example, from OD 1 in 4% glucose only ~ six doublings take place.
11. For extraction of lycopene, it is recommended to optimize the used culture volume in order to minimize the rounds of extractions. Lycopene is more difficult to extract from the cell membrane of *C. glutamicum* than other carotenoids. Here between 200 and 400 μL of culture are optimal for one-round extraction.
12. The dodecane phase might not be fully separated from cell debris, resulting in a viscous interphase. If so, take the upper phase from the medium, and centrifuge it in a 1.5 mL reaction tube at max speed for 5 min.

13. To prevent attrition of the detector, the start of detection should be set such that the measurement of the extraction solvent is prevented which is possible if the product is eluting after the solvent!
14. As a proxy, products can be quantified using a standard of a compound with a similar structure (e.g., β-elemene and germacrene can be quantified with a (+)-valencene standard) when a proper standard is missing; concentrations have to be given as, for example, (+)-valencene equivalents.

Acknowledgments

We acknowledge Sabine A. E. Heider for her work on optimization of HPLC methods, Elizabeth Gingras-Lafleur for her help on carotenoid extraction, and Hironori Taniguchi for support during photometer measurements.

References

1. Kelly M, Jensen SL (1967) Bacterial carotenoids .26. C50-Carotenoids .2. Bacterioruberin. Acta Chem Scand 21(9):2578
2. Pfander H (1994) C-45-carotenoids and C-50-carotenoids. Pure Appl Chem 66(10–11):2369–2374
3. Netzer R et al (2010) Biosynthetic pathway for gamma-cyclic sarcinaxanthin in *Micrococcus luteus*: heterologous expression and evidence for diverse and multiple catalytic functions of C(50) carotenoid cyclases. J Bacteriol 192(21):5688–5699
4. Henke NA et al (2016) Production of the marine carotenoid astaxanthin by metabolically engineered *Corynebacterium glutamicum*. Mar Drugs 14(7):E124
5. Frohwitter J et al (2014) Production of the sesquiterpene (+)-valencene by metabolically engineered *Corynebacterium glutamicum*. J Biotechnol 191:205–213
6. Kinoshita S, Udaka S, Shimono M (1957) Studies on the amino acid fermentation. Production of L-glutamic acid by various microorganisms. J Gen Appl Microbiol 3:193–205
7. Zahoor A, Otten A, Wendisch VF (2014) Metabolic engineering of *Corynebacterium glutamicum* for glycolate production. J Biotechnol 192:366–375
8. Jorge JM et al (2016) Improved fermentative production of gamma-aminobutyric acid via the putrescine route: systems metabolic engineering for production from glucose, amino sugars and xylose. Biotechnol Bioeng 114(4):862–873
9. Jorge JM, Leggewie C, Wendisch VF (2016) A new metabolic route for the production of gamma-aminobutyric acid by *Corynebacterium glutamicum* from glucose. Amino Acids 48(11):2519–2531
10. Pérez-García F, Peters-Wendisch P, Wendisch VF (2016) Engineering *Corynebacterium glutamicum* for fast production of L-lysine and L-pipecolic acid. Appl Microbiol Biotechnol 100(18):8075–8090
11. Rohles CM et al (2016) Systems metabolic engineering of Corynebacterium glutamicum for the production of the carbon-5 platform chemicals 5-aminovalerate and glutarate. Microb Cell Fact 15(1):154
12. Schneider J, Wendisch VF (2010) Putrescine production by engineered *Corynebacterium glutamicum*. Appl Microbiol Biotechnol 88(4):859–868
13. Blombach B, Eikmanns BJ (2011) Current knowledge on isobutanol production with *Escherichia coli*, *Bacillus subtilis* and *Corynebacterium glutamicum*. Bioeng Bugs 2(6):346–350
14. Matsumoto K et al (2011) Production of poly(3-hydroxybutyrate-co-3-hydroxyvalerate) in recombinant *Corynebacterium glutamicum* using propionate as a precursor. J Biotechnol 152(4):144–146

15. Jo SJ et al (2007) Improvement of poly(3-hydroxybutyrate) [P(3HB)] production in *Corynebacterium glutamicum* by codon optimization, point mutation and gene dosage of P(3HB) biosynthetic genes. J Biosci Bioeng 104(6):457–463
16. Kinoshita S, Tanaka K (1972) Glutamic acid. In: Yamada K (ed) The microbial production of amino acids. Halsted Press, New York, NY, pp 263–324
17. Blombach B, Seibold GM (2010) Carbohydrate metabolism in *Corynebacterium glutamicum* and applications for the metabolic engineering of L-lysine production strains. Appl Microbiol Biotechnol 86(5):1313–1322
18. Meiswinkel TM et al (2013) Crude glycerol-based production of amino acids and putrescine by *Corynebacterium glutamicum*. Bioresour Technol 145:254–258
19. Gopinath V et al (2011) Amino acid production from rice straw and wheat bran hydrolysates by recombinant pentose-utilizing *Corynebacterium glutamicum*. Appl Microbiol Biotechnol 92(5):985–996
20. Uhde A et al (2013) Glucosamine as carbon source for amino acid-producing *Corynebacterium glutamicum*. Appl Microbiol Biotechnol 97(4):1679–1687
21. Matano C et al (2014) Engineering of *Corynebacterium glutamicum* for growth and L-lysine and lycopene production from N-acetyl-glucosamine. Appl Microbiol Biotechnol 98(12):5633–5643
22. Tsuchidate T et al (2011) Glutamate production from beta-glucan using endoglucanase-secreting *Corynebacterium glutamicum*. Appl Microbiol Biotechnol 90(3):895–901
23. Kim EM et al (2015) Engineering of *Corynebacterium glutamicum* for growth and succinate production from levoglucosan, a pyrolytic sugar substrate. FEMS Microbiol Lett 362(19)
24. Seibold G et al (2006) Utilization of soluble starch by a recombinant *Corynebacterium glutamicum* strain: growth and lysine production. J Biotechnol 124(2):381–391
25. Heider SA et al (2014) Metabolic engineering for the microbial production of carotenoids and related products with a focus on the rare C50 carotenoids. Appl Microbiol Biotechnol 98(10):4355–4368
26. Heider SA, Peters-Wendisch P, Wendisch VF (2012) Carotenoid biosynthesis and overproduction in *Corynebacterium glutamicum*. BMC Microbiol 12(1):198
27. Heider SA, Wendisch VF (2015) Engineering microbial cell factories: Metabolic engineering of *Corynebacterium glutamicum* with a focus on non-natural products. Biotechnol J 10(8):1170–1184
28. Binder D et al (2016) Light-controlled cell factories: employing photocaged isopropyl-beta-d-thiogalactopyranoside for light-mediated optimization of lac promoter-based gene expression and (+)-valencene biosynthesis in *Corynebacterium glutamicum*. Appl Environ Microbiol 82(20):6141–6149
29. Krubasik P, Kobayashi M, Sandmann G (2001) Expression and functional analysis of a gene cluster involved in the synthesis of decaprenoxanthin reveals the mechanisms for C50 carotenoid formation. Eur J Biochem 268(13):3702–3708
30. Heider SA et al (2014) Optimization of the IPP precursor supply for the production of lycopene, decaprenoxanthin and astaxanthin by *Corynebacterium glutamicum*. Front Bioeng Biotechnol 2:28
31. Heider SA et al (2014) IdsA is the major geranylgeranyl pyrophosphate synthase involved in carotenogenesis in *Corynebacterium glutamicum*. FEBS J 281(21):4906–4920
32. Stansen C et al (2005) Characterization of a *Corynebacterium glutamicum* lactate utilization operon induced during temperature-triggered glutamate production. Appl Environ Microbiol 71(10):5920–5928
33. Peters-Wendisch PG et al (2001) Pyruvate carboxylase is a major bottleneck for glutamate and lysine production by *Corynebacterium glutamicum*. J Mol Microbiol Biotechnol 3(2):295–300
34. Kirchner O, Tauch A (2003) Tools for genetic engineering in the amino acid-producing bacterium *Corynebacterium glutamicum*. J Biotechnol 104(1-3):287–299
35. Schäfer A et al (1994) Small mobilizable multi-purpose cloning vectors derived from the *Escherichia coli* plasmids pK18 and pK19: selection of defined deletions in the chromosome of *Corynebacterium glutamicum*. Gene 145(1):69–73

Chapter 9

Rapid and Selective Screening Method for Isolation and Identification of Carotenoid-Producing Bacteria

Dalal Asker, Tarek S. Awad, Teruhiko Beppu, and Kenji Ueda

Abstract

Carotenoids are naturally occurring yellow to red pigments with many biological activities including antioxidant, anticancer, anti-inflammatory, membrane stabilizers, and precursors for vitamin A. These biological activities are linked with many health benefits (e.g., anticarcinogenic activity, prevention of chronic diseases, etc.), which grew the interest of several industrial sectors especially in food, feed, nutraceuticals, cosmetics, and pharmaceutical industries. The production of natural carotenoids from microbial sources such as bacteria can help meet the growing global market of carotenoids estimated at $1.5 billion in 2014 and is expected to reach 1.8 billion in 2019. This chapter demonstrates, step-by-step, the development of a rapid and selective screening method for isolation and identification of carotenoid-producing microorganisms and their carotenoid analysis. This method involves three main procedures: UV treatment, sequencing analysis of 16S rRNA genes, and carotenoids analysis using rapid and effective HPLC-diode array-MS methods.

Key words Carotenoids, Irradiation, Misasa, Bacteria, Radiotolerant, 16S rRNA genes sequencing analysis, Phylogeny, Diversity, Photodiode array detection, HPLC-DAD, HPLC-MS

1 Introduction

1.1 Structure, Function, and Importance of Carotenoids

Carotenoids are yellow-, orange-, or red-colored natural pigments that provide plants and animals with their attractive colors. Carotenoids are produced by photosynthetic organisms (e.g., plants and algae) and some non-photosynthetic organisms (e.g., yeast, fungi and bacteria) [1]. In photosynthetic organisms, carotenoids serve as accessory pigments in light-capturing complexes by absorbing light and transferring the energy to chlorophyll. In both phototrophic and non-phototrophic organisms, carotenoids protect cells from harmful oxygen radicals [2]. All mammalians including humans, birds, fish, and invertebrates cannot synthesize carotenoids and must thus obtain from the diet [3]. Carotenoids are subdivided into carotenes (hydrocarbons, non-oxygenated such as α- and ß-carotene) and xanthophylls (oxygenated such as

Carlos Barreiro and José-Luis Barredo (eds.), *Microbial Carotenoids: Methods and Protocols*, Methods in Molecular Biology, vol. 1852, https://doi.org/10.1007/978-1-4939-8742-9_9, © Springer Science+Business Media, LLC, part of Springer Nature 2018

zeaxanthin, canthaxanthin, and astaxanthin). Carotenoids especially ß-carotene, canthaxanthin, and astaxanthin have become commercially important as color-enhancing additives to animal feed in the aquaculture, poultry, and swine feed industries. Carotenoids are also used as food colorants [4–7].

Many health benefits are closely linked to the biological activities of carotenoids such as vitamin A precursors and antioxidants [8–10]. The consumption of carotenoids-rich diets enhances the immune system and reduces the risk of degenerative diseases such as cancer, cardiovascular diseases, macular degeneration, and cataract [11–19]. These health benefits diversified the applications of carotenoids in foods, cosmetics, and pharmaceuticals.

Among more than 700 carotenoids so far reported [1], β-carotene, astaxanthin, canthaxanthin, lutein, lycopene, and zeaxanthin (Fig. 1) are the most highest-value carotenoids that are commonly used in food supplements, pharmaceuticals, and feed supplements for proper growth and pigmentation of commercial fish and crustaceans [7]. The global market of carotenoids has increased dramatically from 2004 to 2009 [20], totalled $1.5 billion in 2014 and is expected to reach about $1.8 billion in 2019, a compound annual growth rate (CAGR) of 3.9% [21].

1.2 Microbial Production of Carotenoids

The commercial production of carotenoids is mostly carried out by chemical synthesis [22] or extraction from plant tissue and food industrial wastes [23]. Although chemical synthesis produces carotenoids with high purity and at low cost, they often contain some reaction precursors or by-products of non-biological nature that may have undesired side effects [24, 25] and are not eco-friendly. In addition, consumer awareness of food safety and the health benefits of natural products have increased the demand for biological rather than synthetic production. The major advantages of the biological production of carotenoids are the wide range of their biosynthetic capability and ability to produce only the naturally occurring stereoisomers. Food (i.e., plants or food processing wastes) and microorganisms are examples of carotenoids' biological sources. Unlike microorganisms, the production of carotenoids from food has many disadvantages such as season fluctuation, limited resources, competition with the food industry, and requirement of land beside complicated extraction and purification process. For example, the carotenoids produced from plants contain mixtures of carotenoids, fats, oils, waxes, and unsaponifiable compounds [24, 26]. Nevertheless, the microbial production of carotenoids is still awaiting lot of challenges to reduce cost and simultaneously increase productivity possibly by increasing biomass production and/or carotenoid synthesis. This could be possibly achieved by selecting a strain that grows fast, accumulates high amount of carotenoids, facilitates the extraction and purification processes of carotenoids, and has a lower cost production [27–30]. In addition, the application of modern tools of

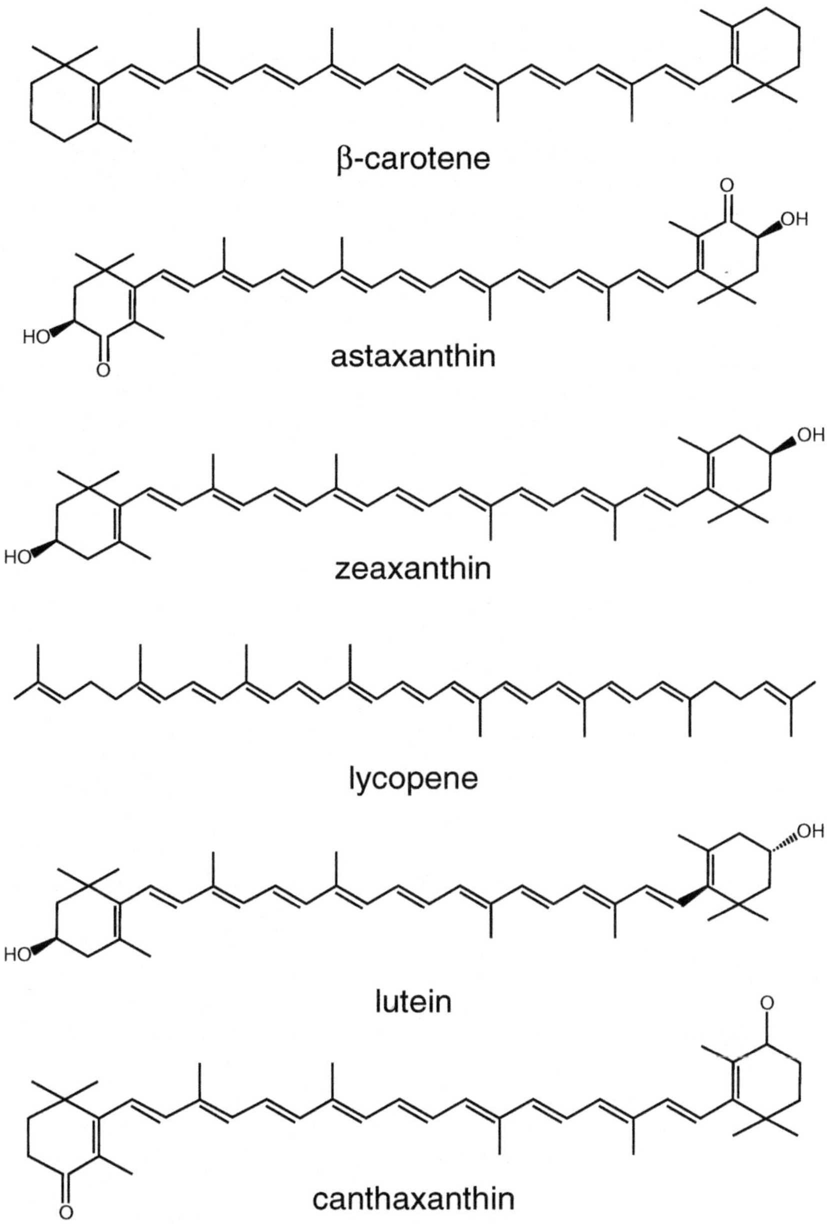

Fig. 1 Chemical structure of high-value carotenoids

bioprocessing and recombinant DNA technology could potentially increase the production of carotenoids.

Table 1 lists the microbial sources of carotenoids. As can be noticed, the commercial production of ß-carotene has been carried out using *Dunaliella salina* [31] and *Blakeslea trispora* [33], while astaxanthin has been commercially produced from *Haematococcus pluvialis* [38] or *Phaffia rhodozyma* [39, 40]. However, there is no effective bacterial production of these carotenoids, with only a few

Table 1
Microorganisms reported for production of high-value carotenoids

Carotenoid	Type of microbe	Microorganism	Yield	Commercial producer	References
β-carotene	Green algae	*Dunaliella salina*[a]	3–5%	Australia, China, India, Israel, Japan, USA	[31]
	Green algae	*Dunaliella bardawil*[a]	10–14%		[32]
	Blue-green algae	*Spirulina*	0.8–1%		
	Mold	*Blakeslea trispora*[a]	30 mg/g	Russia, Ukraine, Spain	[33]
	Mold	*Phycomyces blakesleeanus*	10–35 mg/g		[34, 35]
	Mold	*Mucor circinelloides*			[33]
Lycopene	Mold	*Blakeslea trispora*[a]	24 mg/g		[36]
	Mold	*Fusarium sporotrichioides*	0.5 mg/g		[37]
Astaxanthin	Microalgae	*Haematococcus pluvialis*[a]	1–3%	India, Japan, Sweden, USA	[17, 38]
	Yeast	*Xanthophyllomyces dendrorhous*[a,b]	4–8.1 mg/L	China, Japan, USA	[39, 40]
	Archaea	*Halobacterium salinarium*	265 μg/g		[41]
	Bacteria	*Paracoccus* sp. N81106[c]	0.3 mg/6 L		[42]
	Bacteria	*Paracoccus carotinifaciens* E-396	–		[43]
	Bacteria	*Brevundimonas* sp. SD212[d]	1.5 mg/27 L		[44]
	Bacteria	*Paracoccus* sp. PC1[e]	–		[42]
	Bacteria	*Paracoccus bogoriensis*	0.4 mg/g of Wet cells		[45]
	Bacteria	*Brevundimonas* spp. (wild type)	440–1060 μg/g		[28]
	Bacteria	*Brevundimonas* sp. M7 (mutant)	1.3 mg/g		[46]

(continued)

Table 1
(continued)

Carotenoid	Type of microbe	Microorganism	Yield	Commercial producer	References
Zeaxanthin[f]	Bacteria	*Flavobacterium* sp.	16 mg/g		[47]
	Bacteria	*Sphingobacterium multivorum*[g]	10.6 μg/mL		[48]
	Bacteria	*Synechocystis* sp. PCC 6803	0.98 μg/mL		[49]
	Bacteria	*Erwinia herbicola*			[50]
	Bacteria	*Paracoccus zeaxanthinifaciens*[h]			[51, 52]
	Bacteria	*Nubsella zeaxanthinifaciens*	0.8 mg/g		[53]
	Bacteria	*Mesoflavibacter zeaxanthinifaciens*	0.910 mg/g		[54]
	Bacteria	*Zeaxanthinibacter enoshimensis*			[55]
Canthaxanthin	Bacteria	*Micrococcus roseus*	1.7 μg/mL		[56]
	Bacteria	*Gordonia jacobaea* MV-1	13,373 μg/mL		[57, 58]
	Archaea	*Haloferax alexandrinus* TM	2 mg		[59, 60]
	Bacteria	*Bradyrhizobium* sp.	1.34 mg/g		[61]
β-Cryptoxanthin	Bacteria	*Brevibacterium linens*	0.3 μg/mL		[62]
Lutein	*Chlorophyta*	*Chlorella zofingiensis*	21 μg/mL		[63]
		Chlorella protothecoides CS-41	225 μg/mL		[64]
		Muriellopsis sp.	35 μg/mL		[65]

[a]Effective commercial production
[b]Formerly *Phaffia rhodozyma*
[c]Formally *Agrobacterium aurantiacum* N81106
[d]Formerly *Pseudomonas* sp.
[e]Formally *Alcaligenes* sp. PC-1
[f]Please see (Table 1, Chapter 11) for a complete list of reported zeaxanthin-producing microorganisms
[g]Formerly *Flavobacterium* spp.
[h]Formerly, *Flavobacterium* sp. strain R-1512

bacteria that synthesize astaxanthin, zeaxanthin, canthaxanthin, and cryptoxanthin. To meet the increasing global and industrial demands of natural high-value carotenoids, there are needs to develop screening strategies to enhance the isolation of promising bacterial strains capable of maximizing the production of carotenoids to a commercial level.

1.3 Screening of Carotenoids-Producing Bacteria

In the following sections, we will demonstrate an effective strategy for the screening of carotenoids-producing bacteria. As a result, several potential carotenoids-producing bacteria were isolated from hot spring waters with high radioactivity [66, 67]. Based on 16S rRNA sequencing, we will show how these isolates were identified and classified. We will then describe three HPLC methods for the analysis of the carotenoids produced by the isolates. Finally, we will examine the relationship between radiation resistance and carotenoid production (carotenogenesis).

1.4 Source of Isolation

The natural occurrence of colored microbial mats and dominancy of pigmented microbes in extreme habitats may serve as guides for potential sources of microbial carotenoids. For example, the red coloration of the surface water and soil of salt crystallizer ponds is associated with the growth of extremely halophilic red archaea [59, 68, 69] or bacteria [70]. The yellow and orange colors at Yellowstone hot spring in the USA are carotenoids developed by thermophilic archaea [71–73]. These microbes produce carotenoids to protect their cells against photooxidation induced by the intense sunlight in summer and stabilize their cell membranes at the extremely high temperatures [74–77]. The exposure to solar UV radiations such as type A (320–400 nm) and type B (280–320 nm) may damage biological cells. UV-A radiations generate free radicals that damage DNA, while the exposure to UV-B can lead to direct damage of DNA and proteins. As antioxidants, carotenoids protect cells by quenching free radicals, and some carotenoids have been reported to have the ability of absorbing UV-A [78]. The protective role of carotenoids against UV-B irradiation in some microbial systems has been evidenced [72, 79–81]. In the Gram-negative gliding bacterium *Myxococcus*, carotenoid production occurs in a light-dependent manner, probably to protect the cells from photooxidative damages [82]. Another specific environmental stimulus that is possibly correlated to microbial carotenogenesis is ionizing radiations (e.g., X-ray), which damage living cells mainly by inducing oxidative stress; reactive oxygen species and free radicals attack biological macromolecules such as nucleic acids, proteins, and lipids [83]. Most members of *Deinococcus* spp. are radiotolerant and produce carotenoids. Studies have shown that the colorless mutant of *Deinococcus radiodurans* is more sensitive to ionizing radiation, hydrogen peroxide, and desiccation than the wild type, suggesting an association between carotenoids and tolerance to environmental stresses [84, 85]. However, pigmentation strategies are not obligatory as some microbes have highly efficient DNA repair mechanisms as an alternative to damage prevention [86]. Many carotenoids-producing bacteria have been isolated from various extreme environments such as very low temperatures [75], high salinity [59, 87], strong light [59, 78], acidic and alkaline [88], and thermophilic conditions [71, 74, 77, 89]. One may hypothesize, based on these evidences, that the oxidative stresses in extreme environments are selective factors associated with pigmented microorganisms, which are able to synthesize antioxidants (i.e., carotenoids) to protect their vital molecules (e.g., proteins and nucleic acids). Misasa (Tottori, Japan; Fig. 2) is a radioactive region site that is characterized by high radon activity (^{226}Ra, 0.60 Bq/L; ^{228}Ra, 0.41 Bq/L) [90]. This region was thus selected as a potential source for the isolation of carotenoids-producing bacteria [6]. Our preliminary observations showed many bacterial colonies that exhibited the typical color of carotenoids. The result of an extensive screening study revealed the widespread distribution of carotenoid-producing bacteria in this region as well as their marked taxonomic diversity [6].

1.5 Screening Strategy

Microorganisms produce a wide variety of pigments such as carotenoids, melanins, flavins, quinones, monascins, violacein, phycocyanin, or indigo [91]. In general, there are two types of pigments: cell-bound and non-cell-bound pigments. Cell-bound pigments are nondiffusible pigments (e.g., carotenoids and chlorophyll) that can be recovered from the cells by disruption and solvent extraction. The second class of pigments (i.e., diffusible pigments) is secreted into the broth or agar media and can be recovered by extraction from the media with large quantities of suitable solvents. Carotenoids are cell-bound pigments that are produced inside the cells and do not diffuse in the agar. Carotenoids give the microbial colonies their distinctive color (i.e., colonies on agar may be yellow, orange, pink, or red). Therefore, one can screen environmental water samples for carotenoids-producing microorganisms in two stages (Fig. 3). The first stage includes primary screening and

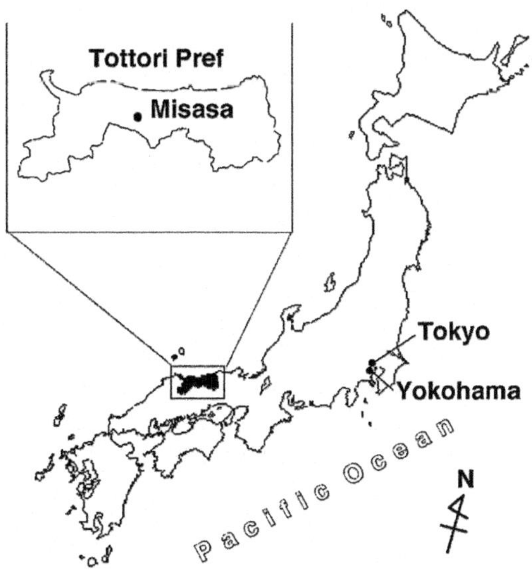

Fig. 2 Geographical location of Misasa (Tottori, Japan) (N35.41, E133.88). Adapted from Asker et al. [6]

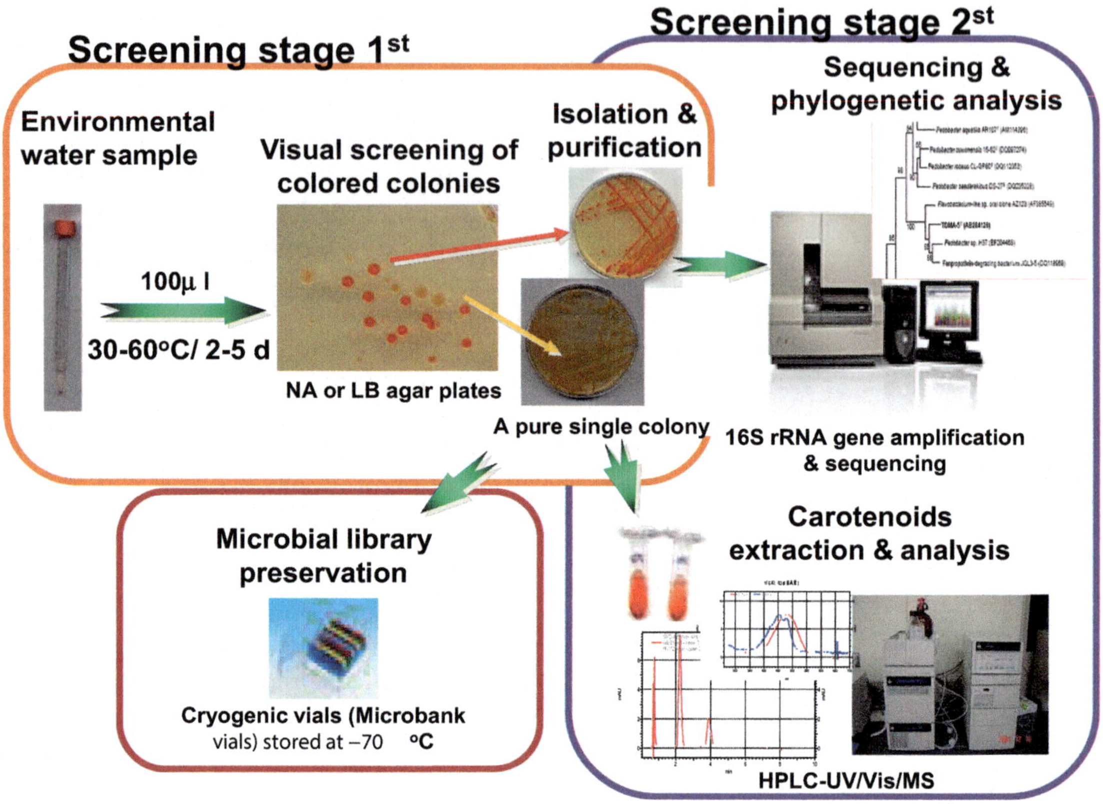

Fig. 3 Screening strategy of carotenoids-producing bacteria

microbial library construction. Primary screening is done visually on cell-bound pigmented colonies (i.e., colonies that produce nondiffusible pigments) that are grown on an appropriate agar media and incubated at various temperatures 30–60 °C (Fig. 3). After incubation, the yellow-, orange-, or red-colored microbial colonies that had formed are isolated, and each organism is purified using the standard microbial manipulation technique. As a selective screening technique, agar plates can be UV-irradiated for 1–3 min before incubation. The UV treatment helps to diminish the uninterested (noncolored) colonies, which have no UV-resistance (Fig. 4). A microbial library that includes the pigmented isolates is then constructed. The second stage involves sequencing of 16S rRNA genes and carotenoids analysis by using a rapid and reliable set of HPLC complementary methods (Fig. 3). By using this screening strategy, we attempted to screen each strain for the three following objectives: (1) discovery of novel microbial sources of carotenoids at level of genus and species, (2) biodiversity of carotenoids-producing bacteria in a specific environment (e.g., radioactive hot spring water), and (3) discovery of novel carotenoids with novel properties.

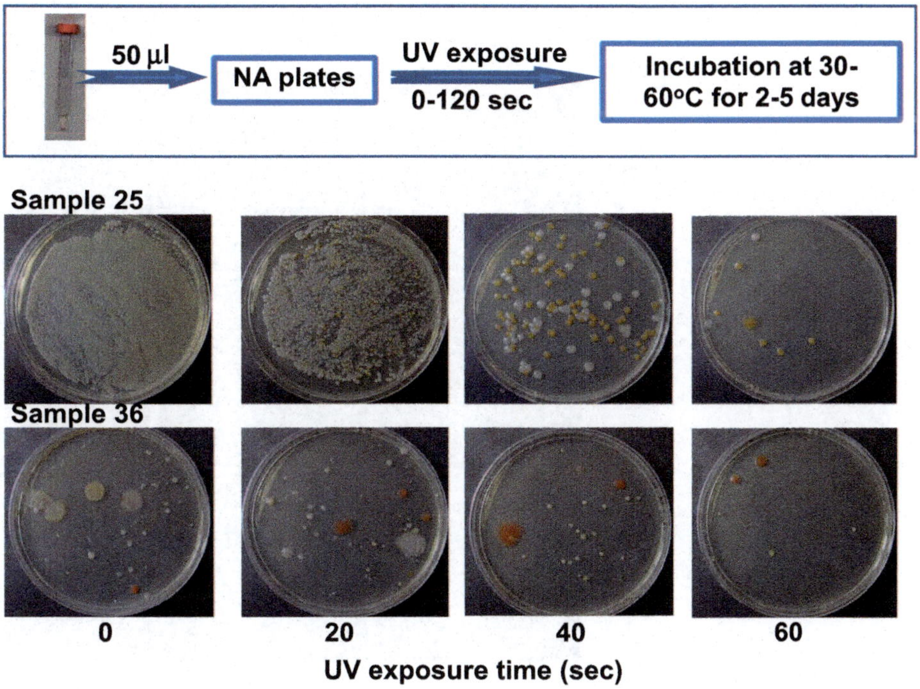

Fig. 4 UV treatment for the screening of carotenoids-producing bacteria from the water samples collected at Misasa. Each sample was diluted 1:10, and 50 μL of each dilution was inoculated onto NA agar plates and incubated at 30–60 °C for 3 days. Prior to incubation, the plates were exposed to UV for the indicated periods

Many yellow-, orange-, pink-, or red-colored colonies were obtained after inoculating and cultivating the water samples collected from Misasa (a representative result is shown in Fig. 5). The frequency at which the colored colonies were obtained from the samples appeared markedly higher than that from the samples collected from normal environments.

1.6 Identification of Microbial Carotenoids

Carotenoids are lipophilic compounds (i.e., insoluble in water while soluble in nonpolar solvents) that are located within the cell attached to cellular membranes. The extraction of carotenoids from microbial cells involves two major steps: cells disruption (i.e., break down or dissolve cell walls) followed by extraction by organic solvents. Disruption of the microbial cells is traditionally done by mechanical, biological, and chemical methods. Mechanical disruption can be achieved by high-pressure homogenization or grinding with glass beads (or silica gel). These methods are however unfavorable because they are slow, require large volumes of cells, have low productivity of carotenoids, and require high energy [92, 93]. Enzymatic method has been widely used for microbial disruption [93–96]; however, it is also time-consuming to break the cells and may simultaneously result in serious degradation of

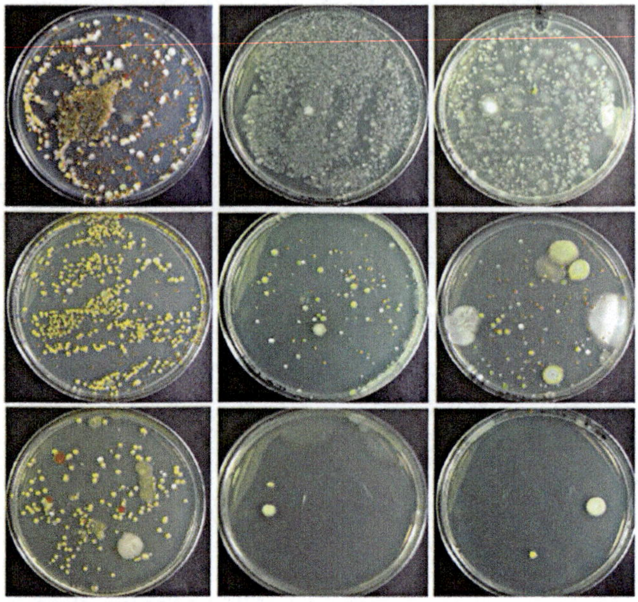

Fig. 5 A representative result of colony isolation from the water samples collected at Misasa. Each sample was diluted 1:10, and 50 μL of each dilution was inoculated onto LB agar plates and incubated at 30 °C for 3 days. Prior to incubation, the plates were exposed to UV for the indicated periods. Typical results of the same treatment using samples collected from normal environments: (**a**) Miya River, Yokohama, Kanagawa. (**b**) Ooka River, Yokohama, Kanagawa. Adapted from Asker et al. [66]

carotenoids. On the other hand, chemical extraction is simple, fast, and low-cost method [97–99]. Nevertheless, it is important to avoid the use of alkaline chemicals as they induce irreversible conversion of astaxanthin to astacene [100]. Other alternative methods for microbial cells disruption including freezing and thawing, ultrasonication, and freeze-drying are also slow and may result in serious degradation of carotenoids. The second step is to extract carotenoids from the microbial cell debris by successive repeated extraction in organic solvents such as acetone, petroleum ether, hexane, chloroform, ethanol, methanol, and dimethyl sulfoxide (DMSO). While a single organic solvent can be used, the use of mixtures of solvents is advantageous as it yields high carotenoids recovery [30, 101]. The simultaneous use of mixtures of some solvents demonstrated synergistic interactions, which resulted in enhanced carotenoids yield [102]. A relatively rapid method for extracting astaxanthin from *Phaffia rhodozyma* was developed by Sedmak et al. (1990) who used hot dimethyl sulfoxide (DMSO) to extract carotenoids and then transferred the dissolved carotenoids into chloroform [102]. However, the later step complicates the

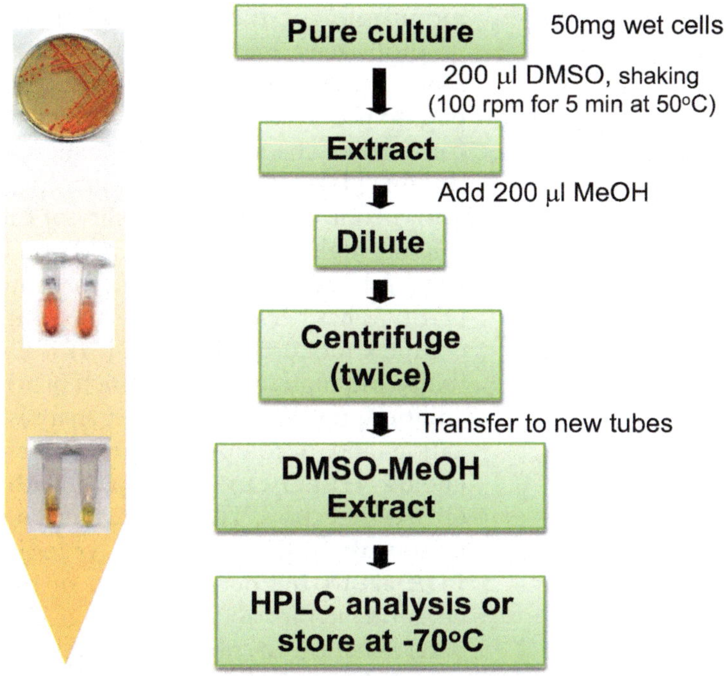

Fig. 6 Outline of a simple and rapid method for bacterial carotenoids extraction

process especially when a large number of samples are extracted. A modification of this method was used by Asker et al. [66] (Fig. 6). In the modified method, the cells (approximately 50 mg) are suspended in 200 μL DMSO and agitated at 50 °C by rotary shaking in the dark (5–10 min) to initiate lysis of the cells. An equal volume of methanol is then mixed with the lysed cells and centrifuged to obtain the carotenoid extract. This method is simple, rapid, and straightforward as it does not involve any cell disruption or carotenoids concentration after extraction, and it excludes the step of carotenoids transfer into chloroform. In addition, it requires a very small amount of the microbial cells for the identification of carotenoids. Importantly, this method enables the extraction of carotenoids from both Gram-negative and Gram-positive bacteria, as well as a wide variety of microorganisms including yeast, fungi, and microalgae (unpublished data). Accordingly, the modified method is ideal for the high-throughput screening of carotenoids-producing microorganisms and their diversity.

Screening for interested carotenoids from microbial sources can be a difficult task possibly due to (1) presence of complex mixture of carotenoids whose structures are slightly different, (2) low content of the targeted carotenoids, (3) large number of samples, and (4) instability of carotenoids to light, oxygen, and heat. Conventional chromatographic methods (e.g., thin layer chromatography (TLC) and column chromatography) can be used for separation and identification of carotenoids as color permits visual monitoring of the separation of carotenoids. However, these methods have many disadvantages such as requirement of large sample volumes, low resolution, time consuming, carotenoids degradation, and low recovery of carotenoids that are needed for further characterization (e.g., spectra analysis and molecular weight) [104]. Gas chromatography (GC) is unsuitable because of low volatility and thermolability of carotenoids [90]. Therefore, a suitable high-performance separation followed by a sensitive detection is required. High-performance liquid chromatography (HPLC) has been utilized for the routine analysis of carotenoids in complex mixtures [104–108]. It is characterized by its reproducibility and high sensitivity, which provide reliable analytical data. In addition, the reasonably short analysis time minimizes the isomerization and decomposition of labile compounds such as carotenoids. HPLC can be coupled with UV/Vis photodiode array (PDA) detection. The HPLC elution profile provides useful information about the classes of carotenoids present in the sample. The PDA allows the determination of the UV/Vis spectrum of each component. In addition, using HPLC/MS system allows the determination of molecular weight of each component. To meet our goals, we designed a rapid and reliable set of HPLC methods for the separation and identification of carotenoids that may have a wide polarity range. These methods (see below) have greatly enabled to detect high-value carotenoids, discover novel carotenoids, and study the diversity of carotenoids produced by unknown bacterial isolates.

In our screening approach, we use three HPLC methods (A, B, and C) (*see* Subheading 3) (Figs. 7, 8, and 9) for the separation and identification of carotenoids from our DMSO/MeOH library extracts. HPLC method A is primarily used to examine the presence of any kind of carotenoid (i.e., polar and apolar) in a short time (Fig. 7a). In method A, a short reversed-phase (RP-C18) HPLC column and moderate polarity mobile phase (acetonitrile/methanol/tetrahydrofuran) are used to separate both high polar carotenoids (e.g., astaxanthin and zeaxanthin) and apolar carotenoids (e.g., β-carotene) within less than 10 min (Fig. 7a). This system is coupled with UV/Vis photodiode array (PDA) detector. The HPLC elution profile provides useful information about the classes of carotenoids present in the sample (Fig. 7a). The PDA allows to online determination of the UV/Vis spectrum of each component (Fig. 7b1–b5). Although method A is very useful in a rapid screening approach, its drawback is the overlapping of slightly similar polar carotenoids such as astaxanthin and zeaxanthin (Fig. 7a). To solve this problem, we use a second method (method B) to separate and identify the overlapping polar carotenoids (Fig. 7b). In method B, using the same column and a high polar solvent such as methanol (90%), it is possible to separate between overlapping polar carotenoids such as zeaxanthin and astaxanthin

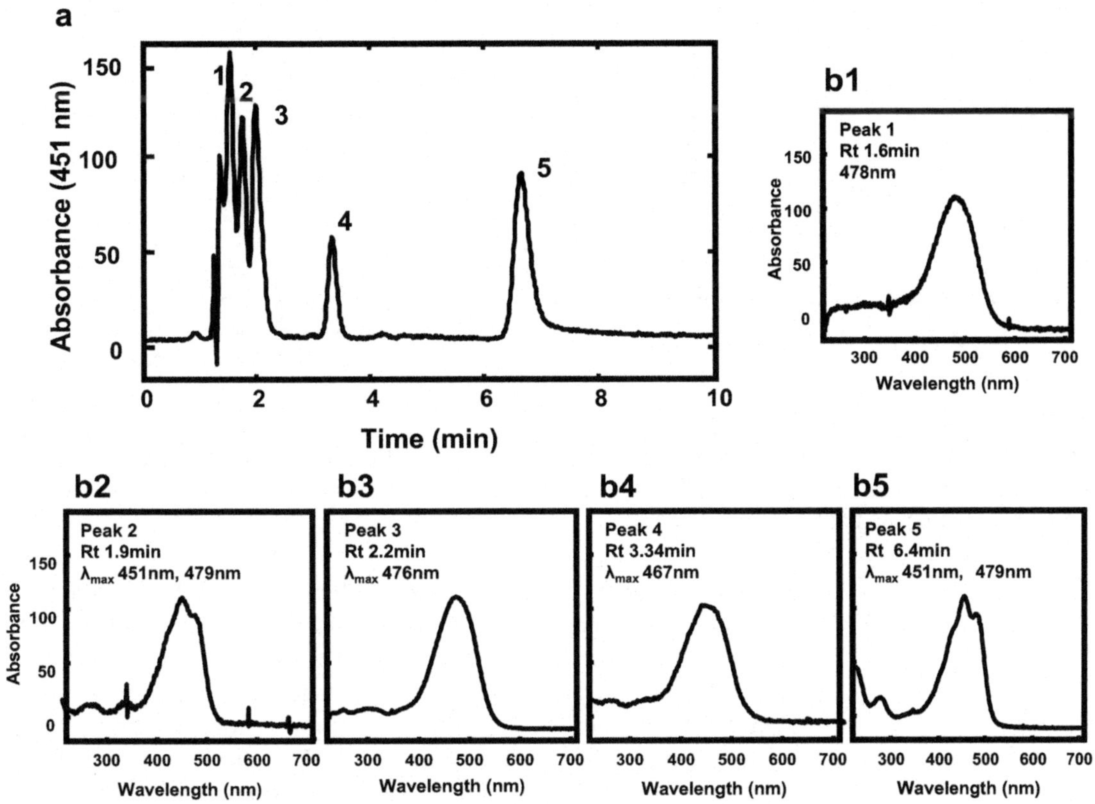

Fig. 7 HPLC method A for carotenoids analysis. Elution profile (λ_{max} 451 nm) of carotenoid standards including astaxanthin, zeaxanthin, canthaxanthin, echinenone, and β-carotene (**a**) and absorption spectrum of each peak (**b**) are shown. The carotenoid fraction was analyzed by reversed-phase liquid chromatography using method A

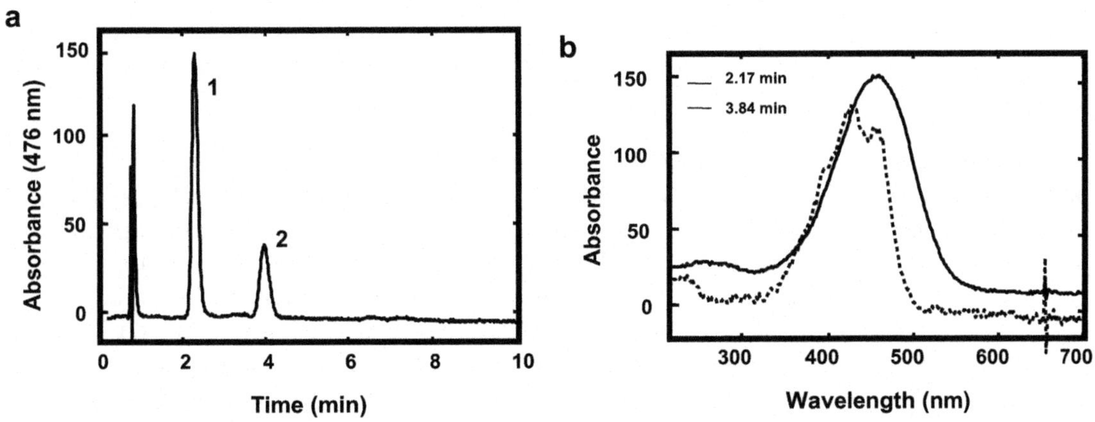

Fig. 8 HPLC method B for carotenoids analysis. Elution profile (λ_{max} 476 nm) of carotenoid standards astaxanthin and zeaxanthin (**a**) and absorption spectrum of each peak (**b**) are shown. The carotenoid fraction was analyzed by reversed-phase liquid chromatography using method B

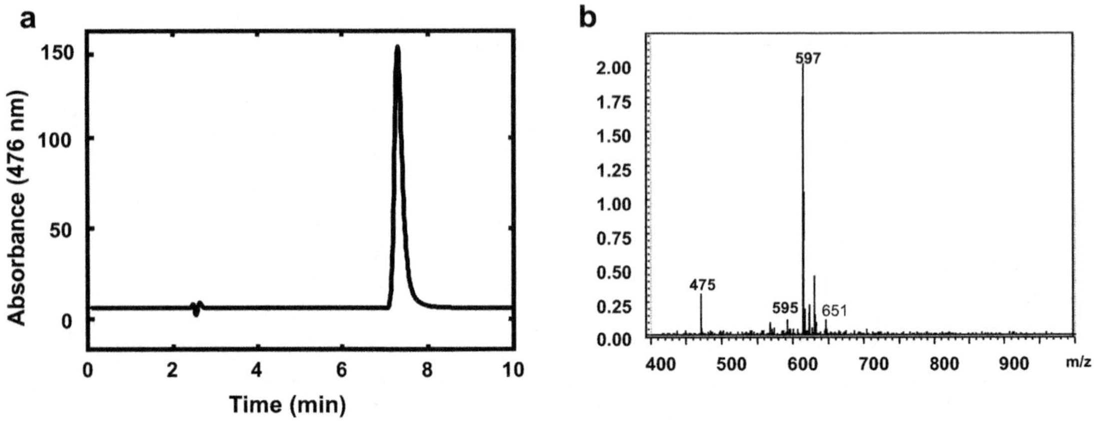

Fig. 9 HPLC method C for carotenoids analysis. Elution profile (λ_{max} 476 nm) of carotenoid standards astaxanthin (**a**) and its mass fragmentation pattern (**b**) is shown. The carotenoid fraction was analyzed by reversed-phase liquid chromatography using method C

in a short time (1–3 min) (Fig. 8a). The same UV/Vis photodiode array (PDA) detection system is also used in method B (Fig. 8b). To determine the molecular weight of the carotenoids, method C is performed on an LC/MS system (see Subheadings 2 and 3) (Fig. 9a,b). In method C, a long column provides more separation between the carotenoids species, while MS allows the determination of molecular weight of individual carotenoids. Therefore, carotenoids can be identified based on retention time (R_t), absorption spectrum, and molecular weight. Commercial β-carotene, astaxanthin, zeaxanthin, canthaxanthin, and echinenone (*see* Subheading 2) are used as standards.

We studied the spectral characteristics of the pigments isolated from the cells of the bacterial isolates. Based on their solubility in organic solvent and spectroscopic properties, it was clear that all the pigments were carotenoids. Table 2 shows the spectroscopic properties of the carotenoids isolated from the representative strains of each species.

1.7 Identification of Carotenoids-Producing Isolates

For interested carotenoids producers, bacterial identification is carried out to determine whether isolates belong to approved taxa (genus or species) or unknown. Traditionally, the identification of unknown bacterial isolates entails numerous morphological, biochemical, cultural, antigenic, and other phenotypic characteristics, which are materials, time, and efforts consuming. For rapid identification of a large number of samples, these methods are thus inappropriate. In contrast, the use of molecular methods based on PCR and DNA sequencing can shorten the identification time. Among these methods, 16S rRNA gene sequencing is used for identification at both the genus and species levels and to study the relatedness of microbial taxa by phylogenetic relationships [109–112]. Recently, high-throughput sequencing and screening methods

Table 2
Spectroscopic and HPLC characteristics of the analyzed carotenoids

		HPLC, retention time (R_t, min)			
Carotenoids	**Absorption (λ_{max}, nm)**	**Method A**	**Method B**	**Method C**	**Mol. wt (M^{+}+1)**
Zeaxanthin	451, 476	0.98	4.27	10.3	569
Nostoxanthin	449, 475	0.89	2.29	6.5	601
Astaxanthin	476	0.81	2.31	7.05	597
Hydroxyastaxanthin	476	0.79	1.79	5.4	613
Dihydroxyastaxanthin	476	0.65	1.34	4.21	629
Canthaxanthin	476	2.04	7.49	20.3	565

have been developed to study microbial diversity and biotechnological potential of various microbial communities [28, 113].

In our screening approach, a conventional 16S rRNA gene-based analysis is used to identify the isolates that produce carotenoids (Fig. 3). After obtaining the sequencing results, 16S rRNA gene sequences are correlated with those in the GenBank/EMBL/DDBJ nucleotide sequence databases by using the BLASTN program (http://www.ncbi.nlm.nih.gov/BLAST/). The strains that share 99% or higher 16S rRNA gene sequence similarity can be grouped into a single species. If the determined sequence of an unknown isolate exhibits high similarity (>97% and ≤100%) with the reference sequence of an approved species, then the unknown isolate may be assigned to this species. When the similarity is <97% and ≥95%, the unknown isolate is assigned to the corresponding genus with possibility of claiming a novel species. When the similarity is <95%, the unknown isolate is assigned to a family with possibility of claiming a novel genus. The next step is to construct a phylogenetic tree based on the complete sequences of 16S rRNA gene for all pigmented isolates. Figure 10 demonstrates a diversity of carotenoids-producing microorganisms (104 isolates), which can be classified into 38 different species belonging to 7 bacterial classes (*Flavobacteria*, *Sphingobacteria*, *α-Proteobacteria*, *γ-Proteobacteria*, *Deinococci*, *Actinobacteria*, and *Bacilli*). The carotenoids produced by these isolates were zeaxanthin (6 strains), dihydroxyastaxanthin (24 strains), astaxanthin (27 strains), canthaxanthin (10 strains), and several unidentified molecular species that were produced by the isolates related to *Deinococcus*, *Exiguobacterium*, and *Flectobacillus* [6]. More details about this study can be also found in the first edition of this book [114].

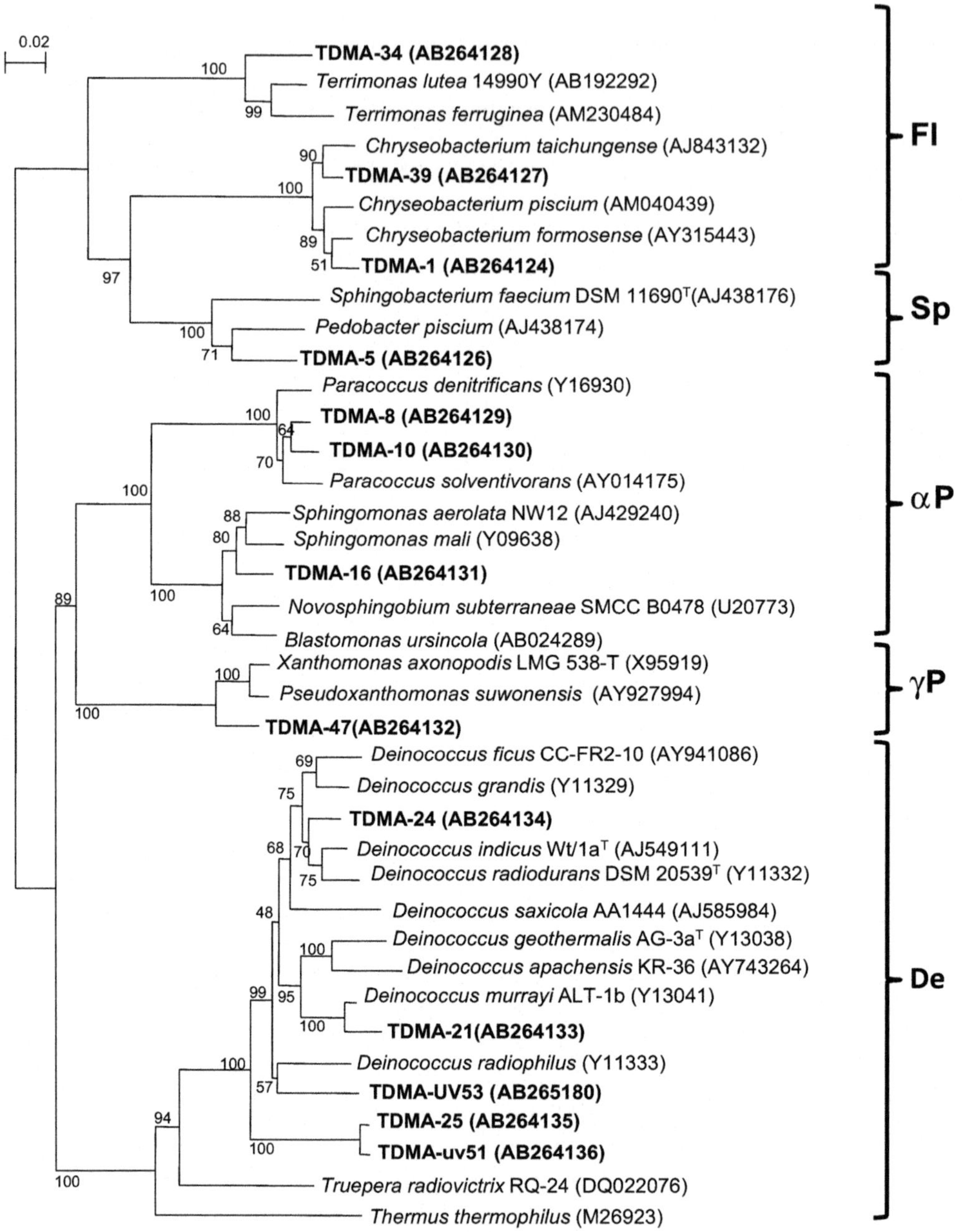

Fig. 10 The 16S rRNA gene-based phylogeny of the taxonomically new carotenoid producers isolated in this study. The tree was constructed using the maximum-likelihood method with ClustalW [115]. The bootstrap value expressed as a percentage of 1000 replicates is given at each node; values above 50% are shown. Nucleotide sequence accession numbers are indicated in parentheses. Fl *Flavobacteria*, Sp *Sphingobacteria*, αP *α-Proteobacteria*, γP *γ-Proteobacteria*, De *Deinococci*. Adapted from Asker et al. [6]

2 Materials

2.1 Water Samples Collection

1. 15 mL sterilized (disposable) polystyrene tubes with cap.
2. Rack.
3. Markers for labelling.
4. Refrigerated storage (4 °C).

2.2 Primary Screening and Microbial Library Construction

1. Analytical balance.
2. 15 mL culture tubes.
3. UV lamp (GL-15 53 μW/cm^2) (Toshiba, Tokyo, Japan).
4. Micropipettes and sterile tips (0–20, 20–200, and 100–1000 μL).
5. Vortexer.
6. Refrigerated microcentrifuge (14,000 × g max speed, rotor holds 1.5 mL tubes).
7. Incubators (28, 30, 37, 40, 50 °C).
8. Sterile loops, sterile toothpicks, and sterile cotton swabs.
9. Clean bench equipped with UV light.
10. Nutrient agar: 3 g/L beef extract, 5 g/L peptone, and 15 g/L agar (Difco Laboratories, Detroit, MI, USA).
11. Luria-Bertani (LB) solid medium: 10 g/L tryptone, 5 g/L yeast extract, 5 g/L NaCl, and 15 g/L agar.
12. Peptone water (Difco Laboratories, Detroit, MI, USA).
13. Dilution tubes with sterile peptone water (9 mL) in each tube.

2.3 Strains Preservation and Library Construction

1. Microbank kit (Pro-Lab Diagnostics, Toronto, Canada).
2. Glycerol 20%.
3. Freezer −70 °C.

2.4 Carotenoids Analysis

1. HPLC grade solvents.
2. β-carotene (Sigma, St. Louis, MO, USA).
3. Astaxanthin (Sigma, St. Louis, MO, USA).
4. Zeaxanthin (DHI; Water & Environment, Hørsholm, Denmark).
5. Canthaxanthin (DHI; Water & Environment, Hørsholm, Denmark).
6. Echinenone (DHI; Water & Environment, Hørsholm, Denmark).

7. BHT (butylhydroxytoluene) (Sigma, St. Louis, MO, USA).
8. Sterile spatula or toothpick.
9. 1.5 mL Eppendorf tubes.
10. Racks.
11. Balance.
12. Eppendorf table top vial shaker.
13. Spectrophotometer.
14. Syringe filter (0.45 μm hydrophobic Teflon membrane).
15. Syringe 1 mL.
16. 96-well plate.
17. HPLC (Class-VP; Shimadzu, Kyoto, Japan) with PDA detector.
18. HPLC/MS system (LCMS-2010EV, Shimadzu).
19. CAPCELL PAK C18 HPLC column (35 mm × 4.6 mm; particle size, 5 μm) (Shiseido, Tokyo, Japan).
20. Shim-Pack FC-ODS HPLC column (150 mm × 4.6 mm; particle size, 5 μm) (Shimadzu, Kyoto, Japan).
21. Acetonitrile/methanol/tetrahydrofuran (5.8:3.5:0.7).
22. Methanol/water (9:1).

2.5 Identification of Bacterial Isolates: 16S rRNA Gene Sequences

1. Sterile material, including Eppendorf tubes, toothpicks, and deionized water.
2. DNA purification kit (Edge BioSystems, Gaithersburg, MD, USA).
3. Primer B8F: 5′-AGAGTTATCCTGGCTCAG.
4. Primer B1492R: 5′-GGTTACCTTGTTACGACTT.
5. Thermocycler.
6. Taq polymerase.
7. Agarose.
8. 10× TBE buffer: 108 g Tris–HCL base, 55 g boric acid, 40 mL 0.5 M EDTA (pH 8.0).
9. 1× TBE buffer: 100 mL of 10× TBE buffer and 900 mL of deionized water.
10. 6× blue loading dye: 30% glycerol in distilled water and 0.25% bromophenol blue.
11. Ethidium bromide (*see* **Note 1**).
12. Agarose gel electrophoresis system.
13. UV safety goggles and UV box.
14. PCR tubes (0.2 mL thin-walled).

15. BigDye Terminator v3.1 Cycle Sequencing Kit (Applied Biosystems, California, USA).
16. Pipettes and tips (0–10, 20–200, 100–1000 μL).
17. GeneClean kit II (Funakoshi, Tokyo, Japan).

3 Methods

3.1 Water Samples Collection

1. Collect water samples from the surface water layer of various places in selected environments.
2. Store samples at 4 °C immediately until screening.

3.2 Primary Screening and Microbial Library Construction

Screening and isolation of carotenoid-producing microorganisms from natural water samples can be done using a direct agar series dilution method. Colored colonies grown on an appropriate media agar plate can be primary isolated by visual screening.

1. Do a series of dilution of 10^{-1}, 10^{-2}, 10^{-3}, and 10^{-4} for each water sample. Spread 0.1 mL of each dilution on LB or nutrient agar in duplicate (*see* **Note 2**).
2. In case of UV-irradiated plates, before incubation and under sterilization condition (inside the clean bench), open the plates, and treat them using a UV lamp (GL-15, 53 μW/cm^2) for 0, 1, 2, 3 min.
3. Incubate aerobically the agar plates at various temperatures (28, 30, 37, 40, and 50 °C).
4. To isolate a single colony, prepare LB or NA agar plates, and then classify each plate into six to eight sections using the marker. Using a sterile inoculating loop (or a sterile toothpick), touch a single-colored colony (yellow, orange, pink, or red), and then streak it into a single section in the appropriate plate (LB or NA) to produce a single colony. Incubate the inoculated plates at an appropriate temperature for 2–3 days.
5. To obtain batches of isolates (biomass), prepare and classify LB or NA agar plates as in **step 4**. Transfer a single colony from each isolate (using a sterile toothpick) into a fresh single section in the appropriate plate (LB or NA), and streak it as a thick batch (*see* **Note 3**).

3.3 Strains Preservation and Library Construction

To preserve isolated strains, use the Microbank kit (Pro-Lab Diagnostics) or the glycerol stock method, and then store at −70 °C (*see* **Note 4**).

1. To preserve isolated strains using the Microbank™ kit, follow the following instructions, which are also provided with the kit manual:

(a) Use a permanent marker to label the vial as desired, one organism per vial to be inoculated.

(b) Under aseptic conditions, inoculate the cryopreservative fluid with pure young colonial growth (18–24 h).

(c) Close vial tightly and invert 4–5 times to emulsify organism (do not vortex).

(d) At this point the microorganisms will be bound to the porous beads. The excess cryopreservative should be well aspirated leaving the inoculated beads as free of liquid as possible.

(e) Record the inoculation coding on the printed Microbank™ Storage Box lid or on any other suitable recording system of choice.

(f) Store the inoculated cryovial at −70 °C for best long-term results.

2. To preserve isolated strains in glycerol, aliquot 1 mL of sterilized glycerol solution (20%) into sterilized Eppendorf tubes. Inoculate the glycerol with pure young bacterial colonial growth (18–24 h). Mix thoroughly and store at −20 or −70 °C.

3.4 Extraction of Carotenoids

1. Using a sterilized spatula or toothpick, transfer about 50 mg of wet microbial cells into the 1.5 mL Eppendorf tube.
2. Add 200 μL of DMSO, and agitate the suspensions at 50 °C by rotary shaking at 100 rpm in the dark until the cells are bleached (5–10 min) (*see* **Note 5**).
3. Add 200 μL of methanol and vortex for 30 s at 15,000 × *g*.
4. Transfer the supernatants to new Eppendorf tubes, and centrifuge at 15,000 × *g* for 10 min.
5. Filter the supernatants (carotenoids extracts) using a 0.45 μm hydrophobic Teflon membrane syringe filter into a 96-well plate for automated HPLC analysis.
6. Proceed with HPLC analysis, or flush carotenoid extracts using N_2 gas, and store at −70 °C until analysis.

3.5 HPLC Method A

1. Analyze carotenoids extracts using HPLC method A to screen for all carotenoids species, ranging from polar carotenoids such as astaxanthin to apolar carotenoids such as β-carotene. You can use an HPLC with automated injection samples for samples loaded into 96-well plate.
2. Analyze standards samples of carotenoids such as β-carotene, astaxanthin, zeaxanthin, and canthaxanthin under the same conditions.

3. Compare HPLC profiles of screened unknown samples with those of the standards.
4. Using PDA, compare the spectral absorption of interested HPLC peaks with those of corresponding standards.
5. Solubilize carotenoids standards in DMSO and dilute with methanol (1:1, v/v).
6. Prepare a standard curve for each carotenoid using HPLC.
7. Place all samples into the 96-well plate. Add carotenoids standards solutions in the beginning wells.
8. Using the automated injection system, inject 10 μL of each DMSO-methanol extract onto the CAPCELL PAK C18 column.
9. Elute with a mixed solvent containing acetonitrile/methanol/tetrahydrofuran (5.8:3.5:0.7), at a flow rate of 1 mL/min.
10. Analyze the HPLC profiles for both R_t and PDA spectra of each peak.
11. Compare both R_t and PDA spectra of each interested peak with those of the carotenoids standards.
12. Select interested carotenoids extract and store at −80 °C for further analysis.
13. Regularly wash HPLC column (*see* **Note 6**).

3.6 HPLC Method B

Subject interested carotenoids extract from method A to method B for screening of interested polar carotenoids such as astaxanthin, zeaxanthin, and carotenoid glucosides such as astaxanthin glucoside.

1. Analyze the standard samples of carotenoids under the same condition.
2. Compare HPLC profiles of screened samples with those of the standards.
3. Use PDA to compare the absorption spectra of interested HPLC peaks with those of the corresponding standards.
4. Apply 10 μL of interested samples onto the same column as above (column temperature 35 °C).
5. Elute with a mixed solvent containing methanol/water (9:1) at a flow rate of 1.0 mL/min.
6. Analyze the HPLC profiles for both R_t and PDA spectra of each peak.
7. Compare both R_t and PDA spectra of each interested peak with those of the carotenoids standards.
8. Select the interested carotenoids extract for more analysis using method C.
9. Regularly wash HPLC column (*see* **Note 6**).

3.7 HPLC Method C

Subject interested carotenoids extracts from methods A and B to method C to determine the molecular weight of interested carotenoids as follows.

1. Inject 10 μL of interested DMSO-methanol extract onto the Shim-Pack FC-ODS.
2. Elute with 95% methanol at a flow rate of 1.0 mL/min.
3. Analyze the HPLC profiles for both R_t, molecular peak, and fragment pattern of interested peak.
4. Compare R_t, molecular peak and fragment pattern of interested peak with those of the corresponding carotenoids standards.
5. Regularly wash HPLC column (*see* **Note 6**).

3.8 16S rRNA Gene Sequences

1. Carefully transfer 50 mg wet cells from each bacterial isolate into Eppendorf tubes.
2. Extract genomic DNA from each bacterial isolate using the bacterial genomic DNA purification kit.
3. Check the quality of extracted genomic DNAs by agarose gel electrophoresis (**steps 4–9**).
4. Prepare 1% gel solution by melting 1 g of agarose in 100 mL of 0.5× TBE buffer in a microwave for approximately 2 min. Allow to cool for 5 min, and then add 2.5 μL ethidium bromide; stir to mix (*see* **Note 7**).
5. Cast the gel using the supplied tray and comb. Allow the gel to set for a minimum of 30–40 min at room temperature.
6. Load into separate wells: 10 μL of 1 kb ladder, diluted samples (5 μL sample + 5 μL water +2 μL 6× loading dye).
7. Run the gel for 30 min at 100 V.
8. Expose the gel to UV light and photograph (wear UV goggles).
9. Confirm DNA quality.
10. Set up PCR reaction mixtures in a final volume of 50 μL: 10 μL 10× PCR-buffer, 1 μL PCR Nucleotide Mix, 50 U Taq DNA polymerase, 1 μL (10 μM) primer B8F, 1 μL (10 μM) primer B1500R, 2 μL DNA extract, and 33 μL deionized and sterilized water.
11. Perform PCR using a protocol that includes an initial denaturation for a period of 5 min at 94 °C; 30 cycles of 94 °C for 1 min, 55 °C for 1 min, and 72 °C for 1 min; 72 °C for 3 min; and incubation at 4 °C until further processing.
12. Purify PCR amplicons using agarose gel electrophoresis (**steps 13–19**).

13. Prepare 2% agarose gel, and let it to solidify at room temperature for 30–40 min.
14. Add 10 μL 6× blue loading dye to each PCR product (50 μL).
15. Carefully load all 60 μL into wide gel well.
16. Run the gel for 30 min at 100 V.
17. Wear UV goggles. Remove gel and place on UV box to identify region with PCR products (*see* **Note 7**).
18. Cut DNA bands from agarose gel and transfer into sterilized Eppendorf tubes.
19. Purify PCR products from the gel using the GeneClean kit.
20. Determine the 16S rRNA gene sequences using the BigDye Terminator v3.1 cycle sequencing kit.

3.9 Phylogenetic Analysis

1. Compare the nearly complete 16S rRNA gene (about 1396–1450 bp) sequences of isolates to those available in GenBank/EMBL/DDBJ nucleotide sequence database using the BLASTN program (http://www.ncbi.nlm.nih.gov/BLAST/).
2. Using ClustalW, align the obtained sequences with those of closed relatives.
3. Construct a neighbor-joining phylogenetic tree (*see* Fig. 10).
4. Use NJ plot to draw and edit the obtained tree.

4 Notes

1. Ethidium bromide is mutagenic and moderately toxic. You must wear gloves and handle the chemical with care.
2. We collected about 40 water samples throughout the spa from the water close to the hot spring wells, the downstream of the water, and the river. Dilution depended on the nature of samples. Samples collected close to the hot spring well (source of radioactivity) are pure, low in organic matter, and free of microbial cells, which did not require dilution. In contrast, samples collected from the water stream on the river are very rich in organic matter and highly rich in microbial cells. Therefore, we do dilution plating of 10^{-1}, 10^{-2}, 10^{-3}, and 10^{-4} serial.
3. You can transfer six to eight colonies on a single plate; each will yield more than 200 mg wet cells. Each section will produce enough biomass for strain preservation, carotenoids extraction, and DNA extraction.
4. You can use 15–20% glycerol stock.
5. Carotenoids are highly unstable and sensitive to light, oxygen, and temperature, which induce their oxidation. To minimize the oxidation of carotenoids, the extraction must be quickly

done without exposure to light, oxygen, and high temperatures. Therefore, all the experimental procedures should be carried out in a dark room. Alternatively, dark tubes or tubes foiled with aluminum foil can be used. Nitrogen gas should be also used for flushing the pigment extracts before fridge storage at −20 °C to avoid oxidation. In addition, the use of an antioxidant such as butylhydroxytoluene (BHT) [116] can significantly reduce the oxidation of carotenoids during extraction.

6. Usually, DMSO-MeOH carotenoids extracts contain impurities such as salts, lipids, fatty compounds, proteins, and other cell compound. These contaminants usually are adsorbed on the stationary phase of HPLC column. Therefore, retention times can shift and tailing can occur. Reversed-phase HPLC columns may become contaminated by the repeated injection of samples. This may cause the column backpressure to build up to intolerably high levels, which overpressures the pump and can cause the column to settle and create a void, depending upon where the blockage occurs. To prevent this problem, we clean the columns on a regular basis throughout our work. Washing the HPLC columns with 100% acetonitrile and then with 100% methanol was sufficient to clean and regenerate the column. We strongly recommend using a pre-column.
7. The PCR product will show up as a single bright glowing band about 1.5 kb. You can also take photographs if needed.

References

1. Britton G, Liaaen-Jensen S, Pfander H (2004) Carotenoids handbook. Birkhäuser, Basel
2. Krinsky NI (2001) Carotenoids as antioxidants. Nutrition 17:815–817
3. Nelis HJ, De Leenheer AP (1989) Profiling and quantitation of bacterial carotenoids by liquid chromatography and photodiode array detection. Appl Environ Microbiol 55:3065–3071
4. Noviendri D, Hasrini RF, Octavianti F (2011) Carotenoids: Sources, medicinal properties and their application in food and nutraceutical industry. J Med Plant Res 5:7119–7131
5. Amaya E, Nickell D (2015) Using feed to enhance the color quality of fish and crustaceans. In: Feed and Feeding Practices in Aquaculture. Woodhead Publishing, Oxford, pp 269–298
6. Francis FJ (2000) Carotenoids as food colorants. Cereal Food World 45:198–203
7. Klaui H, Bauernfeind JC (eds) (1981) Carotenoids as food colors. Academic Press, New York
8. Paiva SA, Russell RM (1999) Beta-carotene and other carotenoids as antioxidants. J Am Coll Nutr 18:426–433
9. Krinsky NI (1989) Antioxidant functions of carotenoids. Free Radic Biol Med 7:617–635
10. Schalch W, Weber P (1994) Vitamins and carotenoids-a promising approach to reducing the risk of coronary heart disease, cancer and eye diseases. Adv Exp Med Biol 366:335–350
11. Agarwal S, Rao AV (2000) Carotenoids and chronic diseases. Drug Metabol Drug Interact 17:189–210
12. Peng YM, Peng YS, Childers JM et al (1998) Concentrations of carotenoids, tocopherols, and retinol in paired plasma and cervical tissue of patients with cervical cancer, precancer, and noncancerous diseases. Cancer Epidemiol Biomarkers Prev 7:347–350
13. Jacques PF, Chylack LT Jr (1991) Epidemiologic evidence of a role for the antioxidant vitamins and carotenoids in cataract prevention. Am J Clin Nutr 53:352S–355S

14. Seddon JM, Ajani UA, Sperduto RD et al (1994) Dietary carotenoids, vitamins A, C, and E, and advanced age-related macular degeneration. Eye disease case-control study group. JAMA 272:1413–1420
15. Wang X, Willen R, Wadstrom T (2000) Astaxanthin-rich algal meal and vitamin C inhibit Helicobacter pylori infection in BALB/cA mice. Antimicrob Agents Chemother 44:2452–2457
16. Jyonouchi H, Sun S, Gross M (1995) Effect of carotenoids on in vitro immunoglobulin production by human peripheral blood mononuclear cells: astaxanthin, a carotenoid without vitamin A activity, enhances in vitro immunoglobulin production in response to a T-dependent stimulant and antigen. Nutr Cancer 23:171–183
17. Guerin M, Huntley ME, Olaizola M (2003) *Haematococcus* astaxanthin: applications for human health and nutrition. Trends Biotechnol 21:210–216
18. Yuan J-P, Peng J, Yin K et al (2011) Potential health-promoting effects of astaxanthin: A high-value carotenoid mostly from microalgae. Molecular nutrition & food research 55:150–165
19. Preobrazhenskii VN, Vasilenko VV, Matveev AA (1997) Beta-carotenoids and their potential use in diseases of the digestive tract. Klin Med (Mosk) 75:43–45
20. März U (2008) FOD025C-The Global Market for Carotenoids In: *BCC Research*
21. März U (2015) FOD025E-The Global Market for Carotenoids In: *BCC Research*
22. Ernst H (2002) Recent advances in industrial carotenoid synthesis. Pure Appl Chem 74:1369–1382
23. Franco-Zavaleta ME, Jimenez-Pichardo R, Tomasini-Campocosio A et al (2010) Astaxanthin extraction from shrimp wastes and its stability in 2 model systems. J Food Sci 75:C394–C399
24. Ausich RL (1997) Commercial opportunities for carotenoid production by biotechnology. Pure Appl Chem 69:2169–2174
25. Bhosale P (2004) Environmental and cultural stimulants in the production of carotenoids from microorganisms. Appl Microbiol Biotechnol 63:351–361
26. Stankovic I (2004) Zeaxanthin chemical and technical assessment (CTA) In: FAO (ed) 63rdJECFA.
27. Reyes LH, Gomez JM, Kao KC (2014) Improving carotenoids production in yeast via adaptive laboratory evolution. Metab Eng 21:26–33
28. Asker D (2018) High throughput screening and profiling of high-value carotenoids from a wide diversity of bacteria in surface seawater. Food Chem 261:103–111
29. Mata-Gómez LC, Montañez JC, Méndez-Zavala A et al (2014) Biotechnological production of carotenoids by yeasts: an overview. Microb Cell Factories 13:12–23
30. Mendes-Pinto MM, Raposo MFJ, Bowen J et al (2001) Evaluation of different cell disruption processes on encysted cells of *Haematococcus pluvialis*: effects on astaxanthin recovery and implications for bio-availability. J Appl Phycol 13:19–24
31. Jahnke LS (1999) Massive carotenoid accumulation in *Dunaliella bardawil* induced by ultraviolet-A radiation. J Photochem Photobiol 48:68–74
32. Ben-Amotz A, Katz A, Avron M (1982) Accumulation of β-carotene in halotolerant algae: purification and characterization of β-carotene rich globules from *Dunaliella bardawil* (*chlorophyceae*). J Phycol 18:529–537
33. Iturriaga EA, Papp T, Breum J et al (2005) Strain and culture conditions improvement for β-carotene production with *Mucor*. Humana Press Inc., Totowa, NJ
34. Murillo FJ, Calderon IL, Lopez-Diaz I et al (1978) Carotene-superproducing strains of *Phycomyces*. Appl Environ Microbiol 36:639–642
35. Mehta BJ, Salgado LM, Bejarano ER et al (1997) New Mutants of *Phycomyces blakesleeanus* for (beta)-Carotene Production. Appl Environ Microbiol 63:3657–3661
36. Pegklidou K, Mantzouridou F, Tsimidou MZ (2008) Lycopene production using *Blakeslea trispora* in the presence of 2-methyl imidazole: yield, selectivity, and safety aspects. J Agric Food Chem 56:4482–4490
37. Jones JD, Hohn TM, Leatherst D (2004) Genetically modified strains of *Fusarium sporotrichioides* for production of lycopene and b-Carotene. In: Society of Industrial Microbiology Annual Meeting, San Diego, p 91
38. Del Rio E, Acien FG, MC G-M et al (2008) Efficiency assessment of the one-step production of astaxanthin by the microalga *Haematococcus pluvialis*. Biotechnol Bioeng 100:397–402
39. Jacobson GK, Jolly SO, Sedmak JJ et al. (1999) Astaxanthin over-producing strains of Phaffia rhodozyma, methods for their cultivation, and their use in animal feeds. United States
40. Liu YS, Wu JY (2006) Hydrogen peroxide-induced astaxanthin biosynthesis and catalase

activity in *Xanthophyllomyces dendrorhous*. Appl Microbiol Biotechnol 73:663–668
41. Calo P, Miguel TD, Sieiro C et al (1995) Ketocarotenoids in halobacteria: 3-hydroxy-echinenone and trans-astaxanthin. J Appl Bacteriol 79:282
42. Yokoyama A, Izumida H, Miki W (1994) Production of astaxanthin and 4-ketozeaxanthin by the marine bacterium, *Agrobacterium aurantiacum*. Biosci Biotech Biochem 58:1842–1844
43. Tsubokura A, Yoneda H, Mizuta H (1999) *Paracoccus carotinifaciens* sp. nov., a new aerobic gram-negative astaxanthin-producing bacterium. Int J Syst Bacteriol 49 Pt 1:277–282
44. Yokoyama A, Miki W, Izumida H et al (1996) New trihydroxy-keto-carotenoids isolated from an astaxanthin-producing marine bacterium. Biosci Biotech Biochem 60:200–203
45. Osanjo GO, Muthike EW, Tsuma L et al. (2009) A salt lake extremophile, *Paracoccus bogoriensis* sp. nov., efficiently produces xanthophyll carotenoids. Afr J Microbiol Res Vol.(8) pp. August, 3:426-433
46. Asker D, Isaka K (2006) Production of astaxanthin by microorganisms. Japan Patent 340676A
47. Shepherd D, Dasek J, Suzanne M et al. (1976) Production of zeaxanthin. US patent 3, 951, 743
48. Bhosale P, Larson AJ, Bernstein PS (2004) Factorial analysis of tricarboxylic acid cycle intermediates for optimization of zeaxanthin production from *Flavobacterium multivorum*. J Appl Microbiol 96:623–629
49. Lagarde D, Beuf L, Vermaas W (2000) Increased production of zeaxanthin and other pigments by application of genetic engineering techniques to *Synechocystis* sp. strain PCC 6803. Appl Environ Microbiol 66:64–72
50. Hundle BS, O'brien DA, Beyer P et al (1993) In vitro expression and activity of lycopene cyclase and beta-carotene hydroxylase from *Erwinia herbicola*. FEBS Lett 315:329–334
51. Berry A, Janssens D, Humbelin M et al (2003) *Paracoccus zeaxanthinifaciens* sp. nov., a zeaxanthin-producing bacterium. Int J Syst Evol Microbiol 53:231–238
52. Mcdermott JC, Britton G, Goodwin TW (1973) Carotenoid biosynthesis in a *Flavobacterium* sp.: stereochemistry of hydrogen elimination in the desaturation of phytoene to lycopene, rubixanthin and zeaxanthin. Biochem J 134:1115–1117
53. Asker D, Beppu T, Ueda K (2008) *Nubsella zeaxanthinifaciens* gen. nov., sp. nov., a zeaxanthin-producing bacterium of the family *Sphingobacteriaceae* isolated from freshwater. Int J Syst Evol Microbiol 58:601–606
54. Asker D, Beppu T, Ueda K (2007) Zeaxanthinibacter enoshimensis gen. nov., sp. nov., a novel zeaxanthin-producing marine bacterium of the family Flavobacteriaceae, isolated from seawater off Enoshima Island, Japan. Int J Syst Evol Microbiol 57:837–843
55. Asker D, Beppu T, Ueda K (2007) *Mesoflavibacter zeaxanthinifaciens* gen. nov., sp. nov., a novel zeaxanthin-producing marine bacterium of the family *Flavobacteriaceae*. Syst Appl Microbiol 30:291–296
56. Cooney JJ, Marks HW, Smith AM (1966) Isolation and Identification of Canthaxanthin from *Micrococcus roseus*. J Bacteriol 92:342–345
57. Veiga-Crespo P, Blasco L, Rosa-Dos-Santos F et al (2005) Influence of culture conditions of *Gordonia jacobaea* MV-26 on canthaxanthin production. Int Microbiol 8:55–58
58. De Miguel T, Sieiro C, Poza M et al (2001) Analysis of canthaxanthin and related pigments from *Gordonia jacobaea* mutants. J Agric Food Chem 49:1200–1202
59. Asker D, Ohta Y (1999) Production of canthaxanthin by extremely halophilic bacteria. J Biosci Bioeng 88:617–621
60. Asker D, Ohta Y (2002) Production of canthaxanthin by *Haloferax alexandrinus* under non-aseptic conditions and a simple, rapid method for its extraction. Appl Microbiol Biotechnol 58:743–750
61. Lorquin J, Molouba F, Dreyfus BL (1997) Identification of the carotenoid pigment canthaxanthin from photosynthetic *bradyrhizobium* strains. Appl Environ Microbiol 63:1151–1154
62. Guyomarch F, Binet A, Dufosse L (2000) Production of carotenoids by *Brevibacterium lines*: variation among strains, kinetic as pects and HPLC profiles. J Ind Microbiol Biotech 24:64–70
63. Del Campo JA, Rodriguez H, Moreno J et al (2004) Accumulation of astaxanthin and lutein in *Chlorella zofingiensis* (*Chlorophyta*). Appl Microbiol Biotechnol 64:848–854
64. Shi XM, Liu HJ, Zhang XW et al (1999) Production of biomass and lutein by *Chlorella protothecoides* at various glucose concentrations in heterotrophic cultures. Process Biochem 34:341–347
65. Del Campo JA, Rodriguez H, Moreno J et al (2001) Lutein production by *Muriellopsis* sp. in an outdoor tubular photobioreactor. J Biotechnol 85:289–295

66. Asker D, Beppu T, Ueda K (2007) Unique diversity of carotenoid-producing bacteria isolated from Misasa, a radioactive site in Japan. Appl Microbiol Biotechnol 77:383–392
67. Asker D (2017) Isolation and characterization of a novel, highly selective astaxanthin-producing marine bacterium. J Agric Food Chem 65:9101–9109
68. Oren A, Rodríguez-Valera F (2001) The contribution of halophilic bacteria to the red coloration of saltern crystallizer ponds. FEMS Microbiol Ecol 36:123–130
69. Jehlička J, Edwards H, Oren A (2013) Bacterioruberin and salinixanthin carotenoids of extremely halophilic Archaea and Bacteria: a Raman spectroscopic study. Spectrochim Acta A Mol Biomol Spectrosc 106:99–103
70. Oren A, Rodriguez-Valera F (2001) The contribution of halophilic bacteria to the red coloration of saltern crystallizer ponds. FEMS Microbiol Ecol 36:123–130
71. Hanada S, Kawase Y, Hiraishi A et al (1997) *Porphyrobacter tepidarius* sp. nov., a moderately thermophilic aerobic photosynthetic bacterium isolated from a hot spring. Int J Syst Bacteriol 47:408–413
72. Kilian O, Steunou AS, Fazeli F et al (2007) Responses of a thermophilic *Synechococcus* isolate from the microbial mat of Octopus Spring to light. Appl Environ Microbiol 73:4268–4278
73. Allewalt JP, Bateson MM, Revsbech NP et al (2006) Effect of temperature and light on growth of and photosynthesis by *Synechococcus* isolates typical of those predominating in the octopus spring microbial mat community of Yellowstone National Park. Appl Environ Microbiol 72:544–550
74. Lutnaes BF, Strand A, Petursdottir SK et al (2004) Carotenoids of thermophilic bacteria - *Rhodothermus marinus* from submarine Icelandic hot springs. Biochem Syst Ecol 32:455–468
75. Strand A, Shivaji S, Liaaenjensen S (1997) Bacterial carotenoids .55. C-50-carotenoids. 25. Revised structures of carotenoids associated with membranes in psychrotrophic *Micrococcus roseus*. Biochem Syst Ecol 25:547–552
76. Lazrak T, Wolff G, Albrecht AM et al (1988) Bacterioruberins reinforce reconstituted *Halobacterium* lipid-membranes. Biochim Biophys Acta 939:160–162
77. Yamano Y, Sakai Y, Hara M et al (2002) Carotenoids and related polyenes. Part 9. Total synthesis of thermozeaxanthin and thermocryptoxanthin and the stabilizing effect of thermozeaxanthin on liposomes. J Chem Soc Perk T 1:2006–2013
78. Stafsnes MH, Josefsen KD, Kildahl-Andersen G et al (2010) Isolation and characterization of marine pigmented bacteria from Norwegian coastal waters and screening for carotenoids with UVA-blue light absorbing properties. J Microbiol 48:16–23
79. Hirabayashi H, Ishii T, Takaichi S et al (2004) The role of carotenoids in the photoadaptation of the brown-colored sulfur bacterium *Chlorobium phaeobacteroides*. Photochem Photobiol 79:280–285
80. Moline M, Libkind D, Dieguez Mdel C et al (2009) Photoprotective role of carotenoids in yeasts: Response to UV-B of pigmented and naturally-occurring albino strains. J Photochem Photobiol B 95:156–161
81. Libkind D, Moline M, Sampaio JP et al (2009) Yeasts from high-altitude lakes: influence of UV radiation. FEMS Microbiol Ecol 69:353–362
82. Martinez-Laborda A, Balsalobre JM, Fontes M et al (1990) Accumulation of carotenoids in structural and regulatory mutants of the bacterium *Myxococcus xanthus*. Mol Gen Genet 223:205–210
83. Halliwell B (1996) Free radicals, proteins and DNA: oxidative damage versus redox regulation. Biochem Soc Trans 24:1023–1027
84. Carbonneau MA, Melin AM, Perromat A et al (1989) The action of free radicals on *Deinococcus radiodurans* carotenoids. Arch Biochem Biophys 275:244–251
85. Tian B, Xu Z, Sun Z et al (2007) Evaluation of the antioxidant effects of carotenoids from *Deinococcus radiodurans* through targeted mutagenesis, chemiluminescence, and DNA damage analyses. Biochim Biophys Acta 1770:902–911
86. Battista JR (1997) Against all odds: the survival strategies of *Deinococcus radiodurans*. Annu Rev Microbiol 51:203–224
87. Anton J, Oren A, Benlloch S et al (2002) *Salinibacter ruber* gen. nov., sp. nov., a novel, extremely halophilic member of the bacteria from saltern crystallizer ponds. Int J Syst Evol Microbiol 52:485–491
88. Takaichi S, Oh-Oka H, Maoka T et al (2003) Novel carotenoid glucoside esters from alkaliphilic heliobacteria. Arch Microbiol 179:95–100
89. Yokoyama A, Sandmann G, Hoshino T et al (1995) Thermozeaxanthins, new carotenoid-glycoside-esters from thermophilic eubacterium

Thermus thermophilus. Tetrahedron Lett 46: 4901–4904

90. Kametani K, Matsumura T (1983) Determination of 238U, 234U, 226Ra and 228Ra in spring waters of sanin district. Radioisotopes 32:18–21
91. Dufossé L (2009) Pigments, Microbial. In: Moselio S (ed) Encyclopedia of Microbiology. Academic Press, Oxford, pp 457–471
92. Tangeras A, Marøy T, Wahlstrøm S et al (1989) Release of astaxanthin from the red yeast *Phaffia rhodozyma* by high pressure homogenization. In: Program 1: Int. Mar. Biotechnol. Conf. (IMB'89), Tokyo
93. Gentles A, Haard NF (1991) Pigmentation of rainbow trout with enzyme-treated and spray-dried *Phaffia rhodozyma*. Prog Fish Cult 53:1–6
94. Johnson EA, Conklin DE, Lewis MJ (1977) The yeast *Phaffia rhodozyma* as a dietary pigment source for salmonids and crustaceans. J Fish Res Board Can 34:2417–2421
95. Johnson EA, Villa TG, Lewis MJ (1980) *Phaffia rhodozyma* as an astaxanthin source in salmonid diets. Aquaculture 20: 123–134
96. Storebakken T, Sørensen M, Bjerkeng B et al (2004) Stability of astaxanthin from red yeast, *Xanthophyllomyces dendrorhous*, during feed processing: effects of enzymatic cell wall disruption and extrusion temperature. Aquaculture 231:489–500
97. Colla E, Pereira A, Hernalsteens S et al (2010) Optimization of trehalose production by *Rhodotorula dairenensis* following a sequential strategy of experimental design. Food Bioprocess Technol 3:265–275
98. Fontana JD, Czeczuga B, Bonfim TMB et al (1996) Bioproduction of carotenoids: The comparative use of raw sugarcane juice and depolymerized bagasse by *Phaffia rhodozyma*. Bioresour Technol 58:121–125
99. Aguilera-Carbo A, Hernández J, Augur C et al (2009) Ellagic acid production from biodegradation of creosote bush ellagitannins by *Aspergillus niger* in solid state culture. Food Bioprocess Technol 2:208–212
100. Johnson EA, An GH (1991) Astaxanthin from Microbial Sources. Crit Rev Biotechnol 11:297–326
101. An GH, Choi ES (2003) Preparation of the red yeast, *Xanthophyllomyces dendrorhous*, as feed additive with increased availability of astaxanthin. Biotechnol Lett 25:767–771
102. Valduga E, Valerio A, Tatsch PO et al (2009) Assessment of cell disruption and carotenoids extraction from *Sporidiobolus salmonicolor* (CBS 2636). Food Bioprocess Technol 2:234–238
103. Sedmak JJ, Weerasinghe DK, Jolly SO (1990) Extraction and quantification of astaxanthin from *Phaffia rhodaozyma*. Biotechnol Tech 4:107–112
104. Ladislav F, Vera P, Karel S et al (2005) Reliability of carotenoid analyses: a review. Curr Anal Chem 1:93–102
105. Khachik F (2009) Analysis of carotenoids in nutritional studies. Birkhäuser Verlag Basel, Basel/Boston/Berlin
106. Pupin AM, Dennis MJ, Toledo MCF (1999) HPLC analysis of carotenoids in orange juice. Food Chem 64:269–275
107. Scott KJ, Finglas PM, Seale R et al (1996) Interlaboratory studies of HPLC procedures for the analysis of carotenoids in foods. Food Chem 57:85–90
108. Hart DJ, Scott KJ (1995) Development and evaluation of an Hplc method for the analysis of carotenoids in foods, and the measurement of the carotenoid content of vegetables and fruits commonly consumed in the UK. Food Chem 54:101–111
109. Stackebrandt E, Goebel BM (1994) A place for DNA-DNA reassociation and 16s ribosomal-Rna sequence-analysis in the present species definition in bacteriology. Int J Syst Bacteriol 44:846–849
110. Schleifer KH, Ludwig W (1989) Phylogenetic relationships of bacteria. Elsevier Science Publishers B. V, Amsterdam
111. Woese CR (1987) Bacterial evolution. Microbiol Rev 51:221–271
112. Rossello-Mora R, Amann R (2001) The species concept for prokaryotes. FEMS Microbiol Rev 25:39–67
113. Egan S, Thomas T, Kjelleberg S (2008) Unlocking the diversity and biotechnological potential of marine surface associated microbial communities. Curr Opin Microbiol 11:219–225
114. Asker D, Awad TS, Beppu T et al (2012) Isolation, characterization, and diversity of novel radiotolerant carotenoid-producing bacteria. Methods Mol Biol 892:21–60
115. Thompson JD, Higgins DG, Gibson TJ (1994) CLUSTAL W: improving the sensitivity of progressive multiple sequence alignment through sequence weighting, position-specific gap penalties and weight matrix choice. Nucleic Acids Res 22:4673–4680
116. Meyers SP, Bligh D (1981) Characterization of astaxanthin pigments from heat-processed crawfish waste. J Agric Food Chem 29:505–508

Chapter 10

Purification and Identification of Astaxanthin and Its Novel Derivative Produced by Radio-tolerant *Sphingomonas astaxanthinifaciens*

Dalal Asker, Tarek S. Awad, Teruhiko Beppu, and Kenji Ueda

Abstract

The red diketocarotenoid, astaxanthin, exhibits extraordinary health-promoting activities such as antioxidant, anti-inflammatory, antitumor, and immune booster, which may potentially protect against many degenerative diseases such as cancers, heart diseases, and exercise-induced fatigue. These numerous health benefits and consumer interest in natural products have therefore increased the market demand of astaxanthin as a nutraceutical and medicinal ingredient in food, aquaculture feed, and pharmaceutical industries. Consequently, many research efforts have been made to discover novel microbial sources with effective biotechnological production of astaxanthin. Using a rapid screening method based on 16S rRNA gene, and effective HPLC-Diode array-MS methods for carotenoids analysis, we isolated a novel astaxanthin-producing bacterium (strain TDMA-17^T) that belongs to the family *Sphingomonadaceae* (Asker et al., FEMS Microbiol Lett 273:140–148, 2007).

In this chapter, we provide a comprehensive description of the methods used for the analysis and identification of carotenoids produced by strain TDMA-17^T. We will also describe the methods of isolation and identification for a novel bacterial carotenoid (an astaxanthin derivative), a major carotenoid that is produced by the novel strain. Finally, the identification methods of the novel strain will be summarized.

Key words Astaxanthin, Carotenoids, Antioxidant, Biotechnology, *Sphingomonadaceae*, *Sphingomonas astaxanthinifaciens*, Screening, 16S rRNA, Phylogenetic, HPLC

1 Introduction

1.1 Structure, Function, and Importance of Astaxanthin

Astaxanthin, a natural red pigment, is the main carotenoid in marine biosystem (e.g., algae and aquatic animals). Astaxanthin like other carotenoids can be synthesized by photosynthetic organisms (plant and microalgae), and some nonphotosynthetic microorganisms such as yeast, fungi, and bacteria, but animals cannot [1–3]. Most wild aquatic animals (shrimp, crawfish, crabs, lobster, and salmon) can accumulate astaxanthin from their food chains, which gives them the beautiful pink to red flesh [4–7]. In the aquaculture industries, astaxanthin has been widely used as a feed

Carlos Barreiro and José-Luis Barredo (eds.), *Microbial Carotenoids: Methods and Protocols*, Methods in Molecular Biology, vol. 1852, https://doi.org/10.1007/978-1-4939-8742-9_10,

supplement for proper growth and survival as well as to provide the favorable pigmentation in commercial fish and crustacean farms [8–12].

Astaxanthin (3) (3,3′-dihydroxy-β,β-carotene-4,4′-dione) is a diketo-dihydroxy-carotenoid that consists of polyene system with 11 conjugated ethylenic double bonds, the hydroxyl and keto endings on each ionone ring. Owing to its unique chemical structure, astaxanthin is characterized by its extraordinary antioxidant activity, high polar configuration, and the ability to be esterified, which increases the solubility and stability of free astaxanthin in the cell and stabilizes the cell membrane [13]. Beyond the activities of other carotenoids (e.g., canthaxanthin, zeaxanthin, and β-carotene), astaxanthin is a super antioxidant that exhibits a strong free radical scavenging, singlet oxygen quenching, and chain-breaking activities, which reduce oxidative damage in the cell. The photoprotective function of astaxanthin for many microorganisms has been well documented [14–17]. It also protects against lipid peroxidation and oxidative damage of low-density lipoprotein cholesterol [18] and against UVA light-induced oxidative stress in cultured fibroblasts [19], prevents tumor formation, and enhances in vitro antibody production [20, 21] and anti-inflammatory agent. Research in health and nutrition have indicated its beneficial roles against many degenerative diseases [22, 23], which signified its demand and market as a nutraceutical, supplement and medicinal ingredient in various industries including aquaculture, medicinal, pharmaceutical, and food industries [22, 24]. In addition, the negative impact of artificial colorants on health (e.g., carcinogenicity, hypersensitivity, and toxicity) increased the demand for safe and natural food colorants such as astaxanthin [25]. As a result, the global market value of astaxanthin increased from $226 million in 2010 to $369 million in 2014 and is expected to reach $423 million in 2019 [26]. Astaxanthin is ranked as the third most important carotenoid on the global market after β-carotene and lutein, with a predicted sales volume of 670 metric tons valued at 1.1 billion US$ in 2020 [26]. Natural astaxanthin is approved as GRAS, and can be sold as a dietary supplement, and is also approved as a food colorant (E161j) by the European Commission, used in the food and beverage industries [27].

1.2 Microbial Production of Astaxanthin

Astaxanthin is commercially produced by extraction from natural sources (shrimp processing waste) [28–30] or by chemical synthesis [31]. Most commercial astaxanthin for aquaculture is produced synthetically. However, synthetic astaxanthin contains a mixture of stereoisomers, which is not eco-friendly and less favorable by the consumer compared to natural astaxanthin. Due to consumer concerns about these undesirable effects, attempts have been made to produce astaxanthin from microbial sources [32–39]. As listed in Table 1, the effective microbial production of astaxanthin was

Table 1
Microorganisms reported for astaxanthin production

Microorganism	Yield	References
Microalgae		
Haematococcus pluvialis	1–3%	[13, 40]
Nannochloropsis gaditana	3 mg/L	[41]
Yeast		
Xanthophyllomyces dendrorhous[a]	2.5–4.7 mg/g	[42–44]
Archaea		
Halobacterium salinarum	265 μg/g	[45]
Bacteria		
Paracoccus sp. N81106[b]	0.3 mg/6 L	[46]
Paracoccus carotinifaciens E-396	–	[47]
Brevundimonas sp. SD212[c]	1.5 mg/27 L	[48]
Paracoccus sp. strain PC1[d]	–	[46]
Paracoccus bogoriensis	0.4 mg/g	[49]
Brevundimonas spp. (wild type)	440–1060 μg/g	[33, 34]
Brevundimonas sp. M7 (mutant)	1.3 mg/g	[39]
Paracoccus strain A-581-1	91.9 mg/L	[50]

[a]Formerly *Phaffia rhodozyma*
[b]Formerly *Agrobacterium aurantiacum* N81106
[c]Formerly *Pseudomonas* sp.
[d]Formerly *Alcaligenes* sp. PC-1

established using only *Haematococcus pluvialis* [13, 51] or *Xanthophyllomyces dendrorhous* (formerly *Phaffia rhodozyma*) [52]. *Haematococcus pluvialis* is known for its high ability to accumulate (3*S*,3′*S*)-astaxanthin, up to 0.2–3.0% (on a dry mass basis) [13]; however, there are few drawbacks in its biotechnological process. Because it grows at mild conditions such as at 20–28 °C, below 15 klux light intensity and at pH 6.8–7.4, a possible contamination by bacteria, fungi, and protozoa may lead to a serious problem. Therefore, the outdoor cultivation of *Haematococcus* represents a challenge and requires control of contamination and environmental conditions such as light and temperature. On the other hand, process involving completely closed photobioreactors with artificial light is costly. In addition, *Haematococcus* accumulates astaxanthin in the aplanospore growth phase, but their thick wall requires physical breakage before the astaxanthin can be extracted or be available to organisms consuming the algae [13].

Xanthophyllomyces dendrorhous is one of the best candidates for the commercial production of astaxanthin with the highest yield reported so far of about 4.7 mg/g cell dry weight [42]. A major drawback in the use of *X. dendrorhous* is its rigid cell wall structure [52]. For efficient intestinal absorption of the pigment, disruption of the cell wall of the yeast biomass is thus required before addition to an animal diet.

Attempts to isolate astaxanthin-producing bacteria have resulted in a few bacterial strains, most of which were from marine water such as *Paracoccus* spp. [46, 47, 53, 54], *Brevundimonas* sp. [33, 34, 39, 48], and *Halobacterium* sp. [45]. Nevertheless, no commercial processes have been developed. Owing to its extraordinary antioxidant activity and its emerging application as nutraceuticals in food, medical, and cosmetic industries [23], and growing demand for natural carotenoids, it is important to maximize the commercial production of astaxanthin by seeking to isolate novel bacterial sources with unique biotechnological properties.

1.3 Isolation of Novel Astaxanthin-Producing Bacteria

For the screening of high-value carotenoid-producing bacteria, we used freshwater samples collected at Misasa (Tottori, Japan), a region known for its high natural radioactivity [55]. Using a rapid method for screening, a comprehensive taxonomic characterization based on 16S rRNA gene sequences and effective HPLC-Diode array-MS methods for carotenoids analysis, we obtained a pigmented bacterial library that contains 104 strains [36, 38] (*see* Chapter 15). Among these strains, strain TDMA-17^T was isolated by its ability to produce red-pigmented colonies on nutrient agar plate after 2 days of cultivation at 40 °C (Fig. 1a). TDMA-17^T exhibited oval- or short rod-shaped cell morphology and occasionally formed a fiber network structure as observed under scanning electron microscope (Fig. 1b). Strain TDMA-17^T exhibited high accumulation of carotenoids including astaxanthin and its derivatives [32, 38]. The 16S rRNA gene sequence revealed that this unique red-pigmented bacterium (strain TDMA-17^T) belongs to the family *Sphingomonadaceae* and that the highest degree of sequence similarity was found with *Sphingomonas* spp. [32, 38]. This was the first description of a red-pigmented *Sphingomonas* spp. and of an astaxanthin-producing *Sphingomonadaceae* bacterium (i.e., strain TDMA-17^T) [32].

1.4 Identification of Carotenoids Produced by Strain TDMA-17^T

For the analysis and identification of the carotenoids produced by strain TDMA-17^T, we used the rapid and effective HPLC methods as described in Chapter 15. Advantageously, the extraction of carotenoids from the wet cells of strain TDMA-17^T was done using only methanol [32, 37], which yielded a significantly high amount of carotenoids (~2.8 mg/g dry cells) [32]. Analysis of the crude pigment isolated from the cells of strain TDMA-17^T using HPLC method A resulted in the detection of a single

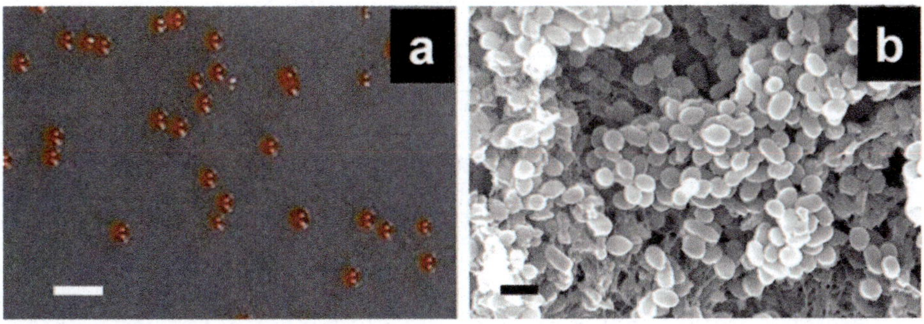

Fig. 1 Morphological characteristics of strain TDMA-17^T. (**a**) Colony appearance and (**b**) image of cells observed under a scanning electron microscope. Cells were grown for 2 days at 40 °C on nutrient agar. Bars, 1.0 mm (**a**) and 1.0 μm (**b**). Adapted from Asker et al. [32]

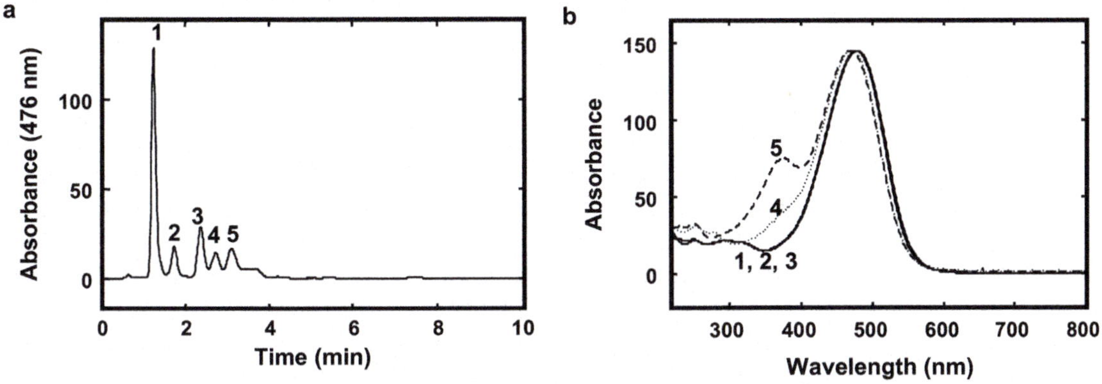

Fig. 2 Absorption profile of the carotenoids produced by strain TDMA-17^T. Elution profile (λ_{max} 476 nm) (**a**) and absorption spectrum of each peak (**b**) are shown. The carotenoid fraction was analyzed by reversed-phase liquid chromatography using method A (*see* Subheadings 2 and 3). The fraction consisted of unidentified carotenoid (peak 1), dihydroxyastaxanthin (peak 2), all-*trans*-astaxanthin (peak 3), 9-*cis*-astaxanthin (peak 4) and 13-*cis*-astaxanthin (peak 5). Adapted from Asker et al. [32]

broad peak at R_t 0.8–1 min. Using HPLC method B, this peak could be resolved into five peaks (Fig. 2a) corresponding to five types of pigments. Table 2 summarizes the spectroscopic data for the HPLC five peaks. The UV-visible absorption spectra of the first three peaks (Fig. 2b) were identical to the spectrum of astaxanthin; they exhibited a single absorption maximum at 476 nm. The astaxanthin and dihydroxy-astaxanthin standards exhibited identical spectroscopic data as those for peaks 3 and 2, respectively. Based on the spectroscopic data in Table 2, we could identify the compounds corresponding to peak 2 (7.5% of the total carotenoids) and peak 3 (24.6% of the total carotenoids) as dihydroxy-astaxanthin and astaxanthin (all-*trans*-astaxanthin), respectively. Astaxanthin was further confirmed by HPLC mass (method C in Chapter 15) retention time (R_t, 7.05 min) and molecular weight [(M^++1), 597] (Table 2). The compounds

Table 2
The spectroscopic data for the HPLC five peaks of carotenoids produced by strain TDMA-17[T]

HPLC peak	Carotenoids	Absorption (λ_{max}, nm)	HPLC, retention time (R_t, min)		Mol. wt (M^{+}+1)
			Method B	Method C	
1	Unknown	476	1.24	3.4	889
2	Dihydroxyastaxanthin	476	1.34	4.21	629
3	Astaxanthin	476	2.31	7.05	597
4	9-*cis*-Astaxanthin	350–370, 464	2.78	8.4	597
5	13-*cis*-Astaxanthin	350–370, 462	3.05	11.0	597

corresponding to peak 4 and 5 represented 3.4% and 6.3% of the total carotenoids, respectively. They exhibited asymmetrical spectra, which were, respectively, assigned as 9-*cis*-astaxanthin and 13-*cis*-astaxanthin by comparing their spectra to literature UV-Vis spectra [56]. However, we were unable to identify the major compound corresponding to peak 1 (~58.4% of the total carotenoids) using these three HPLC methods. Although the spectrum of the unknown compound was similar to that of astaxanthin, the molecular weight and polarity of the compound were much higher than astaxanthin. Therefore, it was primarily identified as "astaxanthin derivative." To identify this novel derivative, further purification steps were required as described in the next section.

1.5 Purification and Identification of a Novel Astaxanthin Derivative

The discovery of the numerous health benefits of carotenoids has increased the demand to explore new carotenoids species with a novel property or activity. More than 700 identified naturally occurring carotenoids [57] differ in the length of the polyene chromophore, nature of the end groups and various substituents. The chemical structure of a carotenoid determines its interaction with other molecules and integration into membranes, which strongly determine its antioxidant activity. Astaxanthin is a diketo and has two hydroxyl groups, which gives it an extraordinary antioxidant activity compared with β-carotene. There are about 12 astaxanthin derivative molecules isolated and identified from marine bacterium, transformed *E. coli* strains, salmon, and microalgae (Table 3). Discovery of astaxanthin-like component that encloses astaxanthin with a modified functional group might increase significantly the polarity and thus may possess a higher antioxidant activity compared to astaxanthin, high solubility, or membrane integrity. All of them can increase or modify the biological function of astaxanthin.

Table 3
Reported natural astaxanthin derivatives

Common name	Formula	Mol. wt	Sources	References
Astaxanthin/ (3*S*,3′*S*)-astaxanthin	$C_{40}H_{52}O_4$	596.84	Salmon, shrimp, krill, etc. (marine animals), *Adonis aestivalis* (flower petals of higher plant), *Haematococcus pluvialis* (green alga) *Agrobacterium aurantiacum* (bacteria)	[5, 46, 47, 58, 59]
(3*R*,3′*R*)-Astaxanthin	$C_{40}H_{52}O_4$	596.84	Shrimps, lobsters, etc. (marine animals) *Xanthophyllomyces dendrorhous* (yeast)	[60, 61]
Meso-astaxanthin/ (3*R*,3′*S*)-astaxanthin	$C_{40}H_{52}O_4$	596.838	Shrimps, lobsters, etc. (marine animals)	[62, 63]
Astaxanthin glucoside	$C_{46}H_{62}O_9$	758.98	Bacteria (*Agrobacterium aurantiacum)*	[64]
Astaxanthin diglucoside	$C_{52}H_{72}O_{14}$	921.12	Bacteria (*Escherichia coli* transformants)	[65]
7,8-Didehydro-astaxanthin	$C_{40}H_{50}O_4$	594.82	Marine animals	[63, 66]
7,8,7′,8′-Tetradehydro-astaxanthin	$C_{40}H_{48}O_4$	592.81	Marine animals	[63, 66]
Astaxanthin esters (mono-, diester)	Average		Marine animals krill	[58, 67]
2-Hydroxy-astaxanthin	$C_{40}H_{52}O_5$	612.838	*Brevundimonas* sp. (bacteria)	[48]
2,2′-Dihydroxy-astaxanthin	$C_{40}H_{52}O_6$	628.837	*Brevundimonas* sp. (bacteria)	[48]
Bastaxanthin D/ bastaxanthol	$C_{40}H_{53}O_8SNa$	716.900	*Ianthella basta* (marine sponge)	[68]
Bastaxanthin C/ bastaxanthin	$C_{40}H_{51}O_8SNa$	714.884	*Ianthella basta* (marine sponge)	[68]

Purification and identification of a novel (i.e., unknown) carotenoid are critical processes due to the instability of carotenoids in the presence of light, heat, or oxygen. Therefore, all procedures must be performed under dim light. Identification of known carotenoid structure in biological samples can be simply done using HPLC-MS system, in which individual carotenoids can be identified by their respective retention time, absorption spectrum, and molecular weight. This requires the use of commercial carotenoid standards or known biological sources as references. On the other hand, identification of unknown carotenoids with novel structure is usually done using HPLC-UV-Vis, NMR, mass, IR, and CD, in which a pure and relatively high amount (at least 10 mg) of the unknown carotenoid is required. In general, individual carotenoids are structurally highly similar, and many compounds interfere in their isolation. It might be easy to isolate carotenoids such as β-carotene and astaxanthin in their natural mixture because of large difference in polarity, i.e., astaxanthin contains extra hydroxyl and keto endings on each ionone ring compared to β-carotene. However, it is a challenge to separate carotenoids with slightly different polarities such as astaxanthin [**3**] and the unknown carotenoid [**1**], as shown in Fig. 2. Therefore, we use open column chromatography (OCC) followed by thin-layer chromatography (TLC) for purification of carotenoid [**1**]. In OCC, silica gel is used as the stationary phase, while a mobile phase (single solvent or a mixture of solvents) is used to separate carotenoid fractions based on polarity. It is also possible to monitor the fractions of carotenoids visually based on color differences.

Figure 3 illustrates a preparative scale for extraction, isolation, and purification of the unknown carotenoid [**1**] from wet cells of strain TDMA-17^T. The crude carotenoid extract of strain TDMA-17^T was subjected to fractionation using OCC with silica gel as the polar stationary phase (Fig. 3). First, astaxanthin [**3**] was eluted with chloroform followed by carotenoid [**1**], which was eluted with $CHCl_3$/methanol (9:1). An additional purification step was done to purify this carotenoid using a preparative TLC (C18-TLC) that was developed with $CHCl_3$/methanol (4:1). By this process, ca. 56 mg carotenoid [**1**] and 4.7 mg astaxanthin [**3**] were obtained. After obtaining a purified unknown carotenoid [**1**], a comprehensive spectroscopy analysis using UV-Vis absorption, NMR, mass, IR, and CD spectra was performed [37]. Properties of the novel carotenoid such as antioxidant activity and solubility were investigated [37].

The purified carotenoid [**1**] was identified using HPLC-UV-Vis, NMR, mass, IR, and CD [35, 37]. As shown in Fig. 4, the structure of the major carotenoid of strain TDMA-17^T was assigned as (3*S*,3′*S*)-3,3′-bis(α-D-rhamnopyranosyloxy)-β,β-carotene-4,4′-dione [**1**]. The detailed analytical data of carotenoid [**1**] are listed in in our previous publications [35, 37].

1.6 Identification of the Novel Astaxanthin-Producing Bacterium

Identification of unknown microorganism is a process that is used to determine whether the unknown belongs to one of the units (species, genus, etc.) and labeled (e.g., nomenclature) [69–71]. This process is very important in academy (e.g., to establish novel taxa) as well as industry (e.g., for obtaining a patent) to claim a novel producer for production of bioactive components at species or genus level. The identification of any taxon involves a polyphasic approach that includes a combination of phenotypic methods (e.g., biochemical testing, physiological analysis, and chemotaxonomic analysis) and genotypic testing methods (e.g., DNA-DNA hybridization, analysis of G+C

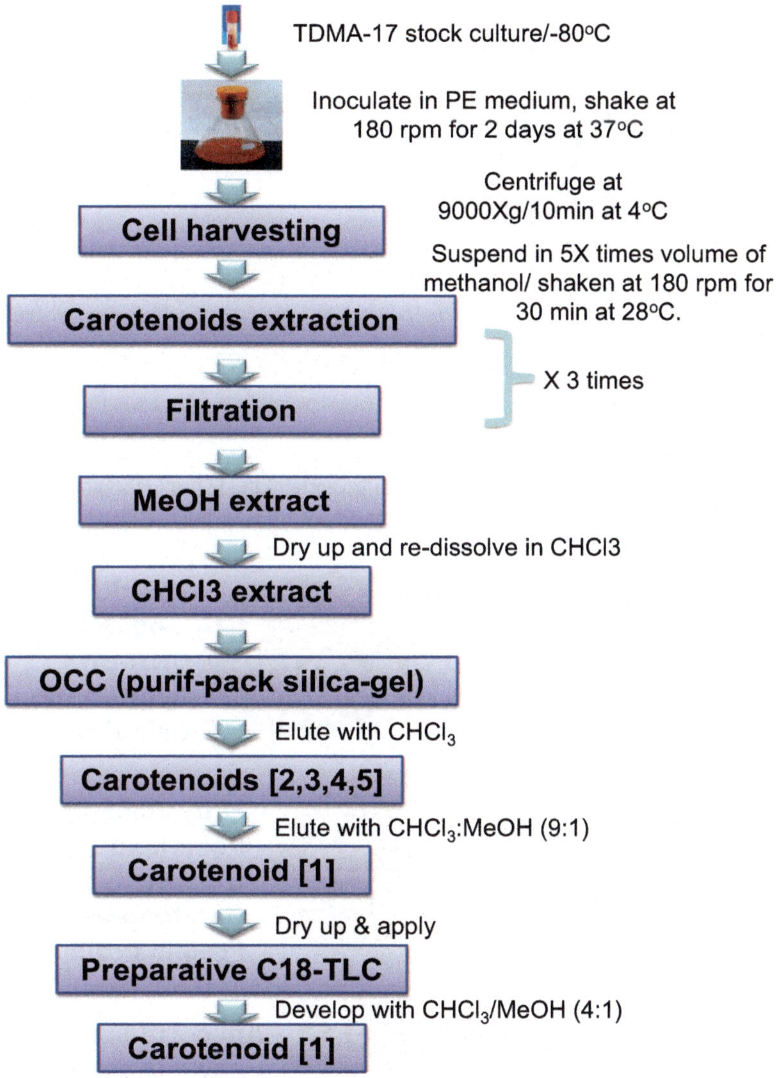

Fig. 3 Outline of isolation and purification process of the novel astaxanthin derivative carotenoid [**1**]

content (in moles percent), 16S rDNA gene sequencing and phylogenetic information) [71–73]. Tindall et al. [74] have mentioned the necessity of a careful selection and use of a wide variety of methodologies for describing a novel taxon [74]. In our screening approach, 16S rRNA gene sequence was used as a fast and reliable molecular tool to determine the phylogenetic delineation (e.g., taxonomic position) of the interested unknown isolates. After classifying the unknown isolates into a group (e.g., species or genus, depending on 16S rRNA gene sequence), appropriate tests that provide all taxonomic information (e.g., phenotypic and genotypic) were carried out to distinguish between unknown isolate and its closely related taxa. Figure 5 summarizes the polyphasic taxonomy approach that we have used for

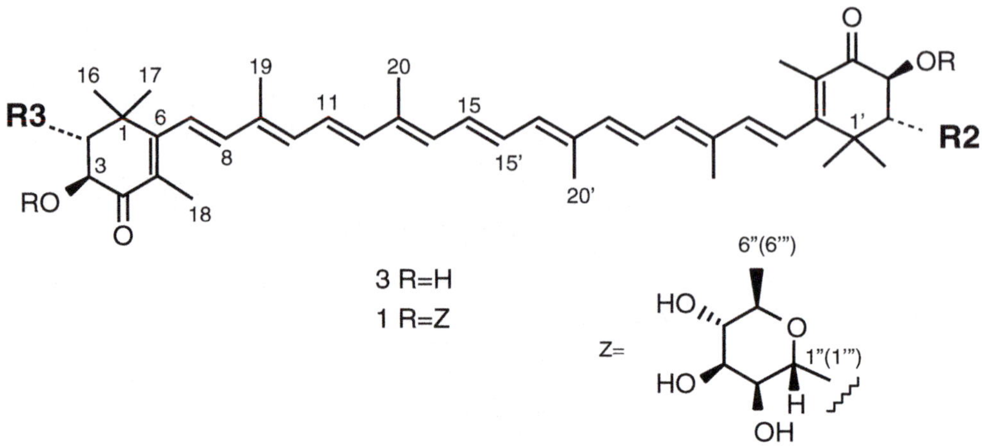

Fig. 4 Chemical structure of astaxanthin dirhamnoside [**1**] and astaxanthin [**2**]. Adapted from Asker et al. [37]

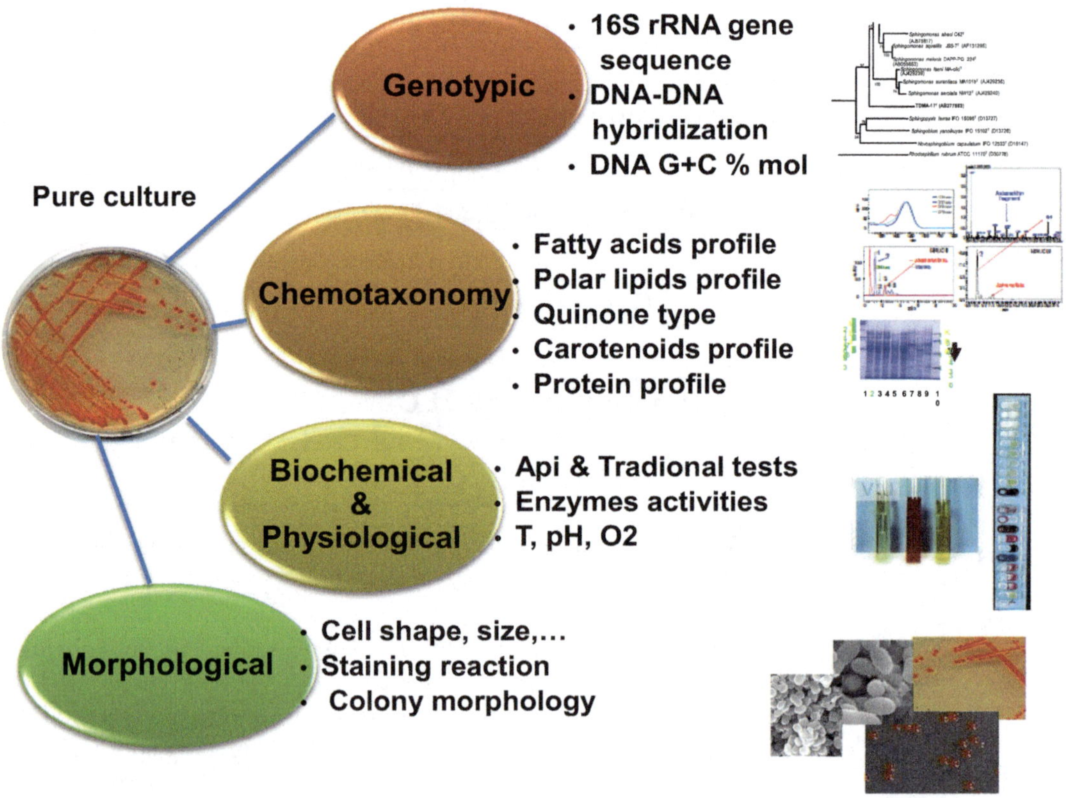

Fig. 5 Summary of the polyphasic taxonomy approach employed for identification of the unknown isolates

identification of the unknown isolates. Nonetheless, specific tests can be added to the above identification approach depending on the taxa involved.

1.7 Phylogenetic Characterization

16S rDNA gene sequencing is a powerful molecular technique presently used for bacterial species identification and to construct phylogenetic trees [75–79]. It is recommended that new species descriptions should include an almost complete 16S rRNA gene sequence (>1300 nt, <0.5% ambiguity) [77]. When the 16S rRNA gene sequence similarity between an unidentified isolate and its closest phylogenetic neighbors is less than 97%, it indicates that a novel species has been isolated [74, 78, 80, 81]. More than 97% similarity indicates a possibility that they belong to the same species, which requires the use of DNA–DNA hybridization, a standard method for defining species and subspecies, to differentiate between them [78]. If the DNA–DNA hybridization between these strains is less than ~70%, it suggests that the two strains belonged to different species [74, 78, 82, 83]. When the unknown isolate exhibits a novel species properties, the phenotypic differences between the novel species and existing species within the genus are reported [74]. At values lower 95% 16S rRNA gene sequence similarity (over full pairwise comparisons), taxa should be tested by other methods to establish whether separate genera are present [74, 84]. To claim a novel producer for production of astaxanthin, we targeted the predictable novel strains that exhibited low sequence homology in the range of 95–97% and 88–94% at species and genus levels, respectively.

A comparison of the 16S rRNA gene sequence of TDMA-17^{T} with those available in GenBank/EMBL/DDBJ nucleotide sequence database revealed that it belonged to the family *Sphingomonadaceae*. Phylogenetic analysis of nearly full-length 16S rRNA sequence of strain TDMA-17^{T} indicated that it is a member of the genus *Sphingomonas* sensu stricto [85], forming a distinct phylogenetic lineage (Fig. 6).

More details about the chemotaxonomic, morphological, physiological, and biochemical characterizations of strain TDMA-17^{T} were described in Asker et al. [32].

2 Materials

2.1 Media and Buffers

1. Nutrient agar (Difco Chemicals, Detroit, MI, USA) containing 3 g/L beef extract, 5 g/L peptone, and 15 g/L agar.
2. PE medium: Bactopeptone (Difco Chemicals, Detroit, MI, USA), 4 g/L; Bacto yeast extract (Difco Chemicals, Detroit, MI, USA), 2 g/L; NaCl, 5 g/L; fish meat extract (Kyokuto, Tokyo, Japan), 2 g/L; $MgSO_4$, 2 g/L; and maltose, 10 g/L (pH 7.2).

2.2 Preparing Media and Cultivation

1. Analytical balance.
2. Flasks for cultivation.

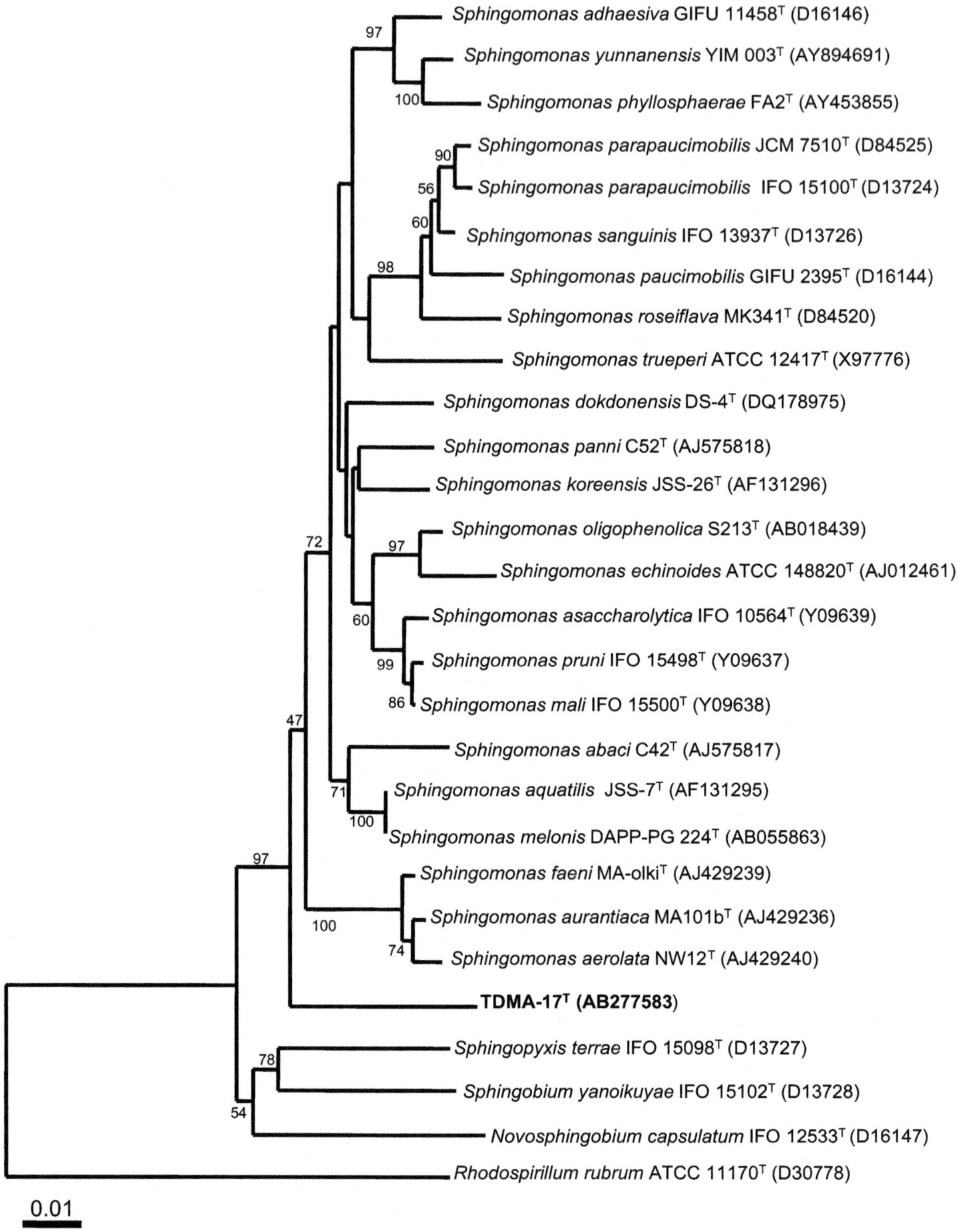

Fig. 6 A phylogenetic tree based on the 16S rRNA gene sequences of strain TDMA-17^{T} and representatives of some related taxa. The tree was constructed by the neighbor-joining method [86]. GenBank accession numbers are shown in parentheses. Numbers at nodes indicate bootstrap values higher than 50%. *Rhodospirillum rubrum* ATCC 11170^{T} was used as an outgroup. Bar, 1 substitution per 100 nucleotide positions. Adapted from Asker et al. [32]

3. Disposable Petri dishes.
4. Micropipettes and sterile tips (0–20, 20–200, and 100–1000 μL).
5. Autoclave.
6. Incubators (28, 30, 37, 40, 50 °C).
7. Sterile loops, sterile toothpicks, and sterile cotton swabs.
8. Clean bench equipped with UV light.

2.3 Carotenoids Analysis

1. HPLC grade solvents including dimethyl sulfoxide (DMSO), methanol, acetonitrile, and tetrahydrofuran.
2. BHT (butylhydroxytoluene) (Sigma, St. Louis, MO, USA) (*see* **Note 1**).
3. Dim light (*see* **Note 1**).
4. Commercial astaxanthin (Sigma, St. Louis, MO, USA).
5. Eppendorf tubes.
6. Racks.
7. Balance.
8. Eppendorf table top vial shaker.
9. Spectrophotometer.
10. 0.45 μm hydrophobic Teflon membrane syringe filter (Advantec, Tokyo, Japan) and syringe (1 mL).
11. 96-well plate.
12. HPLC automated injection with 96-well plate adaptor (Class-VP; Shimadzu, Kyoto, Japan) with PDA detector.
13. HPLC/MS system (LCMS-2010EV, Shimadzu, Kyoto, Japan).
14. HPLC columns: CAPCELLPAC C18 column (35 mm × 4.6 mm; particle size, 5 μm) (Shiseido, Tokyo, Japan) and Shim-Pack FC-ODS (150 mm × 4.6 mm; particle size, 5 μm) (Shimadzu, Kyoto, Japan).

2.4 Purification and Identification of a Novel Astaxanthin Derivative

1. Stock culture of strain TDMA-17^T.
2. PE medium 100 mL in 500 mL flasks.
3. Shaker.
4. Incubators (28, 37, 40, 50 °C).
5. HPLC grade solvents including methanol, acetonitrile, tetrahydrofuran, chloroform, ethyl acetate, hexane.
6. Purif-pack silica-gel chromatography (Moritex, Tokyo), preparative C18-TLC (Whatman).
7. Standards: commercial astaxanthin (Sigma, St. Louis, MO, USA), α-D-rhamnose.

8. HPLC (Class-VP; Shimadzu, Kyoto, Japan) with photodiode array detector (PDA) Shimadzu SPD-M10AVP and Capcell Pak C18 analytical column (20 mm i.d. × 250 mm).
9. NMR (Varian Unity INOVA 500 or JEOL JMN-600 spectrometer) (JEOL, Tokyo, Japan).
10. ESI-TOFMS spectrometer (JEOL JMS-T100LC AccuTOF) (JEOL, Tokyo, Japan).
11. FT/IR 4000 spectrometer (JASCO, Tokyo, Japan).
12. J-720 spectropolarimeter (JASCO, Tokyo, Japan).
13. Lyophilizer.
14. Open column chromatography Dowex-1 (OH-type) column, silica gel column (Wakogel C-200, 40 g).
15. Benzoyl chloride, pyridine, 4-dimethylaminopyridine, NaCl, $NaHCO_3$, anhydrous Na_2SO_4, 4 N HCl.

2.5 Identification of Bacterial Isolates: 16S rRNA Gene Sequences

1. Eppendorf tubes, toothpicks, and deionized water (all sterilized).
2. DNA purification kit (Edge BioSystems, Gaithersburg, MD).
3. Primer B8F (5′-AGAGTTATCCTGGCTCAG).
4. Primer B1492R (5′-GGTTACCTTGTTACGACTT).
5. T1 Thermocycler (Biometra, Goettingen, Germany).
6. Ex Taq polymerase (Takara-shuzo, Kyoto, Japan).
7. Agarose.
8. 10× TBE buffer: 108 g Tris–HCl base, 55 g boric acid, 40 mL 0.5 M EDTA (pH 8.0).
9. 1× TBE buffer: 100 mL of 10× TBE buffer and 900 mL of deionized water.
10. 6× blue loading dye: 30% glycerol in distilled water and 0.25% bromophenol blue.
11. Ethidium bromide (*see* **Note 2**).
12. Agarose gel electrophoresis system (i-Mupid Mini Agarose Gel Electrophoresis Apparatus System Advance Co. Ltd., Tokyo, Japan).
13. UV safety goggles and UV box.
14. 0.2 mL thin-walled PCR tubes.
15. BigDye Terminator v3.1 Cycle Sequencing Kit (Applied Biosystems, California, USA).
16. Pipettes and tips (0–10, 20–200, 100–1000 μL).
17. Gene Clean kit II (Funakoshi, Tokyo, Japan).

3 Methods

3.1 Isolation of the Astaxanthin-Producing Bacterium

1. Collect freshwater samples at Misasa (Tottori, Japan).
2. Do serial dilutions with sterile water.
3. Spread 0.1 mL aliquots on nutrient agar (Difco) and incubate at 40 °C for up to 3 days.
4. Isolate the colony of TDMA-17^T based on its red color.

3.2 Identification of Carotenoids

1. Extract the red pigments by agitating bacterial cells in methanol on the rotary shaker (100 rpm) at 50 °C in the dark until the cells are bleached (*see* **Note 1**).
2. Centrifuge at 5000 × *g* for 5 min, and scan the resultant supernatant for absorbance in the wavelength range (λ = 260–700 nm) at room temperature on a spectrophotometer.
3. Identify each carotenoid using HPLC-DAD (methods B and C) (*see* **Note 3**).
4. Perform HPLC (method B) using a CAPCELLPAC C18 column (35 °C), and elute with methanol/water (9:1, vol/vol). Flow rate: 1.0 mL/min.
5. Carry out HPLC (method C) on an LC/MS system to determine the molecular weights of carotenoids; inject each 10 μL sample into a Shim-Pack FC-ODS (150 mm × 4.6 mm; particle size, 5 μm; column temperature, 35 °C), and elute with 95% methanol at a flow rate 0.8 mL/min. Regularly wash HPLC column (*see* **Note 4**).
6. Use commercial astaxanthin as a standard, and identify astaxanthin by retention time, absorption spectrum, and molecular weight.
7. Use an extract of *Brevundimonas* sp. SD-212 (NBRC 101024) as a reference for dihydroxyastaxanthin (2,3,2′,3′-tetrahydroxy-β,β-carotene-4,4′-dione) [48].

3.3 Purification and Identification of a Novel Astaxanthin Derivative

1. Grow strain TDMA-17^T in PE medium with shaking at 180 rpm for 2 days at 37 °C.
2. Harvest TDMA-17^T cells by centrifugation at 9000 × *g*.
3. Suspend cells in five times volume of methanol, shake at 180 rpm for 30 min at 28 °C, and collect the extract by filtration. Make sure all procedures are performed under dim light.
4. Repeat this extraction process three times (Fig. 3).
5. Combine the filtrates, dry by evaporation, and dissolve in $CHCl_3$.
6. Apply the $CHCl_3$ solution to the purif-pack silica-gel chromatography after washing the column with $CHCl_3$.

7. Elute the pigment with $CHCl_3$/methanol (9:1) and then dry the pigment.
8. Dissolve the dried pigment in $CHCl_3$/methanol (9:1), apply onto a preparative C18-TLC, and develop with $CHCl_3$/methanol (4:1) to isolate the final principle.
9. Identify the purified carotenoid [**1**] using HPLC (method B), and measure its UV-Vis absorption spectrum using the DAD detector.
10. Record its NMR, ESI-TOFMS, IR, and CD spectra.
11. Determine the absolute configuration of rhamnose moieties of [**1**] as described [37].

3.4 Phylogenetic Characterization of the Astaxanthin-Producing Bacterium

1. Carefully transfer 50 mg wet cells from bacterial isolates into Eppendorf tubes.
2. Extract genomic DNA using the bacterial genomic DNA purification kit.
3. Check the quality of the extracted genomic DNA by agarose gel electrophoresis (**steps 4–9**).
4. Prepare 1% gel solution by melting 1 g of agarose in 100 mL of 0.5× TBE buffer in a microwave for approximately 2 min. Allow to cool for 5 min, then add 2.5 μL ethidium bromide, and stir to mix (*see* **Note 2**).
5. Cast the gel using the supplied tray and comb. Allow the gel to set for a minimum of 30–40 min at room temperature.
6. Load into separate wells: 10 μL of 1 kb ladder, diluted samples (5 μL sample + 5 μL water + 2 μL 6× loading dye).
7. Run the gel for 30 min at 100 V.
8. Expose the gel to UV light and photograph (wear UV goggles).
9. Confirm DNA quality.
10. Set up PCR reaction mixtures in a final volume of 50 μL: 10 μL 10× PCR-buffer, 1 μL PCR nucleotide mix, 50 U Taq DNA polymerase, 1 μL (10 μM) primer B8F, 1 μL (10 μM) primer B1500R, 2 μL DNA extract, and 33 μL deionized and sterilized water.
11. Perform PCR using a protocol that includes an initial denaturation for a period of 5 min at 94 °C; 30 cycles of 94 °C for 1 min, 55 °C for 1 min, and 72 °C for 1 min; 72 °C for 3 min; and incubate at 4 °C until further processing.
12. Purify PCR amplicons using agarose gel electrophoresis (**steps 13–19**).
13. Prepare 2% agarose gel, and let it solidify at room temperature for 30–40 min.
14. Add 10 μL 6× blue loading dye to each PCR product (50 μL).

15. Carefully load all 60 μL into wide gel well.
16. Run the gel for 30 min at 100 V.
17. Wear UV goggles. Remove gel and place on UV box to identify region with PCR products.
18. Cut DNA bands from agarose gel and transfer into sterilized Eppendorf tubes.
19. Purify PCR products from the gel using the Gene Clean kit.
20. Determine the 16S rRNA gene sequences using the BigDye Terminator v3.1 cycle sequencing kit.

3.5 Phylogenetic Analysis

1. Compare the nearly complete 16S rRNA gene (about 1396–1450 bp) sequences of isolates to those available in GenBank/EMBL/DDBJ nucleotide sequence database using the BLASTN program (http://www.ncbi.nlm.nih.gov/BLAST/).
2. Using ClustalW, align the obtained sequences with those of closed relatives.
3. Construct a neighbor-joining phylogenetic tree (Fig. 6).
4. Use NJ plot to draw and edit the obtained tree.

3.6 Morphological, Physiological, and Biochemical Characterization

1. Cultivate 17^T cells on NA for 48 h at 40 °C for 2–3 days.
2. Perform Gram staining according to the method described by Smibert and Krieg [87], and observe the bacterial cells under a light microscope.
3. Observe the bacterial cells under scanning electron microscope according to the manufacturer's instructions. The specimen for scanning electron microscopy is prepared using the osmium fixation protocol and critical point drying [88]. For example, scanning electron microscope images for strain TDMA-17 is shown in Fig. 1b.
4. Test the anaerobic growth by incubation of inoculated nutrient agar plates in a GasPak anaerobic system.
5. Perform the following tests as described in each reference: casein hydrolysis [89]; catalase activity, spore formation, and hydrolysis of Tween 80, DNA, and starch [87]; and hydrolysis of chitin, carboxymethyl cellulose (high viscosity; Sigma), and agar [90].
6. Test oxidase activity by using commercial cytochrome oxidase test strips, according to its manual.
7. Examine the other enzyme activities, growth on carbohydrates, acid production from carbohydrates, nitrate reduction, and the production of H_2S, indole, and acetoin using the commercial systems API 20E, API 20NE, and API 50CH, according to the manufacturer's instructions.

8. Determine the optimal growth temperature: cultivate the strain on NA at 4, 10, 20, 30, 37, 40 and 45 °C.
9. Test NaCl tolerance using NB supplemented with different NaCl concentrations.
10. Evaluate growth at various pH values (3–10) in NB adjusted with HCl or NaOH.
11. Determine the survival rate after exposure to gamma radiation:
    (a) Grow cells of TDMA-17^T on NA at 40 °C.
    (b) Suspend the grown cells in 5 mL of NB.
    (c) Prepare aliquots of the cell suspension (1 mL), and irradiate at 2.35 kGy (13 Gy min^{-1}, ^{137}Cs source).
    (d) Dilute serially and plate onto NA plates.
    (e) Aerobically incubate the plates at 40 °C.
    (f) Similarly, treat *E. coli* C600.
    (g) Determine the survival percentage by comparing with unirradiated cultures.

3.7 Chemotaxonomic Characterization

1. Determine the cellular fatty acids of strains TDMA-17^T:
    (a) Grow the cells aerobically on NA at 40 °C.
    (b) Extract and analyze the fatty acid methyl esters according to the standard protocol of the Sherlock Microbial Identification System.
2. Determine the respiratory quinone by an HPLC method [91] using cell extracts of *Sphingopyxis alaskensis* DSMZ 13593^T and *Sphingomonas asaccharolytica* NBRC 15499^T as a reference for Q-10.
3. Analyze glycosphingolipids using TLC as described previously [85, 92]. Determine the polar lipids according to [93].
4. Determine the DNA G+C content by an HPLC method [94].

4 Notes

1. Carotenoids are highly unstable and sensitive to light, oxygen, and temperature, which induce their oxidation. To minimize the oxidation of carotenoids, the extraction must be quickly done without exposure to light, oxygen, and high temperatures. Therefore, all the experimental procedures should be carried out in a dark room. Alternatively, dark tubes or tubes foiled with aluminum foil can be used. Nitrogen gas should be also used for flushing the pigment extracts before fridge storage at −20 °C to avoid oxidation. In addition, the use of an

antioxidant such as butylhydroxytoluene (BHT) [95] can significantly reduce the oxidation of carotenoids during extraction.

2. Ethidium bromide is mutagenic and moderately toxic. You must wear gloves and handle the chemical with care.
3. Details of HPLC methods B and C are in Chapter 15.
4. Usually, DMSO-MeOH carotenoid extracts contain impurities such as salts, lipids, fatty compounds, proteins, and other cell compounds. These contaminants usually are adsorbed on the stationary phase of reversed-phase HPLC column and may become contaminated by the repeated injection of samples. Therefore, retention times can shift (0.5–1 min) and tailing can occur. In addition, this may cause the column backpressure to build up to intolerably high levels, which overpressures the pump and can cause the column to settle and create a void, depending upon where the blockage occurs. To prevent this problem, we clean the columns on a regular basis throughout our work. Washing the HPLC columns with 100% acetonitrile then with 100% methanol was sufficient to clean and regenerate the column. We strongly recommend using a pre-column.

References

1. Johnson EA, An GH (1991) Astaxanthin from microbial sources. Crit Rev Biotechnol 11(4):297–326
2. Nelis HJ, Deleenheer AP (1991) Microbial sources of carotenoid-pigments used in foods and feeds. J Appl Bacteriol 70(3):181–191
3. Britton G, Liaaen-Jensen S, Pfander H (1995) Carotenoids today and challenges for the future. Birkhauser, Basel
4. Rufer CE et al (2008) Bioavailability of astaxanthin stereoisomers from wild (*Oncorhynchus* spp.) and aquacultured (*Salmo salar*) salmon in healthy men: a randomised, double-blind study. Br J Nutr 99(5):1048–1054
5. Page GI, Davies SJ (2006) Tissue astaxanthin and canthaxanthin distribution in rainbow trout (*Oncorhynchus mykiss*) and Atlantic salmon (*Salmo salar*). Comp Biochem Physiol A Mol Integr Physiol 143(1):125–132
6. Fujita T et al (1983) Pigmentation of cultured red sea bream with astaxanthin diester purified from krill oil. Bull Jpn Soc Sci Fish 49:1855–1865
7. Torrissen OJ (1989) Pigmentation of salmonids—interactions of astaxanthin and canthaxanthin on pigment deposition in rainbow-trout. Aquaculture 79(1–4):363–374
8. Bjerkeng B, Berge GM (2000) Apparent digestibility coefficients and accumulation of astaxanthin E/Z isomers in Atlantic salmon (*Salmo salar* L.) and Atlantic halibut (*Hippoglossus hippoglossus* L.). Comp Biochem Phys B 127(3):423–432
9. An GH, Choi ES (2003) Preparation of the red yeast, *Xanthophyllomyces dendrorhous*, as feed additive with increased availability of astaxanthin. Biotechnol Lett 25(10):767–771
10. Christiansen R et al (1995) Antioxidant status and immunity in Atlantic salmon, *Salmo salar* L., fed semi-purified diets with and without astaxanthin supplementation. J Fish Dis 18(4):317–328
11. Christiansen R, Torrissen OJ (1997) Effects of dietary astaxanthin supplementation on fertilization and egg survival in Atlantic salmon (*Salmo salar* L.). Aquaculture 153(1–2):51–62
12. Amaya E, Nickell D (2015) Using feed to enhance the color quality of fish and crustaceans. In: Allen Davis D (ed) Feed and feeding practices in aquaculture. Woodhead Publishing, Oxford, pp 269–298
13. Lorenz RT, Cysewski GR (2000) Commercial potential for *Haematococcus microalgae* as a natural source of astaxanthin. Trends Biotechnol 18(4):160–167

14. Gorton HL, Williams WE, Vogelmann TC (2001) The light environment and cellular optics of the snow alga *Chlamydomonas nivalis* (Bauer) Wille. Photochem Photobiol 73(6):611–620
15. Gorton HL, Vogelmann TC (2003) Ultraviolet radiation and the snow alga *Chlamydomonas nivalis* (Bauer) Wille. Photochem Photobiol 77(6):608–615
16. Kim JH, Chang HI (2006) High-level production of astaxanthin by *Xanthophyllomyces dendrorhous* mutant JH1, using chemical and light induction. J Microbiol Biotechnol 16(3):381–385
17. Fan L et al (1998) Does astaxanthin protect *Haematococcus* against light damage? Z Naturforsch C 53(1–2):93–100
18. Iwamoto T et al (2000) Inhibition of low-density lipoprotein oxidation by astaxanthin. J Atheroscler Thromb 7(4):216–222
19. O'connor I, O'brien N (1998) Modulation of UVA light-induced oxidative stress by beta-carotene, lutein and astaxanthin in cultured fibroblasts. J Dermatol Sci 16(3):226–230
20. Wang X, Willen R, Wadstrom T (2000) Astaxanthin-rich algal meal and vitamin C inhibit *Helicobacter pylori* infection in BALB/cA mice. Antimicrob Agents Chemother 44(9):2452–2457
21. Jyonouchi H, Sun S, Gross M (1995) Effect of carotenoids on in vitro immunoglobulin production by human peripheral blood mononuclear cells: astaxanthin, a carotenoid without vitamin A activity, enhances in vitro immunoglobulin production in response to a T-dependent stimulant and antigen. Nutr Cancer 23(2):171–183
22. Guerin M, Huntley ME, Olaizola M (2003) *Haematococcus* astaxanthin: applications for human health and nutrition. Trends Biotechnol 21(5):210–216
23. Yuan JP et al (2010) Potential health-promoting effects of astaxanthin: a high-value carotenoid mostly from microalgae. Mol Nutr Food Res 55(1):150–165
24. Mortensen A (2009) Supplements. In: Britton G, Liaaen-Jensen S, Pfander H (eds) Nutrition and health. Birkhäuser Verlag, Basel, pp 68–82
25. Vaclavik VA, Christian EW (2014) Fat and oil products. In: Vaclavik VA, Christian EW (eds) Essentials of food science. Springer, New York, pp 233–261
26. März U (2015) FOD025E—the global market for carotenoids. BCC Research, Wellesley, MA
27. Solymosi K et al (2015) Food colour additives of natural origin. In: Scotter MJ (ed) Colour additives for foods and beverages. Elsevier Ltd., Amsterdam, pp 3–34
28. Bi W et al (2010) Task-specific ionic liquid-assisted extraction and separation of astaxanthin from shrimp waste. J Chromatogr B Analyt Technol Biomed Life Sci 878(24):2243–2248
29. Franco-Zavaleta ME et al (2010) Astaxanthin extraction from shrimp wastes and its stability in 2 model systems. J Food Sci 75(5):C394–C399
30. Pacheco N et al (2009) Effect of temperature on chitin and astaxanthin recoveries from shrimp waste using lactic acid bacteria. Bioresour Technol 100(11):2849–2854
31. An GH et al (2004) Pigmentation and delayed oxidation of broiler chickens by the red carotenoid, astaxanthin, from chemical synthesis and the yeast, Xanthophyllomyces dendrorhous. Asian Austral J Anim 17(9):1309–1314
32. Asker D, Beppu T, Ueda K (2007) *Sphingomonas astaxanthinifaciens* sp. nov., a novel astaxanthin-producing bacterium of the family *Sphingomonadaceae* isolated from Misasa, Tottori, Japan. FEMS Microbiol Lett 273(2):140–148
33. Asker D et al (2018) Screening and profiling of natural ketocarotenoids from environmental aquatic bacterial isolates. Food Chem 253:247–254
34. Asker D (2017) Isolation and characterization of a novel, highly selective astaxanthin-producing marine bacterium. J Agric Food Chem 65(41):9101–9109
35. Asker D et al (2012) A novel radio-tolerant astaxanthin-producing bacterium reveals a new astaxanthin derivative: astaxanthin dirhamnoside. In: Barredo J-L (ed) Microbial carotenoids from bacteria and microalgae: methods and protocols. Humana Press, New York, pp 61–97
36. Asker D et al (2012) Isolation, characterization, and diversity of novel radiotolerant carotenoid-producing bacteria. In: Barredo J-L (ed) Microbial carotenoids from bacteria and microalgae: methods and protocols. Humana Press, New York, pp 21–60
37. Asker D et al (2009) Astaxanthin dirhamnoside, a new astaxanthin derivative produced by a radio-tolerant bacterium, *Sphingomonas astaxanthinifaciens*. J Antibiot (Tokyo) 62(7):397–399
38. Asker D, Beppu T, Ueda K (2007) Unique diversity of carotenoid-producing bacteria isolated from Misasa, a radioactive site in Japan. Appl Microbiol Biotechnol 77(2):383–392
39. Asker D, Isaka K (2006) Production of astaxanthin by microorganisms. Office TP, Japan. Patent JP340676A
40. Del Rio E et al (2008) Efficiency assessment of the one-step production of astaxanthin by the

microalga *Haematococcus pluvialis*. Biotechnol Bioeng 100(2):397–402

41. Lubián LM et al (2000) Nannochloropsis (Eustigmatophyceae) as source of commercially valuable pigments. J Appl Phycol 12(3):249–255
42. De La Fuente JL et al (2010) High-titer production of astaxanthin by the semi-industrial fermentation of Xanthophyllomyces dendrorhous. J Biotechnol 148(2–3):144–146
43. Jacobson GK et al. (2003) Astaxanthin overproducing strains of Phaffia rhodozyma, methods for their cultivation, and their use in animal feeds. US patent 20030049241
44. Liu YS, Wu JY (2006) Hydrogen peroxide-induced astaxanthin biosynthesis and catalase activity in *Xanthophyllomyces dendrorhous*. Appl Microbiol Biotechnol 73(3):663–668
45. Calo P et al (1995) Ketocarotenoids in halobacteria: 3-hydroxy-echinenone and trans-astaxanthin. J Appl Bacteriol 79:282
46. Yokoyama A, Izumide H, Miki W (1994) Production of astaxanthin and 4-ketozeaxanthin by the marine bacterium, *Agrobacterium aurantiacum*. Biosci Biotechnol Biochem 58:1842–1844
47. Tsubokura A, Yoneda H, Mizuta H (1999) *Paracoccus carotinifaciens* sp. nov., a new aerobic gram-negative astaxanthin-producing bacterium. Int J Syst Bacteriol 49(Pt 1):277–282
48. Yokoyama A et al (1996) New trihydroxy-keto-carotenoids isolated from an astaxanthin-producing marine bacterium. Biosci Biotechnol Biochem 60:200–203
49. Osanjo GO et al (2009) A salt lake extremophile, *Paracoccus bogoriensis* sp. nov., efficiently produces xanthophyll carotenoids. Afr J Microbiol Res 3(8):426–433
50. Hirasawa K, Tsuborkura A (2014). Method for separating carotenoid. US Patent 8853460B2
51. Bubrick P (1991) Production of astaxanthin from *Haematococcus*. Bioresour Technol 38:237–239
52. Miller MW, Yoneyama M, Soneda M (1976) *Phaffia*, a new yeast genus in the *Deuteromycotina* (*Blastomycetes*). Int J Syst Bacteriol 26:286–291
53. Harker M, Hirschberg J, Oren A (1998) *Paracoccus marcusii* sp. nov., an orange gram-negative coccus. Int J Syst Bacteriol 48(Pt 2):543–548
54. Lee JH et al (2004) *Paracoccus haeundaensis* sp. nov., a Gram-negative, halophilic, astaxanthin-producing bacterium. Int J Syst Evol Microbiol 54(Pt 5):1699–1702
55. Kametani K, Matsumura T (1983) Determination of ^{238}U, ^{234}U, ^{226}Ra and ^{228}Ra in spring waters of sanin district. Radioisotopes 32(1):18–21
56. Nelis HJ, De Leenheer AP (1989) Profiling and quantitation of bacterial carotenoids by liquid chromatography and photodiode array detection. Appl Environ Microbiol 55(12):3065–3071
57. Britton G, Liaaen-Jensen S, Pfander H (2004) Carotenoids handbook. Birkhäuser, Basel
58. Takaichi S et al (2003) Fatty acids of astaxanthin esters in krill determined by mild mass spectrometry. Comp Biochem Physiol B Biochem Mol Biol 136(2):317–322
59. Breithaupt DE (2004) Identification and quantification of astaxanthin esters in shrimp (*Pandalus borealis*) and in a microalga (*Haematococcus pluvialis*) by liquid chromatography-mass spectrometry using negative ion atmospheric pressure chemical ionization. J Agric Food Chem 52(12):3870–3875
60. Matsuno T et al (1984) The occurence of enantiomeric and meso-astaxanthin in aquatic animals. Bull Jpn Soc Sci Fish 50:1589–1592
61. Johnson EA, Schroeder W (1995) Astaxanthin from the yeast *Phaffia rhodozyma*. Stud Mycol 38:81–90
62. Rønneberg H et al (1980) Natural occurrence of enatiomeric and meso-astaxanthin 1. Ex lobster eggs (*Homarus gammarus*). Helv Chim Acta 63:711–715
63. Bernhard K et al (1982) Carotenoids of the carotenoprotein asteriarubin. Optical purity of asterinic acid. Helv Chim Acta 65:2224–2229
64. Yokoyama A, Adachi K, Shizuri Y (1995) New carotenoid glucosides, astaxanthin glucoside and adonixanthin glucoside, isolated from the astaxanthin-producing marine bacterium, *Agrobacterium aurantiacum*. J Nat Prod 58:1929–1933
65. Misawa N, Shimada H (1998) Metabolic engineering for the production of carotenoids in non-carotenogenic bacteria and yeasts. J Biotechnol 59(3):169–181
66. Maoka T, Tsushima M, Matsuno T (1989) New acetylenic carotenoids from the starfishes *Asterina pectinifera* and *Asterias amurensis*. Comp Biochem Physiol 93:829–834
67. Maoka T et al (1985) Stereochemical investigation of the carotenoids in the antarctic krill *Euphausia superba*. Bull Jpn Soc Sci Fish 51:1671–1673
68. Hertzberg S et al (1983) Carotenoid sulfates: 2. Structural elucidation of bastaxanthin. Acta Chem Scand 37:267–280
69. Cowan ST (1968) A dictionary of microbial taxonomic usage. Oliver & Boyd, Edinburgh
70. Staley JT, Krieg NJ (1984) Classification of prokaryotic organisms: an overview. The Williams & Wilkins Co., Baltimore
71. Vandamme P et al (1996) Polyphasic taxonomy, a consensus approach to bacterial systematics. Microbiol Rev 60(2):407–438

72. Colwell RR (1970) Polyphasic taxonomy of the genus *Vibrio*: numerical taxonomy of *Vibrio cholerae*, *Vibrio parahaemolyticus*, and related *Vibrio* species. J Bacteriol 104:410–433
73. Murray RGE et al (1990) Report of the Ad-Hoc committee on approaches to taxonomy within the *Proteobacteria*. Int J Syst Bacteriol 40(2):213–215
74. Tindall BJ et al (2010) Notes on the characterization of prokaryote strains for taxonomic purposes. Int J Syst Evol Microbiol 60(Pt 1):249–266
75. Egan S, Thomas T, Kjelleberg S (2008) Unlocking the diversity and biotechnological potential of marine surface associated microbial communities. Curr Opin Microbiol 11(3):219–225
76. Schleifer KH, Ludwig W (1989) Phylogenetic relationships of bacteria. Elsevier Science Publishers B.V., Amsterdam
77. Stackebrandt E et al (2002) Report of the ad hoc committee for the re-evaluation of the species definition in bacteriology. Int J Syst Evol Microbiol 52:1043–1047
78. Stackebrandt E, Goebel BM (1994) A place for DNA–DNA reassociation and 16s ribosomal-RNA sequence-analysis in the present species definition in bacteriology. Int J Syst Bacteriol 44(4):846–849
79. Woese CR (1987) Bacterial evolution. Microbiol Rev 51(2):221–271
80. Fox GE, Wisotzkey JD, Jurtshuk P Jr (1992) How close is close: 16S rRNA sequence identity may not be sufficient to guarantee species identity. Int J Syst Bacteriol 42(1):166–170
81. Martinez-Murcia AJ, Collins MD (1990) A phylogenetic analysis of the genus Leuconostoc based on reverse transcriptase sequencing of 16S rRNA. FEMS Microbiol Lett 70(1):73–83
82. Wayne LG et al (1987) Report of the ad hoc committee on reconciliation of approaches to bacterial systematics. Int J Syst Bacteriol 37(4):463–464
83. Johnson JL (1984) Nucleic acids in bacterial classification. Williams & Wilkins, Baltimore, MD
84. Yarza P et al (2008) The All-Species Living Tree project: a 16S rRNA-based phylogenetic tree of all sequenced type strains. Syst Appl Microbiol 31(4):241–250
85. Takeuchi M, Hamana K, Hiraishi A (2001) Proposal of the genus *Sphingomonas sensu stricto* and three new genera, *Sphingobium*, *Novosphingobium* and *Sphingopyxis*, on the basis of phylogenetic and chemotaxonomic analyses. Int J Syst Evol Microbiol 51(Pt 4):1405–1417
86. Saitou N, Nei M (1987) The neighbor-joining method: a new method for reconstructing phylogenetic trees. Mol Biol Evol 4(4):406–425
87. Smibert RM, Krieg NR (1994) Phenotypic characterization. In: Gerhardt P, Murray RGE, Wood WA, Krieg NR (eds) Methods for general and molecular bacteriology. American Society for Microbiology, Washington, DC, pp 607–654
88. Beveridge TJ, Popkin TJ, Cole RM (1994) Electron microscopy. In P. Gerhardt (ed.), Methods for general molecular bacteriology. American Society for Microbiology, Washington, D.C., pp 42–71
89. Norris JR, Ribbons DW, Varma AK (1985) Methods in microbiology. Academic Press, London
90. Cowan ST, Steel KJ (1993) Manual for the identification of medical bacteria. Cambridge University Press, London
91. Collins MD (1994) Isoprenoid quinones. In: O'Donnell MGAG (ed) Chemical methods in prokaryotic systematics. Wiley, Chichester, pp 265–310
92. Kawahara K et al (1991) Chemical structure of glycosphingolipids isolated from *Sphingomonas paucimobilis*. FEBS Lett 292(1–2):107–110
93. Tindall BJ (1990) Lipid composition of *Halobacterium lacusprofundi*. FEMS Microbiol Lett 66:199–202
94. Mesbah M, Whitman WB (1989) Measurement of deoxyguanosine/thymidine ratios in complex mixtures by high-performance liquid chromatography for determination of the mole percentage guanine + cytosine of DNA. J Chromatogr 479(2):297–306
95. Meyers SP, Bligh D (1981) Characterization of astaxanthin pigments from heat-processed crawfish waste. J Agric Food Chem 29(3):505–508

Chapter 11

Screening, Isolation, and Identification of Zeaxanthin-Producing Bacteria

Dalal Asker, Tarek S. Awad, Teruhiko Beppu, and Kenji Ueda

Abstract

Zeaxanthin is a yellow xanthophyll, dihydroxy-carotenoid, that is naturally found in some of the green, orange, and yellow vegetables and fruits and has a powerful antioxidant activity. Epidemiological evidences suggest that increasing the consumption of zeaxanthin in the diet is associated with a lower risk of age-related macular degeneration (ARMD) and cataracts, two of the leading causes of blindness in the world. Zeaxanthin is a promising nutraceutical/colorant with many applications in feed, food, and pharmaceutical industries. Currently, the commercial production of zeaxanthin is dependent on synthetic routes with limitation in production from biological sources. However, the biotechnological production of natural zeaxanthin is favored due to its safety, potential large-scale production and consumers' preference for natural additives. In this chapter, we describe a rapid screening method based on 16S rRNA gene sequencing and effective HPLC with diode array detector/MS methods for the isolation and identification of zeaxanthin-producing bacteria and their carotenoid analysis.

Key words Zeaxanthin, Nutraceuticals, Antioxidants, Colorants, Ketocarotenoids, Bacteria, Fresh water, 16S rDNA phylogeny, HPLC-DAD-MS, HPLC-MS

1 Introduction

1.1 Structure, Function, and Importance of Zeaxanthin

Zeaxanthin is a yellow xanthophyll, dihydroxy-carotenoid (3,3′-dihydroxy-β-carotene), that is naturally found in some of the green, orange, and yellow vegetables and fruits such as spinach, collard greens, lettuce, broccoli, zucchini, kiwifruit, garden peas, alfalfa, and marigold flowers [1–6]. Its chemical structure includes 40 carbon atoms with 11 conjugated ethylenic double bonds and two rings on both (Fig. 1). Zeaxanthin and lutein (another carotenoid) are isomers that differ from each other just by the shift of a single double bond. They are important antioxidants that accumulate in the macula of the human eye and constitute the macular pigment, where they act as UV filters to protect the eye against photooxidative damage from sun UV light, which reduces the risk of age-related macular degeneration (AMD) and cataracts, two of the

Carlos Barreiro and José-Luis Barredo (eds.), *Microbial Carotenoids: Methods and Protocols*, Methods in Molecular Biology, vol. 1852, https://doi.org/10.1007/978-1-4939-8742-9_11,

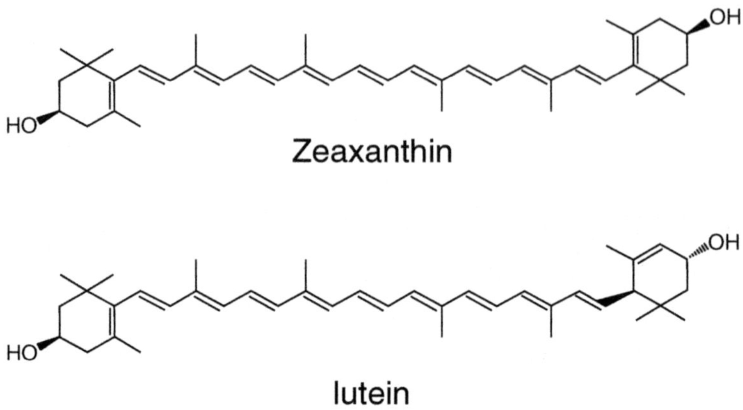

Fig. 1 Chemical structure of zeaxanthin and lutein

leading causes of blindness in the world [7–11]. They also increase brain efficiently, reduce bone fracture risk, protect stem cells, and can be used in medical diagnostics [12–16]. They may also play a critical role in the prevention of cancer [17, 18]. The macular pigments can be increased in primates either by increasing the intake of foods rich in lutein and zeaxanthin or supplementation.

Zeaxanthin, like other carotenoids, can be synthesized by photosynthetic organisms (e.g., plant and microalgae) and some non-photosynthetic microorganisms (e.g., yeast, fungi, and bacteria) [19–23], where it serves as an accessory light-gathering pigment as well as a protectant against the toxic effects of ultraviolet radiation and oxygen radicals. Animals and humans cannot synthesize zeaxanthin and thus must be obtained through diet. To enhance the pigmentation of animal flesh and egg yolk, zeaxanthin is supplemented as an ingredient in feed for fish, birds, swine, and poultry [5]. It is also suitable for use as a colorant in cosmetics and food industries [24]. Increasing the industrial applications of zeaxanthin has thus increased the commercial demand for its production. The global market for zeaxanthin reached $10 million and is expected to increase to $18 million in 2018 with a CAGR of 1.5% [25].

1.2 Microbial Production of Zeaxanthin

At present, the commercial production of zeaxanthin is mostly carried out either by the extraction from plant tissues such as marigolds and alfalfa [26] or chemical synthesis. Synthetic zeaxanthin is prepared by the Wittig reaction, which yields 96–98% of *trans*-(3*R*,3*R*)-zeaxanthin and minor quantities of impurities [27]. A biological route of zeaxanthin extraction is from marigold (*Tagetes erecta*) flowers; however, the produced zeaxanthin has low purity as the extract comprises only about 20% of *trans*-zeaxanthin; various amounts of other carotenoids, fats, oils, and waxes; and other organic solvent-extractable, mainly unsaponifiable, compounds originating from plant material [27]. A comparative study has shown that the bioavailability of zeaxanthin produced from a microbial source (*Flavobacterium multivorum*) is 2–3 times higher than that produced from marigold, which also had low stability [28]. Although various taxa including microalgae, yeast, and bacteria (Table 1) have been reported to produce zeaxanthin, its microbial production at the commercial level has not yet been established. A demand for natural zeaxanthin is now emerging in the pharmaceutical industry owing for its health benefits as antioxidant for reducing the risk of AMD, cataracts, and cancer-preventive properties.

1.3 Screening of Novel Zeaxanthin-Producing Bacteria

We screened for high-value carotenoid-producing bacteria from freshwater samples that were collected at Misasa (Tottori, Japan), a region known for the high natural radioactivity content [46] (*see* Chap. 9). Using a rapid screening method, comprehensive taxonomic characterization based on 16S rRNA gene sequences and rapid and effective HPLC with diode array/MS methods for carotenoids analysis, we discovered a unique diversity of carotenoid producers including six yellow-pigmented strains (TDMA-2, 5, 12, 14, 15, and 16) that produce zeaxanthin and its derivatives [44]. These strains were isolated based on their abilities to form yellow-pigmented colonies on LB or NA agar media. From 16S rRNA gene sequences and phylogenetic analysis, we found these strains belong to three different bacterial classes (α-*Proteobacteria*, *Flavobacteria*, *Sphingobacteria*) and may represent novel taxa at the species or genus level [44]. Importantly, the HPLC analysis of carotenoids revealed that the strains with noteworthy production efficiency of zeaxanthin were TDMA-5 (*Flavobacteria*) and TDMA-16 (*Sphingobacteria*), which represented novel taxa that belong to family *Sphingobacteriaceae* and family *Sphingomonadaceae*, respectively [44, 47].

This chapter deals with methodology of isolation and identification of the novel zeaxanthin-producing bacteria (strains TDMA-5 and 16) and also identification of their carotenoids [45, 48].

1.4 Identification of Zeaxanthin Production by Novel Isolates

In Chap. 9 we summarized rapid and effective HPLC methods with diode array or MS spectroscopy for the analysis and identification of carotenoids in their natural biological mixtures. These methods were applied for the analysis and identification of the two zeaxanthin-producing strains (TDMA-5^T and TDMA-1) [45, 48]. Extraction of carotenoids from the wet cells of both strains was done as described in Chap. 9. The yellow pigments produced by the two strains were identified as carotenoids. The amount of carotenoids produced by strains TDMA-5^T was slightly high (0.8 mg/g dry cells of zeaxanthin). Analysis of the crude pigment isolated from the cells of strain TDMA-5^T using HPLC method A

Table 1
Reported microorganisms for production of zeaxanthin

Microorganism	Yield	References
Cyanobacteria		
Microcystis aeruginosa	–	[29]
Spirulina sp.	–	[30]
Synechococcus sp. PCC 7942[a] (*Anacystis nidulans*)	–	[31]
Prochloron sp.	–	[32]
Green algae		
Neospongiococcum excentricum	0.65%	[33]
Dunaliella salina zea1 (mutant)	6 mg/g	[34]
Dunaliella parva	–	[35]
Dunaliella bardawil	–	[36]
Blue green algae		
Cyanophora paradoxa	–	[37]
Glaucocystis nostochinearum		
Yellow-green algae		
Pleurochloris commutata	–	[37]
Red algae		
Corallina officinalis	–	[38]
C. elongate and *Jania* sp.		
Yeast		
Phaffia rhodozyma	–	[22]
Bacteria		
Sphingobacterium multivorum ATCC 55238[b]	10.6 μg/mL	[39]
Synechocystis sp. PCC 6803	0.98 μg/mL	[40]
Erwinia herbicola	–	[41]
Paracoccus zeaxanthinifaciens[c]	13.76 ± 0.14 mg/L	[42]
E. coli (metabolic engineered strain)	1.6 mg/g	[43]
Zeaxanthinibacter enoshimensis	–	[19]
Mesoflavibacter zeaxanthinifaciens	0.91 mg/g	[20]
Nubsella zeaxanthinifaciens	0.8 mg/g	[44, 45]

[a]Formerly *Anacystis nidulans* R2
[b]Formerly *Flavobacterium*
[c]Formerly *Flavobacterium* sp. R-1512

showed a sharp peak at $R_t = 0.98$ min, which is identical to zeaxanthin standard (Fig. 2a). In addition, the UV-visible absorption spectrum of this peak was also identical to that of zeaxanthin, which exhibited absorption maximum at 451 and 476 nm (Fig. 2b). These results thus indicated that zeaxanthin was the predominant carotenoid produced by TDMA-5^T.

TDMA-16^T produced significantly large amount of carotenoids (1.7 mg/g dry cells), which comprised a mixture of zeaxanthin, nostoxanthin, and an unknown carotenoid. Analysis of its crude pigment using HPLC (method A) showed the presence of a broad peak at R_t = 0.8–1.1 min. However, HPLC method B showed the presence of four pigments (corresponding to peaks 1–4 in Fig. 3a). Table 2 summarizes the spectroscopic data for the HPLC four peaks of carotenoids produced by strain TDMA-16^T. The UV-visible absorption spectra of the first and third peaks (Fig. 3b) were identical to the spectrum of zeaxanthin standard as they exhibited absorption maxima at 451 and 476 nm. The zeaxanthin standard showed identical R_t and spectroscopic data as those for peak 3. Therefore, the compound corresponding to peak 3 (27.5% of the total carotenoids) was identified as zeaxanthin. The UV-visible absorption spectrum of peak 2 (2.5% of the total carotenoids) (Fig. 3b) was at 449 and 475 nm, which is identical

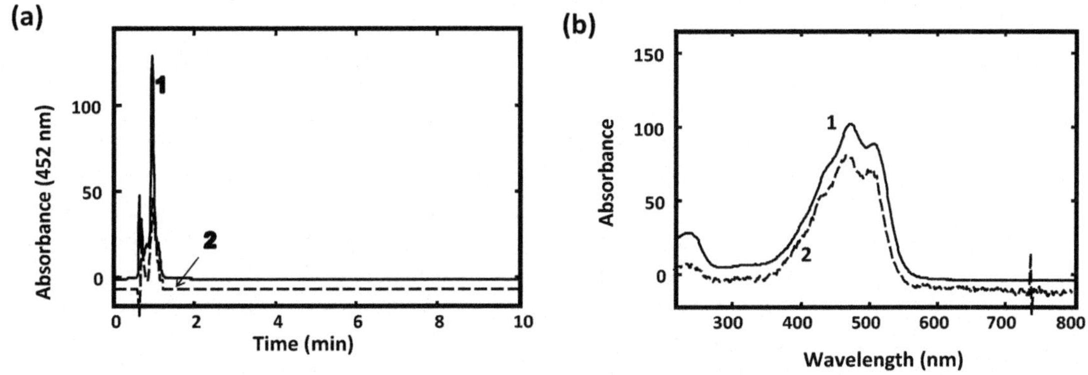

Fig. 2 Absorption profile of the carotenoids produced by strain TDMA-5^T. Elution profile (λ_{max} 452 nm) (**a**) and absorption spectrum for each peak (**b**) are shown. The carotenoid fraction was analyzed by reverse-phase liquid chromatography using method A (*see* Subheadings 2 and 3). The fraction consisted of zeaxanthin as a major carotenoid

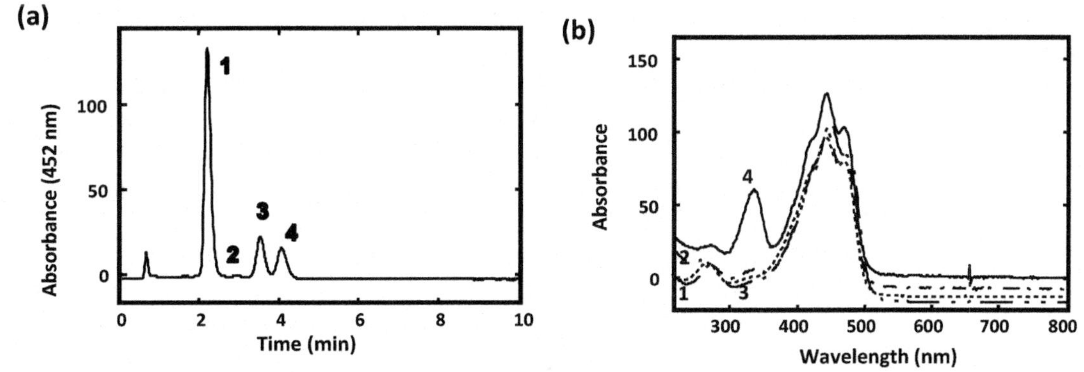

Fig. 3 Absorption profile of the carotenoids produced by strain TDMA-16^T. Elution profile (λ_{max} 452 nm) (**a**) and absorption spectrum of each peak (**b**) are shown. The carotenoid fraction was analyzed by reverse-phase liquid chromatography using method B (*see* Subheadings 2 and 3). The fraction consisted of unidentified carotenoid (peak 1), dihydroxy-zeaxanthin (peak 2), zeaxanthin (peak 3), 9-*cis*-zeaxanthin (peak 4) and 13-*cis*-zeaxanthin (peak 5)

Table 2
The spectroscopic data for the HPLC five peaks of carotenoids produced by strain TDMA-16^T

HPLC peak	Carotenoids	Absorption (λ_{max}, nm)	HPLC, retention time (R_t, min)		Mol. wt (M^++1)
			Method B	Method C	
1	Unknown	451–476	1.24	3.4	861
2	Nostoxanthin (di-hydroxy zeaxanthin)	449–475	2.29	6.5	601
3	Zeaxanthin	451–476	3.85	10.3	569
4	13-*cis*-zeaxanthin	350–370, 451–476	4.27	11.04	569

to nostoxanthin that was extracted from cells *Sphingomonas asaccharolytica* NBRC 15499^T. Zeaxanthin was further confirmed by HPLC-Mass (method C) [R_t at 10.3 min; and molecular weight (M^++1), 569] (Table 2). The compounds corresponding to peak 4 (15.7% of the total carotenoids), which showed asymmetrical spectrum, were predicted to be 13-*cis*-zeaxanthin by comparing their UV/VIS spectra to those published [49]. Peak 1 (54.3% of the total carotenoids) could not be identified although its spectrum was typical of zeaxanthin but with higher molecular weight and polarity. Therefore, this unknown was primarily identified as zeaxanthin derivative. In addition, the presence of nostoxanthin in TDMA-16^T is consistent with its presence in *S. paucimobilis* and *S. echinoides* [50, 51].

1.5 16S rRNA Gene-Based Phylogenetic Analysis by Novel Isolates

The phylogenetic position of the isolates was studied by the standard analysis based on 16S rRNA gene sequence. A sequence similarity search in the GenBank/EMBL/DDBJ nucleotide sequence database performed using the BLASTN program (http://www.ncbi.nlm.nih.gov/BLAST/) revealed that TDMA-5^T belonged to the family *Sphingobacteriaceae*. For example, the phylogenetic analysis of strain TDMA-5^T showed that it formed a clade together with several uncharacterized bacterial strains with sequence similarities of 96–97%. This clade did not cluster robustly with any recognized species or genus (Fig. 4). It is evident from the phylogenetic tree shown in Fig. 4 that strain TDMA-5^T belonged to a distinct lineage in the family *Sphingobacteriaceae*, sharing 89.4–93.4% sequence similarity with members of the nearest genus *Pedobacter*. This result thus suggests that TDMA-5^T represents a novel genus within the family *Sphingobacteriaceae*.

More detailed information about the morphological, physiological, and biochemical characterization and taxonomic positions of these novel zeaxanthin-producing isolates are in our previous publications [45, 48, 53].

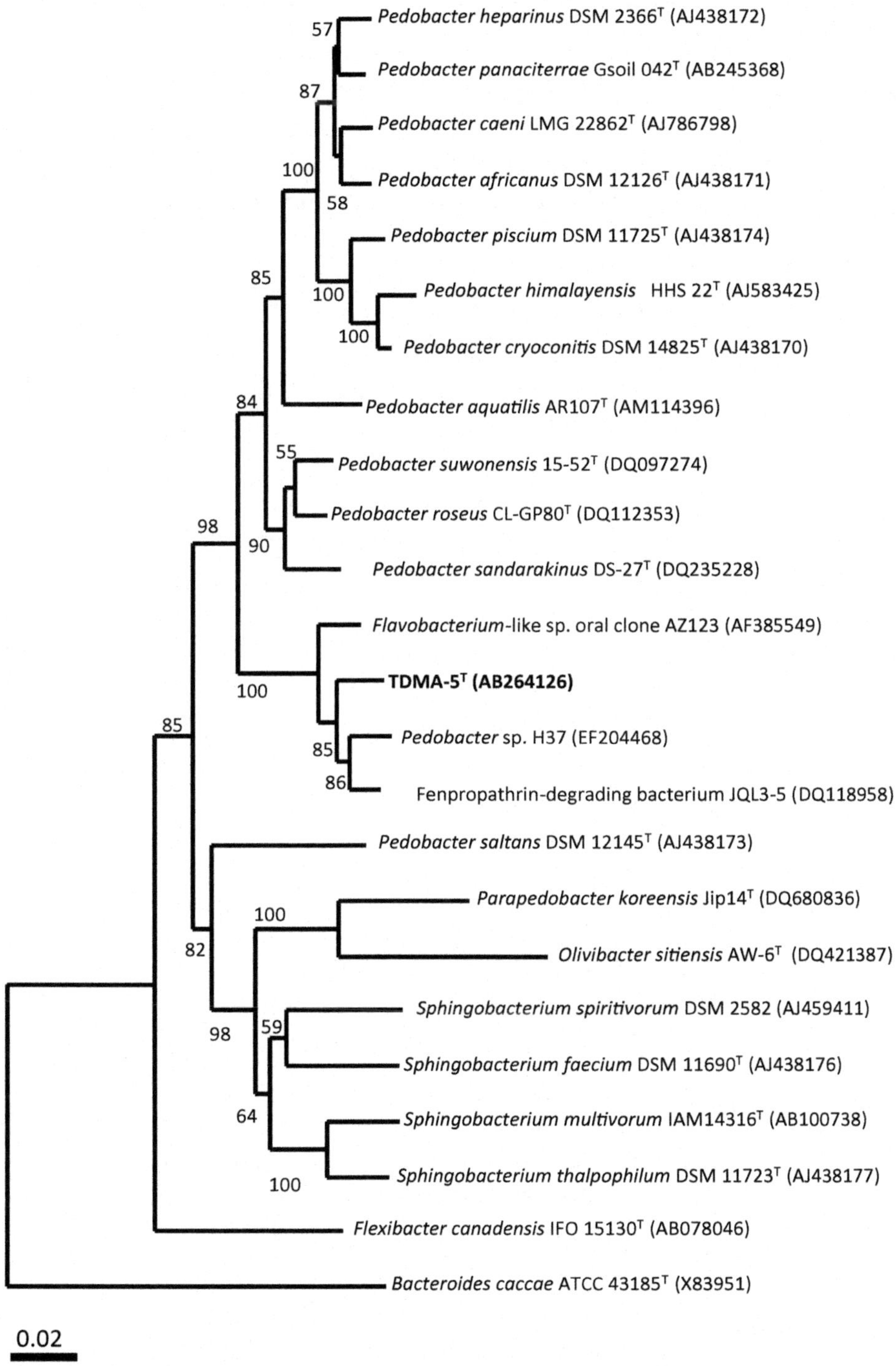

Fig. 4 A phylogenetic tree based on the 16S rRNA gene sequences of strain TDMA-5^T and representative members of related genera of the family *Sphingobacteriaceae*. The tree was constructed by the neighbour-joining method [52]. Bar, 2 substitutions per 100 nucleotide positions. Bootstrap values, expressed as a percentage of 1000 replicates, are given at the branching points. Values above 50% are shown. 16S rRNA gene sequence from *Bacteroides caccae* ATCC 43185^T (AB078046) was used as outgroup. GenBank accession numbers are shown in parentheses. Adapted from Asker et al. [45]

2 Materials

2.1 Media and Buffers

1. Nutrient agar: 3 g/L beef extract, 5 g/L peptone, and 15 g/L agar (Difco Laboratories, Detroit, MI, USA).
2. Luria–Bertani (LB) solid medium: 10 g/L tryptone, 5 g/L yeast extract, 5 g/L NaCl, and 15 g/L agar.

2.2 Preparing Media and Cultivation

1. Analytical balance.
2. Flasks for cultivation.
3. Disposable Petri dishes.
4. Micropipettes and sterile tips (0–20, 20–200, and 100–1000 μL).
5. Autoclave.
6. Incubators (28, 30, 37, 40, and 50 °C).
7. Sterile loops, sterile toothpicks and sterile cotton swabs.
8. Clean bench equipped with UV light.

2.3 Carotenoids Analysis

1. Butylhydroxytoluene (BHT) (Sigma, St. Louis, MO, USA) (*see* **Note 1**).
2. Dim light (*see* **Note 1**).
3. Zeaxanthin (DHI; Water & Environment, Hørsholm, Denmark).
4. Eppendorf tubes.
5. Racks.
6. Balance.
7. Eppendorf table top vial shaker.
8. Spectrophotometer.
9. 0.45 μm hydrophobic Teflon membrane syringe filter (Advantec, Tokyo, Japan) and syringe (1 mL).
10. 96-well plate.
11. HPLC automated injector with 96-well plate adaptor Class-VP (Shimadzu, Kyoto, Japan) with PDA detector.
12. HPLC/MS system LCMS-2010EV (Shimadzu, Kyoto, Japan).
13. HPLC column CAPCELLPAC C18 (35 mm × 4.6 mm; particle size, 5 μm) (Shiseido, Tokyo, Japan).
14. HPLC column Shim-Pack FC-ODS (150 mm × 4.6 mm; particle size, 5 μm) (Shimadzu, Kyoto, Japan).

2.4 Identification of Bacterial Isolates: 16S rRNA Gene Sequences

1. Eppendorf tubes, toothpicks, and deionized water (all sterilized).
2. DNA purification kit (Edge BioSystems, Gaithersburg, MD, USA).
3. Primer B8F: 5′-AGAGTTATCCTGGCTCAG-3′.
4. Primer B1492R: 5′-GGTTACCTTGTTACGACTT-3′.
5. Thermocycler.
6. Ex Taq polymerase (Takara-shuzo, Kyoto, Japan).
7. 10× TBE buffer: 108 g Tris base, 55 g boric acid, and 40 mL 0.5 M EDTA (pH 8.0).
8. 1× TBE buffer: 100 mL of 10× TBE buffer and 900 mL of deionized water.
9. 6× blue loading dye: 30% glycerol in distilled water and 0.25% bromophenol blue.
10. Ethidium bromide (*see* **Note 2**).
11. Agarose gel electrophoresis system.
12. UV safety goggles and UV box.
13. 0.2 mL thin-walled PCR tubes.
14. BigDye Terminator v3.1 Cycle Sequencing Kit (Applied Biosystems, Foster City, CA, USA).
15. Pipettes and tips (0–10, 20–200, 100–1000 μL).
16. Gene Clean kit II (Funakoshi, Tokyo, Japan).

2.5 Morphological, Physiological, and Biochemical Characterization

1. Nutrient agar plates and nutrient broth (5 mL).
2. Incubators (4, 10, 20, 30, 37, 40, and 45 °C).
3. BD™ Gram staining kit (BD Biosciences, Franklin Lakes, NJ, USA).
4. Zeiss Axioskop 2 microscope (Carl Zeiss MicroImaging LLC, Thornwood, NY, USA).
5. Scanning electron microscope VE-8800 (Keyence, Tokyo, Japan).
6. GasPak anaerobic system BBL (BD Biosciences, Franklin Lakes, NJ, USA).
7. Cytochrome oxidase test strips (Fluka Biochemika, Sigma-Aldrich GmbH, Taufkirchen, Germany).
8. API 20E, API 20NE and API 50CH (bioMérieux, Marcy l'Etoile, France).
9. *E. coli* C600 (The Coli Genetic Stock Center CGSC, New Haven, CT, USA).
10. Irradiation at 2.35 kGy (13 Gy min^{-1}, ^{137}Cs source).

11. 0.1 M HCl
12. 0.1 M NaOH.

2.6 Chemotaxonomic Characterization

1. Sherlock Microbial Identification System Version 5.0 (MIDI, Inc., Newark, DE, USA).
2. HPLC automated injector with 96-well plate adaptor Class-VP with PDA detector (Shimadzu, Kyoto, Japan).
3. *Sphingopyxis alaskensis* DSMZ 13593^T (DSMZ culture collection, Braunschweig, Germany).
4. *Sphingomonas asaccharolytica* NBRC 15499^T (NBRC, Chiba, Japan).
5. Thin-layer chromatography (TLC) plates size 20 cm × 10 cm (Merck, Whitehouse Station, NJ, USA).

3 Methods

3.1 Isolation of the Zeaxanthin-Producing Bacteria

1. Collect water samples from the surface water layer (our samples were collected at Misasa radioactive site in Tottori, Japan).
2. Do serial dilutions with sterile water.
3. Spread 0.1 mL aliquots on nutrient agar or LB agar.
4. Incubate at 30 and 40 °C for up to 3 days.
5. Isolate and purify yellow colonies to a single colony. For example, we isolated the colonies of TDMA-5^T and TDMA-16^T based on their ability to form yellow colonies on LB medium.
6. For all characteristics, grow TDMA-5^T cells on LB for 3 days at 30 °C and TDMA-16^T cells on NA at 40 °C.

3.2 Identification of Carotenoids

1. Extract the yellow pigments by agitating the bacterial cells in 200 μL of DMSO on a rotary shaker (100 rpm) at 50 °C in the dark until the cells are bleached (*see* **Note 1**).
2. Add 100 μl of methanol, mix and centrifuge at 5000 × *g* for 5 min.
3. Scan the resultant supernatant for absorbance in the wavelength range (λ = 260–700 nm) at room temperature on a spectrophotometer.
4. Identify each carotenoid on a high-performance liquid chromatography (HPLC) system containing a photodiode array detector using three methods (A, B, and C) (*see* **Note 3**).
5. Perform HPLC method A using a CAPCELLPAC C18 column, and elute with a mixed solvent containing acetonitrile/methanol/tetrahydrofuran (5.8:3.5:0.7), at a flow rate of 1 mL/min (more detailed information in Chap. 9).

6. Perform HPLC method B using a CAPCELLPAC C18 column (35 mm × 4.6 mm; particle size, 5 μm; column temperature, 35 °C), and elute with methanol/water (9:1, vol/vol) at a flow rate of 1.0 mL/min.
7. Perform HPLC method C on a LC/MS system to determine the molecular weight of carotenoids. Apply each 10 μL sample onto a Shim-Pack FC-ODS (150 mm × 4.6 mm, particle size, 5 μm; column temperature, 35 °C), and elute with 95% methanol at a flow rate 0.8 mL/min. Wash the column regularly (*see* **Note 4**).
8. Analyze the HPLC profiles for R_t, PDA, and mass spectra of each peak, and compare with those of the zeaxanthin standard.
9. Identify zeaxanthin by retention times, absorption spectrum, and molecular weight. Use an extract of *Sphingomonas paucimobilis* and *S. echinoides* as a reference for nostoxanthin [50, 51].

3.3 Phylogenetic Characterization of the Zeaxanthin-Producing Bacteria

1. Carefully transfer 50 mg wet cells from bacterial isolates into Eppendorf tubes.
2. Extract genomic DNA using the bacterial genomic DNA purification kit.
3. Check the quality of the extracted genomic DNA by agarose gel electrophoresis (**steps 4–9**).
4. Prepare 1% gel solution by melting 1 g of agarose in 100 mL of 0.5× TBE buffer in a microwave for approximately 2 min. Allow to cool for 5 min, then add 2.5 μL ethidium bromide, and stir to mix (*see* **Note 2**).
5. Cast the gel using the supplied tray and comb. Allow the gel to set for a minimum of 30–40 min at room temperature.
6. Load into separate wells: 10 μL of 1 kb ladder, diluted samples (5 μL sample + 5 μL water + 2 μL 6× loading dye).
7. Run the gel for 30 min at 100 V.
8. Expose the gel to UV light and photograph (wear UV goggles).
9. Confirm DNA quality.
10. Set up PCR reaction mixtures in a final volume of 50 μL: 10 μL 10× PCR-buffer, 1 μL PCR Nucleotide Mix, 50 U Taq DNA polymerase, 1 μL (10 μM) primer B8F, 1 μL (10 μM) primer B1500R, 2 μL DNA extract, and 33 μL deionized and sterilized water.
11. Perform PCR using a protocol that includes an initial denaturation for a period of 5 min at 94 °C; 30 cycles of 94 °C for 1 min, 55 °C for 1 min, and 72 °C for 1 min; 72 °C for 3 min; and incubation at 4 °C until further processing.

12. Purify PCR amplicons using agarose gel electrophoresis (**steps 13–19**).
13. Prepare 2% agarose gel, and let it solidify at room temperature for 30–40 min.
14. Add 10 μL 6× blue loading dye to each PCR product (50 μL).
15. Carefully load all 60 μL into wide gel well.
16. Run the gel for 30 min at 100 V.
17. Wear UV goggles. Remove gel and place on UV box to identify region with PCR products.
18. Cut DNA bands from agarose gel and transfer into sterilized Eppendorf tubes.
19. Purify PCR products from the gel using the Gene Clean kit.
20. Determine the 16S rRNA gene sequences using the BigDye Terminator v3.1 cycle sequencing kit.

3.4 Phylogenetic Analysis

1. Compare the nearly complete 16S rRNA gene (about 1396–1450 bp) sequences of isolates to those available in GenBank/EMBL/DDBJ nucleotide sequence database using the BLASTN program (http://www.ncbi.nlm.nih.gov/BLAST/).
2. Using ClustalW, align the obtained sequences with those of closed relatives.
3. Construct a neighbor-joining phylogenetic tree (Fig. 4).
4. Use NJ plot to draw and edit the obtained tree.

3.5 Morphological, Physiological, and Biochemical Characterization

1. Cultivate TDMA-5^T and TDMA-16^T cells on LB and nutrient agar for 48 h at 40 and 30 °C for 2–3 days.
2. Perform Gram staining according to the method described by Smibert and Krieg [54], and observe the bacterial cells under a microscope.
3. Observe the bacterial cells under scanning electron microscope according to the manufacturer's instructions. The specimen for scanning electron microscopy was prepared using the osmium fixation protocol and critical point drying [55]. For example, scanning electron microscope images for strain TDMA-5 and TDMA-16 are shown in Fig. 5a, b, respectively.
4. Test the anaerobic growth of both strains by incubation of inoculated nutrient agar plates in a GasPak anaerobic system.
5. Perform the following tests as described in each reference: casein hydrolysis [56]; catalase activity, spore formation, and hydrolysis of Tween 80, DNA, and starch [54]; and hydrolysis of chitin, carboxymethyl cellulose (high viscosity; Sigma), and agar [57].

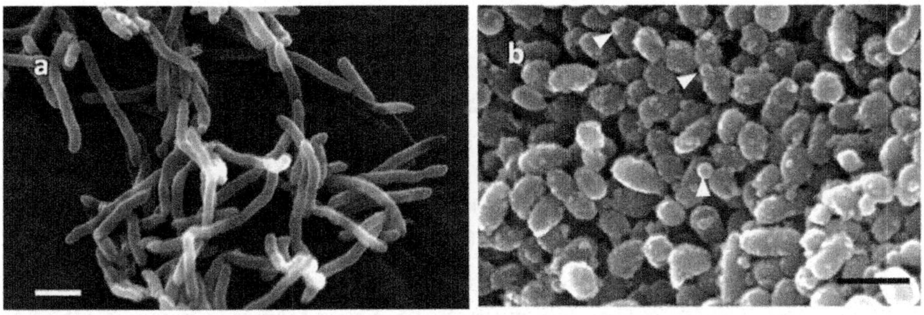

Fig. 5 Scanning electron micrographs of novel strains. (**a**) Strain TDMA-5^T grown on LB agar at 30 °C. Cells were long rods after 3 days cultivation. Bar, 1 μm. Adapted from Asker et al. [45]. (**b**) SEM micrograph of the TDMA-16^T cells. Pleomorphic-shaped morphology and budding formation (indicated by arrow heads) on mature cells were observed. Cells were grown for 4 days at 37 °C on Nutrient agar. Bar, 1.0 μm. Adapted from Asker et al. [48]

6. Test oxidase activity by using commercial cytochrome oxidase test strips, according to its manual.
7. Examine the other enzyme activities, growth on carbohydrates, acid production from carbohydrates, nitrate reduction, and the production of H_2S, indole, and acetoin using the commercial systems API 20E, API 20NE, and API 50CH, according to the manufacturer's instructions.
8. Determine the optimal growth temperature: cultivate the strain on NA at 4, 10, 20, 30, 37, 40, and 45 °C.
9. Test the NaCl tolerance using nutrient broth supplemented with different concentrations of NaCl.
10. Evaluate the growth at various pH values (pH 3.0–10.0) in NB adjusted with HCl or NaOH.
11. Determine the survival rate after exposure to gamma radiation.
12. Grow TDMA-5^T cells and grow TDMA-16^T cells on NA 40 °C:
    (a) Grow cells of TDMA-5^T on LB agar at 30 °C, and TDMA-16^T on NA at 40 °C for 3 days.
    (b) Suspend the grown cells in 5 mL of LB (TDMA-5^T) or NB (TDMA-16^T).
    (c) Prepare aliquots of the cell suspension (1 mL), and irradiate at 2.35 kGy (13 Gy min^{-1}, ^{137}Cs source).
    (d) Dilute serially and plate onto nutrient agar or LB agar plates.
    (e) Aerobically incubate the plates incubated at 30 or 40 °C.
    (f) Similarly treat *E. coli* C600.
    (g) Determine survival percentage by comparing with unirradiated cultures.

3.6 Chemotaxonomic Characterization

1. Determine the cellular fatty acids of strains TDMA-5^T and TDMA-16^T:
   (a) Grow the cells aerobically on Nutrient agar at 40 or 30 °C.
   (b) Extract and analyze the fatty acid methyl esters of strains TDMA-5^T and TDMA-16^T according to the standard protocol of the Sherlock Microbial Identification System (Version 5.0).
2. Determine the respiratory quinone by an HPLC method [58] using cell extracts of *Sphingopyxis alaskensis* DSMZ 13593^T and *Sphingomonas asaccharolytica* NBRC 15499^T as a reference for Q-10.
3. Analyze glycosphingolipids using TLC as described previously [59, 60]. Determine the polar lipids according to [61].
4. Determine the DNA G+C content by an HPLC method [62].

4 Notes

1. Carotenoids are highly unstable and sensitive to light, oxygen, and temperature, which induce their oxidation. To minimize the oxidation of carotenoids, the extraction must be quickly done without exposure to light, oxygen, and high temperatures. Therefore, all the experimental procedures should be carried out in a dark room. Alternatively, dark tubes or tubes foiled with aluminum foil can be used. Nitrogen gas should be also used for flushing the pigment extracts before fridge storage at −20 °C to avoid oxidation. In addition, the use of an antioxidant such as butylhydroxytoluene (BHT) [23] can significantly reduce the oxidation of carotenoids during extraction.
2. Ethidium bromide is mutagenic and moderately toxic. You must wear gloves and handle the chemical with care.
3. For details, see HPLC methods B and C in Chap. 9.
4. Usually, DMSO-MeOH carotenoid extracts contain impurities such as salts, lipids, fatty compounds, proteins, and other cell compounds. These contaminants usually are adsorbed on the stationary phase of reversed-phase HPLC column, which may become contaminated by the repeated injection of samples. This causes shifts in retention times (0.5–1 min) and tailing may occur. In addition, this may cause the column backpressure to build up to intolerably high levels, which overpressures the pump and can cause the column to settle and create a void, depending upon where the blockage occurs. To prevent this problem, we clean the columns on a regular basis throughout our work. Washing the HPLC columns with 100% acetonitrile and then with 100% methanol is sufficient to clean and regenerate the column. We strongly recommend using a pre-column.

References

1. Sommerburg O et al (1998) Fruits and vegetables that are sources for lutein and zeaxanthin: the macular pigment in human eyes. Br J Ophthalmol 82(8):907–910
2. Nelis HJ, Deleenheer AP (1991) Microbial sources of carotenoid-pigments used in foods and feeds. J Appl Bacteriol 70(3):181–191
3. Handelman GJ et al (1999) Antioxidant capacity of oat (*Avena sativa* L.) extracts. 1. Inhibition of low-density lipoprotein oxidation and oxygen radical absorbance capacity. J Agric Food Chem 47(12):4888–4893
4. Humphries JM, Khachik F (2003) Distribution of lutein, zeaxanthin, and related geometrical isomers in fruit, vegetables, wheat, and pasta products. J Agric Food Chem 51(5):1322–1327
5. Hadden WL et al (1999) Carotenoid composition of marigold (*Tagetes erecta*) flower extract used as nutritional supplement. J Agric Food Chem 47(10):4189–4194
6. Perry A, Rasmussen H, Johnson EJ (2009) Xanthophyll (lutein, zeaxanthin) content in fruits, vegetables and corn and egg products. J Food Compos Anal 22(1):9–15
7. Khachik F et al (2002) Transformations of selected carotenoids in plasma, liver, and ocular tissues of humans and in nonprimate animal models. Invest Ophthalmol Vis Sci 43(11):3383–3392
8. Snodderly DM (1995) Evidence for protection against age-related macular degeneration by carotenoids and antioxidant vitamins. Am J Clin Nutr 62(6):S1448–S1461
9. Schweitzer D et al (2010) Suppelementation with lutein and zeaxanthin—a possible protection against age-related macular degeneration. Spektrum Augenheilkd 24(4):242–247
10. Moeller SM et al (2008) Associations between age-related nuclear cataract and lutein and zeaxanthin in the diet and serum in the carotenoids in the age-related eye disease study, an ancillary study of the women's health initiative. Arch Ophthalmol 126(3):354–364
11. Keri L et al (1997) The carotenoids β-carotene, canthaxanthin and zeaxanthin inhibit macrophage-mediated LDL oxidation. FEBS Lett 401:262–266
12. Scripsema NK, Hu D-N, Rosen RB (2015) Lutein, zeaxanthin, and meso-zeaxanthin in the clinical management of eye disease. J Ophthalmol 865179:1–13
13. Lindbergh CA et al (2017) Relationship of lutein and zeaxanthin levels to neurocognitive functioning: an fMRI study of older adults. J Int Neuropsychol Soc 23(1):11–22
14. Xu J et al (2017) Carotenoids and risk of fracture: a meta-analysis of observational studies. Oncotarget 8(2):2391
15. Liu Y et al (2017) Precise regulation of miR-210 is critical for the cellular homeostasis maintenance and transplantation efficacy enhancement of Mesenchymal stem cells in acute liver failure therapy. Cell Transplant 26(5):805–820
16. Chen F et al (2017) Carotenoid intake and risk of non-Hodgkin lymphoma: a systematic review and dose-response meta-analysis of observational studies. Ann Hematol 96(6):957–965
17. Nishino H et al (2002) Carotenoids in cancer chemoprevention. Cancer Metastasis Rev 21(3–4):257–264
18. Nishino H et al (2000) Cancer prevention by natural carotenoids. Biofactors 13(1–4):89–94
19. Asker D, Beppu T, Ueda K (2007) *Zeaxanthinibacter enoshimensis* gen. nov., sp. nov., a novel zeaxanthin-producing marine bacterium of the family *Flavobacteriaceae*, isolated from seawater off Enoshima Island, Japan. Int J Syst Evol Microbiol 57(Pt 4):837–843
20. Asker D, Beppu T, Ueda K (2007) *Mesoflavibacter zeaxanthinifaciens* gen. nov., sp. nov., a novel zeaxanthin-producing marine bacterium of the family *Flavobacteriaceae*. Syst Appl Microbiol 30(4):291–296
21. Berry A et al (2003) *Paracoccus zeaxanthinifaciens* sp. nov., a zeaxanthin-producing bacterium. Int J Syst Evol Microbiol 53(Pt 1):231–238
22. Hoshino T, Ojima K, Setoguchi Y (2004) Production of zeaxanthin by recombinant *Phaffia rhodozyma* strain. PCT Int Appl US 20030923
23. Bhosale P, Bernstein PS (2005) Microbial xanthophylls. Appl Microbiol Biotechnol 68(4):445–455
24. Dufossé L (ed) (2009) Microbial and microalgal carotenoids as colourants and supplements. Birkhäuser Verlag, Basel
25. März U (2011) FOD025C—the global market for carotenoids. BCC Research, Wellesley, MA
26. Bosma TL, Dole JM, Maness NO (2003) Optimizing marigold (*Tagetes erecta* L.) petal and pigment yield. Crop Sci 43(6):2118–2124

27. Stankovic I (2004) Zeaxanthin chemical and technical assessment (CTA). In: FAO (ed) 63rd JECFA
28. Gierhart DL (1994) Production of zeaxanthin and zeaxanthin-containing compositions. Applied Food Biotechnology, Inc. US 5,308,759
29. Chen F et al (2005) Isolation and purification of the bioactive carotenoid zeaxanthin from the microalga *Microcystis aeruginosa* by high-speed counter-current chromatography. J Chromatogr A 1064(2):183–186
30. Liao WL et al (1993) Pigmentation of cultured black tiger prawn by feeding with a *Spirulina* supplemented diet. Nippon Suisan Gakkaishi/ Bull Jpn Soc Sci Fish 59:165–169
31. Murata N et al (1981) Separation and characterization of thylakoid and cell envelope of the blue-green alga (cyanobacterium) *Anacystis nidulans*. Plant Cell Physiol 22(5):855–866
32. Withers NW et al (1978) Photosynthetic unit size, carotenoids, and chlorophyll-protein composition of *Prochloron* sp., a prokaryotic green alga. Proc Natl Acad Sci U S A 75(5):2301–2305
33. Liao HH et al. (1995) Carotenoid producing culture using *Nespongiococcum excentricum*. US Patent 5,437,997
34. Jin ES, Feth B, Melis A (2003) A mutant of the green alga *Dunaliella salina* constitutively accumulates zeaxanthin under all growth conditions. Biotechnol Bioeng 81(1):115–124
35. Andrew Y, Britton G (1990) Photobleaching in the unicellular green alga *Dunaliella parva* 19/9. Photosynth Res 25:129–136
36. Ben-Amotz A, Katz A, Avron M (1982) Accumulation of β-carotene in halotolerant algae: purification and characterization of β-carotene rich globules from *Dunaliella bardawil* (*chlorophyceae*). J Phycol 18:529–537
37. Goodwin TW (1976) Distribution of carotenoids. In: Goodwin TW (ed) Chemistry and biochemistry of plant pigments. Academic Press Inc., New York
38. Palermo JA, Gros EG, Seldes AM (1991) Carotenoids from three red algae of the *Corallinaceae*. Phytochemistry 30(9):2983–2986
39. Bhosale P et al (2004) Identification and characterization of a pi isoform of glutathione S-transferase (GSTP1) as a zeaxanthin-binding protein in the macula of the human eye. J Biol Chem 279(47):49447–49454
40. Lagarde D, Beuf L, Vermaas W (2000) Increased production of zeaxanthin and other pigments by application of genetic engineering techniques to *Synechocystis* sp. strain PCC 6803. Appl Environ Microbiol 66(1):64–72
41. Hundle BS et al (1993) In vitro expression and activity of lycopene cyclase and beta-carotene hydroxylase from *Erwinia herbicola*. FEBS Lett 315(3):329–334
42. Joshi C, Singhal RS (2018) Zeaxanthin production by *Paracoccus zeaxanthinifaciens* ATCC 21588 in a lab-scale bubble column reactor: artificial intelligence modelling for determination of optimal operational parameters and energy requirements. Korean J Chem Eng 35(1):195–203
43. Albrecht M, Misawa N, Sandmann G (1999) Metabolic engineering of the terpenoid biosynthetic pathway of *Escherichia coli* for production of the carotenoids beta-carotene and zeaxanthin. Biotechnol Lett 21(9):791–795
44. Asker D, Beppu T, Ueda K (2007) Unique diversity of carotenoid-producing bacteria isolated from Misasa, a radioactive site in Japan. Appl Microbiol Biotechnol 77(2):383–392
45. Asker D, Beppu T, Ueda K (2008) *Nubsella zeaxanthinifaciens* gen. nov., sp. nov., a zeaxanthin-producing bacterium of the family *Sphingobacteriaceae* isolated from freshwater. Int J Syst Evol Microbiol 58(Pt 3):601–606
46. Kametani K, Matsumura T (1983) Determination of ^{238}U, ^{234}U, ^{226}Ra and ^{228}Ra in spring waters of sanin district. Radioisotopes 32(1):18–21
47. Asker D et al (2012) Isolation, characterization, and diversity of novel radiotolerant carotenoid-producing bacteria. In: Barredo J-L (ed) Microbial carotenoids from bacteria and microalgae: methods and protocols. Humana Press, New York, pp 21–60
48. Asker D, Beppu T, Ueda K (2007) *Sphingomonas jaspsi* sp. nov., a novel carotenoid-producing bacterium isolated from Misasa, Tottori, Japan. Int J Syst Evol Microbiol 57(Pt 7):1435–1441
49. Sajilata MG, Singhal RS, Kamat MY (2008) The carotenoid pigment Zeaxanthin—a review. Compr Rev Food Sci Food Saf 7(1):29–49
50. Jenkins CL et al (1979) The pigment of *Pseudomonas paucimobilis* is a carotenoid (nostoxanthin), rather than a brominated aryl-polyene (xanthomonadin). Curr Microbiol 3(1):1–4
51. Rowe NJ et al (2000) Lipid composition and taxonomy of [*Pseudomonas*] *echinoides*: transfer to the genus *Sphingomonas*. Microbiology 146(Pt 11):3007–3012
52. Saitou N, Nei M (1987) The neighbor-joining method: a new method for reconstructing phylogenetic trees. Mol Biol Evol 4(4):406–425

53. Asker D et al (2012) Novel zeaxanthin-producing bacteria isolated from a radioactive hot spring water. In: Barredo J-L (ed) Microbial carotenoids from bacteria and microalgae: methods and protocols. Humana Press, New York, pp 99–131
54. Smibert RM, Krieg NR (1994) Phenotypic characterization. In: Gerhardt P, Murray RGE, Wood WA, Krieg NR (eds) Methods for general and molecular bacteriology. American Society for Microbiology, Washington, DC, pp 607–654
55. Beveridge TJ, Popkin TJ, Cole RM (1994) Electron microscopy. In P. Gerhardt (ed.), Methods for general molecular bacteriology. American Society for Microbiology, Washington, D.C., pp 42–71
56. Norris JR, Ribbons DW, Varma AK (1985) Methods in microbiology. Academic Press, London
57. Cowan ST, Steel KJ (1993) Manual for the identification of medical bacteria. Cambridge University Press, London
58. Collins MD (1994) Isoprenoid quinones. In: O'Donnell MGAG (ed) Chemical methods in prokaryotic systematics. Wiley, Chichester, pp 265–310
59. Kawahara K et al (1991) Chemical structure of glycosphingolipids isolated from *Sphingomonas paucimobilis*. FEBS Lett 292(1–2):107–110
60. Takeuchi M, Hamana K, Hiraishi A (2001) Proposal of the genus *Sphingomonas sensu stricto* and three new genera, *Sphingobium*, *Novosphingobium* and *Sphingopyxis*, on the basis of phylogenetic and chemotaxonomic analyses. Int J Syst Evol Microbiol 51(Pt 4):1405–1417
61. Tindall BJ (1990) Lipid composition of *Halobacterium lacusprofundi*. FEMS Microbiol Lett 66:199–202
62. Mesbah M, Whitman WB (1989) Measurement of deoxyguanosine/thymidine ratios in complex mixtures by high-performance liquid chromatography for determination of the mole percentage guanine + cytosine of DNA. J Chromatogr 479(2):297–306

Chapter 12

Synthesis of Carotenoids of Industrial Interest in the Photosynthetic Bacterium *Rhodopseudomonas palustris*: Bioengineering and Growth Conditions

Eric Giraud, Laure Hannibal, Clémence Chaintreuil, Joël Fardoux, and André Verméglio

Abstract

Rhodopseudomonas palustris is a purple photosynthetic bacterium that accumulates in the inner membrane the photosynthetic pigment spirilloxanthin, formed from lycopene. Here, we describe the procedures used to successfully engineer *Rps. palustris* strains to reroute the production of lycopene toward the synthesis of ß-carotene or canthaxanthin. The *crtCD* genes specifically involved in spirilloxanthin were replaced by *crtY* and *crtW* genes from *Bradyrhizobium* ORS278 to synthesize ß-carotene and (or) canthaxanthin, two pigments of industrial interest. Since the synthesis of canthaxanthin depends on the presence of oxygen, the procedure to optimize their production is also proposed. By modulating the light and oxygen during the growth process, a single species of photosynthetic bacteria, with an efficient growth rate, produces various carotenoids of economical interest.

Key words Canthaxanthin, Spirilloxanthin, ß-Carotene, Lycopene, Carotenoid, *crt* genes, Photosynthetic *Bradyrhizobium*, *Rhodopseudomonas palustris*, Light regulation, Oxygen regulation, Genetic engineering

1 Introduction

Carotenoids are a large class of pigments widely distributed in living organisms. They are present in all photosynthetic organisms, from bacteria to plants, where they play three essential roles: accessory light harvesting pigments, stabilization of the light-harvesting complexes, photoprotection by direct quenching of both triplet states of (bacterio)chlorophylls and singlet oxygen [1].

The carotenoid synthesis requires the functioning of up to ten enzymes. Depending upon the considered photosynthetic organism, the biosynthetic pathway diverges after the formation of lycopene, leading to a variety of carotenoids (Fig. 1). Carotenoids play also important vital functions in animals. Since, at the exception of aphids [2], animals are unable to synthetize carotenoids, they

Carlos Barreiro and José-Luis Barredo (eds.), *Microbial Carotenoids: Methods and Protocols*, Methods in Molecular Biology, vol. 1852, https://doi.org/10.1007/978-1-4939-8742-9_12, © Springer Science+Business Media, LLC, part of Springer Nature 2018

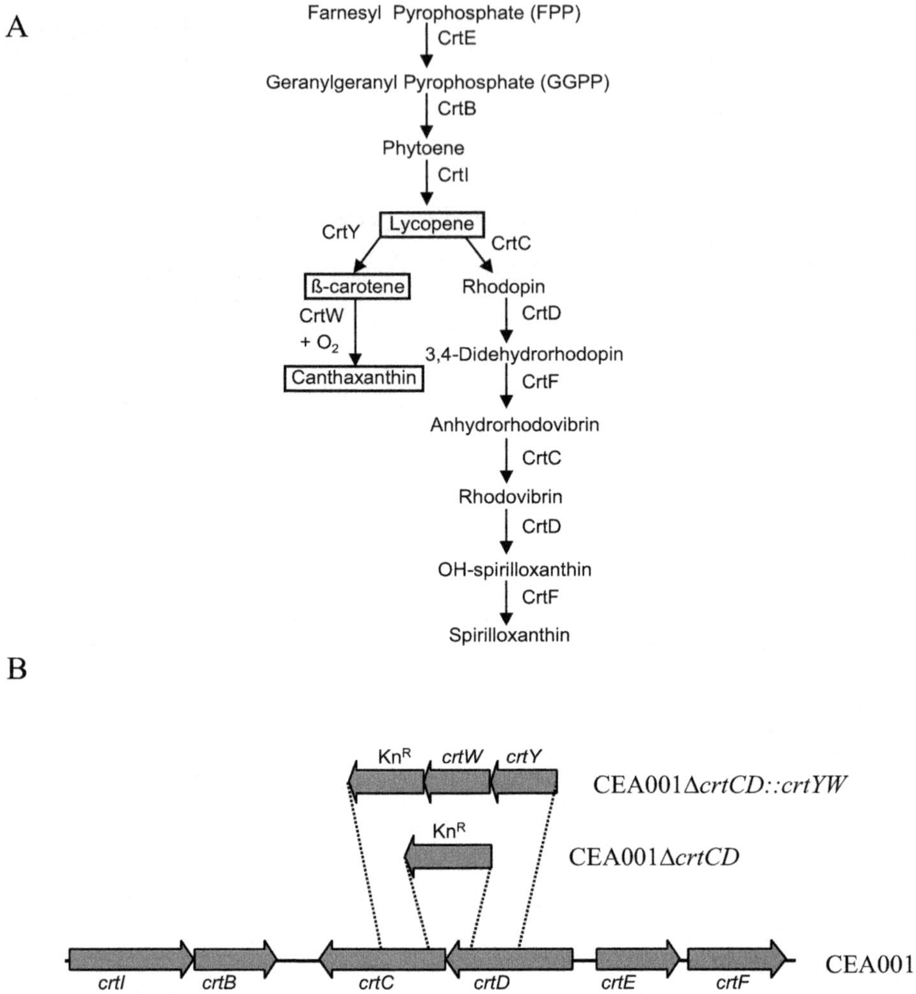

Fig. 1 General strategy to produce carotenoids of interests in ***Rhodopseudomonas palustris*** CEA001 strain. (**a**) Biosynthesis pathway of carotenoids considered in this paper. (**b**) Organization of the carotenoid gene cluster in the wild-type strain ***Rps. palustris*** and in its derived mutants

obtain these pigments via their diets. For example, ß-carotene is converted into retinal, which is a key molecule involved in vision and carotenoids like lycopene act as antioxidants due to their ability to quench singlet oxygen [3]. Therefore, carotenoids have been utilized for several decades in the food-processing, pharmaceutical and cosmetic industries [4]. The antioxidant and coloring properties of ß-carotene and lycopene are used in sunscreen and tanning products. Canthaxanthin is used as coloring agent in salmon, trout, or poultry farming [5]. Today the synthesis of these carotenoids is essentially carried out chemically. A potential market for carotenoids of natural origin, plant or microbial, does however exist. There are nevertheless several hindrances for an optimal natural production of carotenoids. In the case of bacteria, carotenoids producing species accumulate only small amounts of products and have very often a limited growth rate. To overcome these difficulties, one possibility is to use a photosynthetic bacterium as a producer. Indeed these bacteria synthesize various types of carotenoids, present an efficient growth rate, and develop large amount of internal membrane, where the hydrophobic carotenoids can be stored. Among them, *Rhodopseudomonas (Rps.) palustris* presents several advantages. This bacterium grows anaerobically in the light and aerobically in the dark, degrades a variety of aromatic compounds, and presents a strong resistance to toxic metals [6, 7]. The formation of the photosynthetic apparatus, and consequently the synthesis of the photosynthetic carotenoid (spirilloxanthin), is regulated by both light quality and oxygen concentration [8]. Based on our present knowledge of this regulation and the genetic of this bacterium, we describe in the present article a method to produce several carotenoids of commercial interest in *Rps. palustris.* The general strategy is to divert the spirilloxanthin endogenous synthesis pathway to the production of lycopene, β-carotene, or canthaxanthin by deleting some specific spirilloxanthin genes and introduce in place other exogenous carotenoid (*crt*) genes.

2 Materials

2.1 Engineering of *Rps. palustris* to Produce Carotenoids of Industrial Interest

1. *Rps. palustris* strain CEA001 [9].
2. *Bradyrhizobium* sp. strain ORS278 (LMG P-12187) [10].
3. crtD.f primer: 5′-GAGCTCGAGTTCGCCGGCATCGGCCTGAACTTCTC-3′.
4. crtC.r primer: 5′-CTGCTCGAGAGGAGTATTACGGACTGATCGAAC-3′.
5. crtY.f primer: 5′-TGAGATCTGGAGGCTGTCGTCATGAGTCGAGATGCCGA-CGTCATCGTC-3′.
6. crtY.r primer: 5′-GTTGAATTCCTGGTACCTCATGGGGTCTTGAAGGCGC-TCGCCTCA-3′.
7. crtW.f primer: 5′-CGGTACCGGGAGCTTTGCCAATGCATGCAGCAACCG-CCAAGGCTAC-3′.
8. crtW.r primer: 5′-GTGAATTCCATGCTCGAGCGGGTTTAGTCACGCCTTTC-CAG-3′.
9. puc4K (GE Healthcare).
10. pJQ200mp18 [11].
11. pGEM-T vector (Promega, Madison, WI, USA).
12. *E. coli* S17-1 λpir (Biomedal, Sevilla, Spain).
13. Luria-Bertani (LB) broth: 10 g/L tryptone, 5 g/L yeast extract and 10 g/L NaCl.

14. LB agar: LB and 16 g/L agar.
15. Metals 44: Dissolve 250 mg of EDTA, 1.1 mg of $ZnSO_4 \cdot 7H_2O$, 0.5 mg of $FeSO_4 \cdot 7H_2O$, 154 mg of $MnSO_4 \cdot H_2O$, 39 mg of $CuSO_4 \cdot 5H_2O$, 18 mg of $NaBO_4 \cdot H_2O$, 25 mg of $Co(NO_3)_2 \cdot 6H_2O$ and 0.15 mg of H_2SO_4 (6 N) in 100 mL distilled water.
16. Concentrated base: Dissolve 5 g of nitriloacetic acid in 250 mL distilled water. Add KOH at 10% to reach pH 5. Then add one by one 14.5 g $MgSO_4 \cdot 7H_2O$, 1.25 g $CaCl_2$, 4.6 mg ammonium molybdate, 50 mg $FeSO_4$ and 25 mL of metals 44. Add about 12.5 mL of KOH 10% to obtain a final pH of 6.2. Complete to 500 mL with distilled water.
17. 1 M Phosphate buffer: Dissolve separately 114 g of K_2HPO_4 in 500 mL distilled water and 68 g KH_2PO_4 in 400 mL distilled water. Add the desired volume (about 250 mL) of the KH_2PO_4 solution into the K_2HPO_4 solution to obtain pH 7. Complete with distilled water to 1 L.
18. 0.5 M Ammonium malate: Dissolve 67 g of malic acid in 200 mL distilled water. Add 70 mL of concentrated ammonia water and adjust to pH 6.7 by using a solution of 10% ammonia water. Complete to 500 mL with distilled water.
19. Glutamate acetate solution: Dissolve 60 g of monosodium glutamate and 50 g sodium acetate anhydrous in 500 mL distilled water.
20. Growth factors: Dissolve 2 mg of biotin, 50 mg of $NaHCO_3$, 100 mg of nicotinic acid, 50 mg of thiamine hydrochloride and 100 mg of *para*-aminobenzoic acid in 100 mL distilled water.
21. Hutner medium [12]: Add to 1 L of distilled water 20 mL of concentrated base, 20 mL of 1 M phosphate buffer, 20 mL of 0.5 M ammonium malate, 20 mL of glutamate acetate, 2 mL of growth factor and 1 g yeast extract.
22. Hutner agar medium: Add 16 g agar per 1 L Hutner medium.
23. Kanamycin stock solution (50 mg/mL).
24. Carbenicillin stock solution (50 mg/mL).

2.2 Light and Oxygen Control of Carotenoids Biosynthesis

1. Hutner medium [12].
2. Airtight bottle of 100 mL.
3. Erlenmeyer flask of 250 mL with baffles.
4. Vacuum pump.
5. Dinitrogen gas.
6. 100 W incandescent bulb.
7. Acetone/methanol: 7 mL acetone and 2 mL methanol.

8. Hypersil C18 column: 5 μm, 250 by 4.6 mm (Phenomenex, Torrance, CA).
9. Acetonitrile/methanol/isopropanol: 40 mL acetonitrile, 50 mL methanol, and 10 mL isopropanol.

3 Methods

3.1 Engineering of Rps. palustris to Produce Carotenoids of Industrial Interest

The photosynthetic *Rps. palustris* CEA001 strain contains six *crt* genes that are involved in spirilloxanthin synthesis: the *crtE*, *crtB*, and *crtI* genes permit the production of lycopene from farnesyl pyrophosphate, while the *crtC*, *crtD*, and *crtF* genes are involved downstream in the biosynthesis pathway and permit the conversion of lycopene into spirilloxanthin (Fig. 1a). These six genes are found inside the photosynthetic gene cluster and are organized in three operons as described in Fig. 1b. Most of the carotenoids of industrial interest are derived from lycopene. The synthesis of ß-carotene needs one enzyme, CrtY, which catalyzes cyclization of lycopene into ß-carotene; canthaxanthin requires one additional enzyme CrtW to oxygenate ß-carotene (*see* Fig. 1a). To overproduce carotenoids of interest in *Rps. palustris* CEA1001 strain, two engineered strains were constructed: (1) a mutant (named CEA001Δ*crtCD*) accumulating lycopene which was obtained by deleting the endogenous *crtC* and *crtD* genes (*see* **Note 1**) and (2) a mutant (named CEA001Δ*crtCD::crtYW*) accumulating ß-carotene and/or canthaxanthin—this mutant was obtained by introducing the *crtY* and *crtW* genes from *Bradyrhizobium* ORS278 strain into the *crtC* and *crtD* genes (*see* **Note 2**).

The following protocol describes the different strategies used to construct these two mutant strains.

1. Amplify by polymerase chain reaction (PCR) the DNA region (2.7-kb) encompassing the *crtC* and *crtD* genes of *Rps. palustris* CEA001 using the primers crtD.f and crtC.r (*see* **Note 3**). Use a touchdown PCR program as follows: initial denaturation at 94 °C for 5 min followed by 20 cycles consisting of a 30 s denaturation at 94 °C, 30 s annealing temperature from 60 to 50 °C, and 2 min primer extension at 72 °C, followed by 15 cycles consisting of a 30 s denaturation at 94 °C, 30 s annealing temperature at 50 °C, and 2 min primer extension at 72 °C.
2. Digest the PCR product by *Xho*I, and clone it into the *Sal*I site of pJQ200mp18 suicide vector [11] (*see* **Note 4**). The resulting plasmid is named pJQ200mp18::*crtCD* (*see* **Note 5**).
3. For the construction of the mutant CEA001Δ*crtCD*, delete a 0.6-kb *Sal*I fragment of pJQ200mp18::*crtCD* corresponding

to the 5′-end of *crtD* and to the 3′-end of *crtC* and insert in place the 1.2-kb *Sal*I Kn^r cartridge of puc4K. This leads to the plasmid pJQ200mp18::*crtCD*::Kn^r (*see* **Note 6**).

4. For the construction of the mutant CEA001Δ*crtCD::crtYW*, amplify firstly the *crtY* and *crtW* genes of ORS278 strain by using the couple of primers crtY.f/crtY.r and crtW.f/crtW.r, respectively (*see* **Note 7**). After cloning the PCR products independently into the pGEM-T vector, insert the 1.2-b *Sal*I Kn^r cartridge of puc4K into the *Xho*I site of pGEMT::*crtW* and subsequently liberate the tandem *crtW*/Kn^r by *Eco*RI/*Kpn*I digestion and clone it into pGEMT::*crtY* after linearization by the same enzymes. From the resulting plasmid (pGEMT::*crtYW*::KnR), liberate the *crtY*/*crtW*/Kn^r genes by *Bgl*II/*Eco*RI digestion and insert them into the plasmid pJQ200mp18::*crtCD* previously obtained in **step 2** and digested by the same enzymes. This leads to the plasmid pJQ200mp18::*crtCD::crtYW*::Kn^r (*see* **Note 8**).
5. Transfer by electroporation the two derived pJQ200mp18 plasmids obtained in **steps 3** and **4** into *E. coli* S17-1 λpir to permit their mobilization by bipartite conjugation in the CEA1000 strain.
6. Grow the *E. coli* S17-1 λpir strains containing the various derived pJQ200mp18 plasmids overnight at 37 °C in LB medium supplemented with kanamycin (50 μg/mL). Grow *Rps. palustris* CEA001 strain for 3 days at 34 °C in Hutner medium [12].
7. For each construction, harvest 2.5 mL of *E. coli* S17-1 and 10 mL of *Rps. palustris* cells by centrifugation (3200 × *g*, 30 min, 4 °C).
8. Wash the pellets twice with NaCl 9 g/L.
9. Resuspend each pellet in 200 μL of NaCl 9 g/L.
10. Gently mix the two pellets and transfer on a Hutner plate. Incubate 2 days at 34 °C.
11. Resuspend the mating in 1 mL of Hutner medium, make serial dilutions until 10^{-4} into NaCl 9 g/L, and spread 100 μL of each dilution on Hutner plates containing kanamycin (50 μg/mL) and carbenicillin (50 μg/mL) (*see* **Note 9**).
12. Incubate plates at 34 °C during 1 week.
13. Colonies are then screened by PCR to check that the first event of recombination occurred.
14. One recombinant (or positive strain) is grown for 3 days at 34 °C in Hutner medium supplemented with kanamycin (50 μg/mL) and carbenicillin (50 μg/mL).

15. Make serial dilutions until 10^{-4} into NaCl 9 g/L, and spread 100 μL of each dilution onto Hutner plates containing sucrose (7%), kanamycin (50 μg/mL) and carbenicillin (50 μg/mL) (*see* **Note 10**).
16. Incubate plates at 34 °C during 1 week.
17. Colonies are then screened by PCR to check that the second event of recombination occurred.

3.2 Light and Oxygen Control of Carotenoids Biosynthesis

The carotenoid synthesis is modulated by both light and oxygen tension in *Rps. palustris* [8]. At oxygen tension higher than 8%, the synthesis of both the photosynthetic system and the carotenoids is inhibited. This inhibition is however suppressed by near infrared light via the action of a bacteriophytochrome [13]. While the synthesis of lycopene and ß-carotene occurs in the light under anaerobic conditions, the enzyme responsible for the synthesis of canthaxanthin, CrtW (Fig. 1a), requires the presence of O_2. Addition of oxygen during growth under illumination has however two important effects. The cell density is decreased by a factor of 2 from anaerobic to aerobic conditions (21% O_2) and the amount of total carotenoids by a factor of 30. To overcome this problem, it is essential to develop different growth conditions, described below, to optimize the production of a given carotenoid.

3.2.1 Production of Lycopene or ß-Carotene

1. Prepare a fresh culture of the mutant CEA001Δ*crtCD* for lycopene production or the mutant CEA001Δ*crtCD::crtYW* for ß-carotene production (*see* **Note 11**).
2. Place the bacteria in an airtight bottle of 100 mL filled with degassed Hutner medium (*see* **Note 12**).
3. Grow the culture in the light at 30 °C for 72 h (*see* **Note 13**).
4. Harvest the cells by centrifugation (3200 × *g* for 30 min at 4 °C) and resuspend the pellets with 1 mL of cold acetone/methanol to extract carotenoids.
5. Place the samples in a rotary agitator and incubate for 30 min at 4 °C in the dark.
6. Centrifuge (3200 × *g* for 30 min at 4 °C), keep the supernatant and repeat twice the extraction procedure on the pellets.
7. Pool the supernatants and analyze the amount of the different carotenoids produced by HPLC using the following conditions: Hypersil C18 column, acetonitrile/methanol/isopropanol as eluent, and a flow rate of 0.8 mL/min to determine the amount of the different carotenoids produced (*see* **Note 14**).

3.2.2 Production of Canthaxanthin

1. Prepare a fresh culture of mutant CEA001Δ*crtCD::crtYW*.
2. Place the bacteria in an airtight bottle of 100 mL filled with degassed Hutner medium (*see* **Note 12**).

3. Grow the culture in the light at 30 °C for 48 h (*see* **Note 13**).
4. Transfer 50 mL of the cells in a 250 mL Erlenmeyer with baffles shaked at 170 rpm in the dark for 1 day (*see* **Note 15**).
5. Harvest the cells by centrifugation (3200 × *g* for 30 min at 4 °C), and resuspend the pellets with 1 mL of cold acetone/methanol to extract carotenoids.
6. Place the samples in a rotary agitator and incubate for 30 min at 4 °C in the dark.
7. Centrifuge (3200 × *g* for 30 min at 4 °C), keep the supernatant and repeat twice the extraction procedure on the pellets.
8. Pool the supernatants and analyze the carotenoids content by HPLC to determine the amount of the different carotenoids produced (*see* **Note 16**).

4 Notes

1. As shown in Fig. 1b, the deletion of the *crtCD* genes cannot impact on the expression of *crtEB* and *crtI* genes necessary for the synthesis of lycopene because these genes are transcribed in the opposite direction.
2. We used preferentially the *crtY* and *crtW* genes originating from the photosynthetic *Bradyrhizobium* ORS278 strain because this strain is phylogenetically close to *Rps. palustris*, and therefore the usage codon of these genes should be well adapted to *Rps. palustris*. However, other origin of the *crtY* and *crtW* genes could be considered.
3. The general strategy for the construction of these two mutants is designed so that the flanking regions of the genes of interest to insert in the recipient strain correspond to the 5′-end of *crtD* and 3′-end of *crtC*. These flanking regions have a size of at least 600-bp in order to promote the double crossing-over event.
4. The use of the pJQ200mp18 suicide plasmid, which does not replicate in CEA001 strain, permits to integrate the DNA region containing the deleted *crtC* and *crtD* genes and the genes of interest into the chromosomal *crtC* and *crtD* genes due to the similarity between one of the flanking regions of the insert and the genomic DNA.
5. The plasmid pJQ200::*crtCD* is used for the construction of the two mutants.
6. The kanamycin resistance cartridge is used to facilitate the selection of the single and double recombinants in the following steps.

7. The primers were designed in order to introduce a ribosome binding site upstream the start codon of *crtY* and *crtW* genes. In addition, they contain appropriate restriction sites to permit the introduction of the Kn^r cartridge and their cloning in tandem. It is to highlight that after introduction in *Rps. palustris* CEA001 strain, the expression of the *crtY* and *crtW* genes is under the control of the *crtD* promoter.
8. The *EcoRI*/*BglII* digestion of pJQ200::*crtCD* led to a deletion of a 1200 bp fragment corresponding to the 5′-end of *crtD* and to the 3′-end of *crtC*. However it remains a part of the *crtC* and *crtD* genes flanking the *crtY*/*crtW*/*Kn^r* genes to permit the subsequent double crossing over in CEA001 strain (Fig. 1b).
9. The *Rps. palustris* CEA001 strain is resistant to carbenicillin. Its addition in the medium permits to inhibit the growth of *E. coli* cells. The addition of kanamycin permits in return to select the *Rps. palustris* recombinant cells in which the pJQ200mp18 vector has been integrated in the chromosome.
10. Clones in which double events of recombination occurred are selected by plating on sucrose-containing medium. Cells retaining the vector backbone are lost since the *sacB* gene on the pJQ200mp18 backbone converts sucrose to toxic levan.
11. In anaerobic conditions, despite the presence of the *crtW* gene, the CEA001Δ*crtCD::crtYW* mutant produces mainly a cocktail of lycopene and ß-carotene but not canthaxanthin because the CrtW enzyme is not active in the absence of oxygen.
12. Degassed Hutner medium could be prepared either by bubbling pure N_2 or placing the medium under vacuum setting to evacuate O_2.
13. Illumination is provided by a 100 W incandescent bulb.
14. In such condition, the CEA001Δ*crtCD* mutant produced essentially lycopene as carotenoid at a concentration of 7.2 mg/L, while the CEA001Δ*crtCD::crtYW* mutant produces a mixture of lycopene (2.7 mg/L) and ß-carotene (4.7 mg/L).
15. This additional aerobic phase permits the conversion of a part of the ß-carotene into canthaxanthin by the CrtW enzyme that is functional in such conditions.
16. In such condition, the CEA001Δ*crtCD::crtYW* mutant produces a mixture of lycopene (2.4 mg/L), ß-carotene (4.7 mg/L) and canthaxanthin (0.8 mg/L).

References

1. Takaichi S (1999) Carotenoids and carotenogenesis in anoxygenic photosynthetic bacteria. In: Frank HA, Young AJ, Britton G, Cogdell RJ (eds) The photochemistry of carotenoids. Kluwer Academic Publishers, Dordrecht
2. Moran NA, Jarkiv T (2010) Lateral transfer of genes from fungi underlines carotenoid production in aphids. Science 328:624–627
3. Di Mascio P, Kaiser S, Sies H (1989) Lycopene as the most efficient biological carotenoid singlet oxygen quencher. Arch Biochem Biophys 274:532–538
4. Kläui H (1982) Industrial and commercial uses of carotenoids. In: Britton G, Goodwin TW (eds) Carotenoid chemistry and biochemistry. Pergamon Press, Inc., Oxford
5. Simpson KL, Katayama T, Chichester CO (1981) Carotenoids in fish feeds. In: Bauernfeind JC (ed) Carotenoids as colorants and vitamin A precursors. Academic Press, Inc., New York
6. Harwood CS, Gibson J (1988) Anaerobic and aerobic metabolism of diverse aromatic compounds by the photosynthetic bacterium *Rhodopseudomonas palustris*. Appl Environ Microbiol 54:712–717
7. Larimer FW, Chain P, Hauser L et al (2004) Complete genome sequence of the metabolically versatile photosynthetic bacterium *Rhodopseudomonas palustris*. Nat Biotechnol 22:55–61
8. Giraud E, Fardoux J, Fourrier N et al (2002) Phytochrome controls the photosystem synthesis in anoxygenic bacteria. Nature 417:202–205
9. Giraud E, Zappa S, Jaubert M et al (2004) Bacteriophytochrome and regulation of the synthesis of the photosynthetic apparatus in *Rhodopseudomonas palustris*: pitfalls of using laboratory strains. Photochem Photobiol Sci 3:587–591
10. Lorquin J, Molouba F, Dreyfus BL (1997) Identification of the carotenoid canthaxanthin from photosynthetic *Bradyrhizobium* strains. Appl Environ Microbiol 63:1151–1154
11. Quandt J, Hynes MF (1993) Versatile suicide vectors which allow direct selection for gene replacement in gram-negative bacteria. Gene 127:15–21
12. Clayton RK (1963) Towards the isolation of a photochemical reaction center in *Rhodopseudomonas sphaeroides*. Biochim Biophys Acta 75(312–318):1963
13. Kojadinovic M, Laugraud A, Vuillet L et al (2008) Dual role for a bacteriophytochrome in the bioenergetic control of *Rhodopsdeudomonas palustris*: enhancement of photosystem synthesis and limitation of respiration. Biochim Biophys Acta 1777:163–172

Chapter 13

Molecular Tools for Carotenogenesis Analysis in the Mucoral *Mucor circinelloides*

Francisco E. Nicolás, María Isabel Navarro-Mendoza, Carlos Pérez-Arques, Sergio López-García, Eusebio Navarro, Santiago Torres-Martínez, and Victoriano Garre

Abstract

The carotene producer *Mucor circinelloides* is the fungus within the Mucoromycota phylum with the widest repertoire of molecular tools to manipulate its genome. The initial development of an effective procedure for genetic transformation and later improvements have resulted in an expansion of available tools, which include gene replacement, inactivation of gene expression by RNA silencing, gene overexpression, and functional genomics. Moreover, sequencing of its genome has given a definitive boost to these techniques making attainable the study of genes involved in many physiological or developmental processes, including carotenoid biosynthesis. Here, we describe in detail the latest molecular techniques currently used in *M. circinelloides* that have made it a valuable model for studying gene function within its phylum.

Key words Transformation, Gene replacement, Gene overexpression, Gene silencing, RNAi, Functional genomics

1 Introduction

Early divergent lineages of fungi, also termed basal fungi, are receiving a growing attention since they can contribute to understand fungal evolution [1]. Moreover, these fungal groups are evolutionary distant to the so-called higher fungi, *Ascomycetes* and *Basidiomycetes*, their physiology, development, metabolism, and ecology being less characterized [2]. *M. circinelloides*, informally referred to herein as *Mucor*, belongs to Mucorales, the largest order of the early divergent phylum Mucoromycota [1]. *Mucor* is an emblematic species that has contributed importantly to a better knowledge of the biology of this phylum because of the development of several molecular tools to manipulate its genome, including genetic transformation [3, 4], gene replacement [5], RNA-mediated gene silencing [5], and overexpression of genes [6]. The potential

Carlos Barreiro and José-Luis Barredo (eds.), *Microbial Carotenoids: Methods and Protocols*, Methods in Molecular Biology, vol. 1852, https://doi.org/10.1007/978-1-4939-8742-9_13,

of these procedures has been amplified by the availability of *Mucor* genome sequence [7]. Main aspects of *Mucor* biology that are currently receiving special attention and have been beneficiated by the existing molecular tools include responses to light [8, 9], sex determination and its evolution [10, 11], gene silencing or RNA interference (RNAi) [12], regulation of gene expression by endogenous small RNAs (esRNAs) [13], pathogenesis [14], lipid accumulation [15], and biosynthesis of carotenoids [9, 16, 17].

Mucorales are characterized for a fast growth and short lifespan which facilitates successful exploitation for the production of lipids and carotenoids. In fact, *Mucor* was the first microorganism industrially used for the production of oil [18], while the mucoral *Blakeslea trispora* is the microorganism of choice for the production of carotenes on the industrial scale [19]. Most of the information about the biosynthesis of carotenoids in Mucorales derived from *B. trispora* but also from the model *Phycomyces blakesleeanus* [20]. However, both fungi exhibit the same the problem, the lack of molecular genetics tools to manipulate their genomes. This is in part compensated by the availability of such tools in Mucor, which also have contributed to understand the regulation of carotenoid biosynthesis in Mucorales [8, 9, 17]. Moreover, manipulation of genes involved in regulation of carotenoid biosynthesis and/or structural genes has produced strains that overaccumulate lycopene or β-carotene [16, 17, 21].

Molecular techniques that are of routine use in *Mucor* were gathered in a chapter of a previous book of this series [5], which should be consulted by researchers working with *Mucor*. Here, we complement it by updating some procedures and introducing novel techniques, such as gene overexpression and functional genomic screening.

2 Materials

1. *M. circinelloides* CBS277.49 (CBS-KNAW Fungal Biodiversity Center, Utrecht, The Netherlands).
2. *M. circinelloides* R7B [22] (*see* **Note 1**).
3. *M. circinelloides* MU402 [23] (*see* **Note 2**).
4. SPB (0.1 M sodium phosphate buffer pH 6.5): 1.42 g of Na_2HPO_4 in a final volume of 100 mL double-distilled water (solution 1) and 1.38 g of NaH_2PO_4 monohydrate in a final volume of 100 mL double-distilled water (solution 2). Take 100 mL of solution 2 and add solution 1 slowly until pH 6.5 is reached (about 53 mL).
5. PS buffer: Mix 18.22 g of sorbitol and 20 mL of SPB. Adjust to 200 mL with double-distilled water (final sorbitol concentration: 0.5 M).

6. Lysing enzymes (L-1412, Sigma-Aldrich, St. Louis, MO, USA).
7. Chitosanase (C-0794, Sigma-Aldrich, St. Louis, MO, USA).
8. 0.5 M Sorbitol: Dissolve 22.77 g of sorbitol in 200 mL of double-distilled water. Adjust to 250 mL with double-distilled water.
9. Gene Pulser Cuvette (0.2-cm electrode gap) (Bio-Rad, Hercules, CA, USA).
10. Bio-Rad Gene Pulser Xcell (Bio-Rad, Hercules, CA, USA).
11. YPG: 3 g/L of yeast extract, 10 g/L of peptone, and 20 g/L of glucose (*see* **Note 3**).
12. YPGS: Add 91.1 g of sorbitol to the YPG medium before adjusting the volume to 1000 mL with distilled water (final sorbitol concentration: 0.5 M) (*see* **Note 3**).
13. YPG-agar and YPGS-agar: YPG or YPGS with 15 g/L agar (*see* **Note 3**).
14. YNB: 1.5 g/L of ammonium sulfate, 1.5 g/L of glutamic acid, 0.5 g/L of yeast nitrogen base (w/o ammonium sulfate and amino acids), and 10 g/L of glucose. After autoclaving add thiamine and niacin at a final concentration of 1 μg/mL (*see* **Note 3**).
15. YNBS: Add 91.1 g/L of sorbitol to the YNB medium before adjusting the volume to 1000 mL with distilled water (final sorbitol concentration: 0.5 M) (*see* **Note 3**).
16. YNB-agar and YNBS-agar: YNB or YNBS with 20 g/L agar (*see* **Note 3**).
17. MMC: 10 g/L of casaminoacids, 0.5 g/L of yeast nitrogen base (w/o ammonium sulfate and amino acids), and 20 g/L of glucose. After autoclaving add thiamine and niacin at a final concentration of 1 μg/mL (*see* **Note 3**).
18. MMCS: Add 91.1 g/L of sorbitol to the MMC medium before adjusting the volume to 1000 mL with distilled water (final sorbitol concentration: 0.5 M) (*see* **Note 3**).
19. MMC-agar and MMCS-agar: MMC or MMCS with 15 g/L agar (*see* **Note 3**).
20. GeneJET Gel Extraction Kit (ThermoFisher Scientific).
21. DMSO (dimethyl sulfoxide).
22. Herculase II Fusion DNA Polymerase (Agilent Technologies, Santa Clara, CA, USA).
23. *carRP*-F1: 5′-CCGGGCGCATTGTAGATAAACT-3′.
24. *carRP*-R1: 5′-GGGCATGTGTAACAGTGCATTGG-3′.
25. *pyrG*fwd: 5′-TGCCTCAGCATTGGTACTTG-3′.

26. *pyrG*rev: 5′-GTACACTGGCCATGCTATCG-3′.
27. *pyrG*ext: 5′-ATCCCACCAGAAGGAGTACATGG-3′.
28. *peuka-1*: 5′-CATGAAGTGTGAGACATTGCG-3′.
29. *pzrt1*: 5′-GGAGATGTGCCTTGATGATATGCTCT-3′.
30. pMAT1476 (*see* **Note 4**).
31. pMAT1477 (*see* **Note 5**).
32. pMAT1552 (*see* **Note 6**).
33. pMAT1702 (*see* **Note 7**).
34. Mixture for overlapping PCR: Mix 1 μL of Herculase II Fusion DNA Polymerase, 10 μL of 5× Herculase II reaction buffer, 0.5 μL of a 100 mM solution of dNTPs (25 mM each), 2 μL of a 10 μM solution of each primer, 100 ng of each upstream and downstream fragments, and 200 ng of *pyrG* marker fragment, and adjust to 50 μL of double-distilled water.
35. Lysis buffer: To prepare 50 mL of buffer, dissolve 23.6 g of guanidine thiocyanate (118.16 g/L) in 25 mL of double-distilled water. Once dissolved add 2.5 mL of 1 M Tris–HCl pH 7.0 (121.14 g/L), 2 mL of 0.5 M EDTA pH 8.0 (186.12 g/L), and 0.05 mL of Triton X-100. Add double-distilled water to 50 mL (final concentrations, 4 M guanidine thiocyanate, 50 mM Tris–HCl pH 7.0, 20 mM EDTA, and 0.1% Triton X-100).
36. DNA binding solution: First, wash diatomaceous earth (silica particles) by resuspending 5 g in 50 mL of water and allowing them to settle. Decant the supernatant after 30 s and repeat this step twice keeping the supernatant. Centrifuge at 1400 × *g* for 1 min, discard the supernatant and resuspend in 25 mL of acetone. Centrifuge at 1400 × *g* for 1 min, discard the supernatant, and dry the silica particles at 37 °C for 24 h. To prepare 50 mL of buffer, dissolve 23.6 g of guanidine thiocyanate in 25 mL of double-distilled water. Once dissolved add 2.5 mL of 1 M Tris–HCl pH 7.0, 2 mL of 0.5 M EDTA pH 8.0, and 1 g of diatomaceous earth previously filtrated. Add double-distilled water to 50 mL (final concentrations, 4 M guanidine thiocyanate, 50 mM Tris–HCl pH 7.0, 20 mM EDTA, and 20 g/L diatomaceous earth).
37. Wash buffer: To prepare 50 mL of buffer, mix 2.5 mL of 4 M NaCl, 1 mL of 0.5 M EDTA pH 8.0, 2.5 mL of 1 M Tris–HCl pH 7.4, and 25 mL of absolute ethanol. Add double-distilled water to 50 mL (final concentrations, 200 mM NaCl, 10 mM EDTA, 50 mM Tris–HCl pH 7.4, and 50% ethanol).
38. Elution buffer: Add 0.5 mL of Tris–HCl pH 8.5 to 45 mL of double-distilled water (final concentration, 10 mM Tris–HCl pH 8.5).

3 Methods

3.1 Transformation

Genetic transformation of *Mucor* is based on the use of auxotrophies since this fungus is naturally resistant to antibiotics and antifungals commonly used in *Ascomycetes* and *Basidiomycetes*. Three different auxotrophies are used for genetic transformation of *Mucor*, methionine [24], leucine [3], and uridine [25]. However, the most widely used strains are auxotrophs for leucine and uridine that are complemented with *leuA1* and *pyrG* genes, respectively. All plasmids that complement auxotrophies in *Mucor* are self-replicative, making necessary the use of linear fragments in transformation to favor chromosomal integration. The procedure described below works for both self-replicative plasmids and linear integrative DNA fragments.

1. Collect fresh spores from mycelium grown in YPGS pH 4.5 for 4–6 days at 26 °C in the light. Release spores from the sporangia by pouring 10 mL/plate of sterile distilled water on the mycelium and spreading it with a Drigalski spatula. Take the water with the spores by using a pipette and put in a conical tube.
2. Wash the spore solution by centrifugation for 5 min at 440 × *g*. Resuspend the spores in 10 mL of sterile distilled water and repeat the centrifugation. Resuspend the pellet in 5 mL of sterile distilled water and store the spore solution at 4 °C. Count the spores using a hemocytometer and light microscope.
3. Take an aliquot of spores containing 2.5×10^8 (no more than 1 week old) and centrifuge for 5 min at 440 × *g*. Resuspend the pellet in 25 mL of YPG medium pH 4.5 (final spore concentration to 10^7 spores/mL), supplemented with 200 μg/mL of uridine when a uridine auxotroph is going to be transformed.
4. Incubate overnight at 4 °C, without shaking.
5. Incubate the spores at 26 °C with shaking (300 rpm) until most of them are germinated having a germ tube length shorter than the swollen spore diameter (Fig. 1). This usually takes 3–4 h.
6. Wash the cells twice by centrifugation in 5 mL of PS buffer pH 6.5 at 91 × *g* for 5 min.
7. Resuspend the pellet in 4 mL of PS buffer. Transfer the germinated spore solution to a 50-mL Erlenmeyer flask.
8. Add 5 mg of lysing enzymes dissolved in 1 mL PS buffer and 0.3 μL (15 U/mg) of chitosanase previously diluted in 50 μL PS buffer. Incubate at 30 °C with gentle shaking (60 rpm) for about 90 min (*see* **Note 8**).

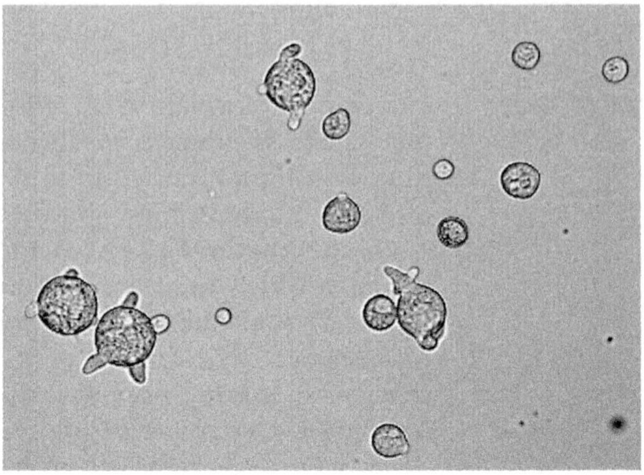

Fig. 1 Spore preparation at the moment to be treated with lysing enzymes and chitosanase to produce protoplasts

9. Transfer the 5 mL solution to a screw cap centrifuge tube and fill the tube with cold 0.5 M sorbitol. Wash twice by centrifugation in cold 0.5 M sorbitol at 91 × *g* for 5 min.
10. Resuspend the pellet gently in 800 μL of cold 0.5 M sorbitol. This 800 μL solution allows four or eight different transformation experiments depending on the type of DNA used (see next step).
11. Each tube of transformation mixture must contain 100 μL protoplast solution and 10 μL DNA sample (1–3 μg total DNA in double-distilled water) for transformation experiments with circular plasmid or 200 μL protoplast solution and 20 μL DNA sample (3–6 μg total DNA) for linear fragments. Include a negative control with either 10 or 20 μL of double-distilled water instead of DNA, respectively.
12. Mix and transfer to the electroporation cuvette.
13. Apply an electrical pulse using the following conditions: field strength of 4 kV/cm, capacitance of 25 μF, and constant resistance of 400 Ω.
14. Immediately after the pulse, remove the cuvette and add 1 mL cold YPGS 4.5. Keep on ice until all cuvettes have been pulsed.
15. Transfer the liquid of each cuvette to 2.0-mL microcentrifuge tubes.
16. Incubate for 1 h at 26 °C and 100 rpm.
17. Centrifuge at 91 × *g* for 3 min, and gently resuspend the pellet in a final volume of 400–600 μL YNBS 4.5.
18. Inoculate plates of the adequate medium containing 0.5 M sorbitol with 200 μL protoplast solution (*see* **Note 9**).
19. Incubate in the dark at 26 °C for 3–4 days to obtain the transformants (*see* **Note 10**).

3.2 Gene Replacement

Generation of null mutants by gene replacement requires the designing of replacement fragment (RF) that disrupts the target gene after homologous recombination (HR). This RF must contain a selectable marker gene flanked by upstream and downstream DNA sequences (approximately 1 kb long each) of the target gene (Fig. 2). The construction of the RF based on *pyrG* marker, which is the preferred marker for disruption, and selection of transformants are described below. The same procedure can be applied for RFs based on *leuA1* gene (*see* **Note 11**).

1. Set up independent PCR reactions to amplify 1-kb sequence of both upstream and downstream flanking regions of the target gene using genomic DNA (gDNA) as template. Upstream sequence is amplified using primers Ufwd and Urev*pyrG*, whereas downstream sequence is amplified using primers Dfwd*pyrG* and Drev (*see* **Note 12**).
2. Amplify the marker gene and its promoter sequence by PCR from either gDNA or any appropriate vector containing the *pyrG* gene, such as pMAT1702 [14], using primers *pyrG*fwd and *pyrG*rev.
3. Check correct amplification of all fragments by agarose gel electrophoresis. Purify the PCR fragments from agarose slabs using a commercial kit, such as GeneJET Gel Extraction Kit.
4. PCR fragments from the upstream region, selectable marker, and downstream region are mixed in 1:1:1 molar proportion and join together by overlapping PCR using primers Ufwd and Drev (Fig. 2) (*see* **Note 13**).
5. Check correct amplification of the RF by agarose gel electrophoresis and purify it as described in **step 4**.
6. The RF thus obtained is used to transform *Mucor* using the protocol described in Subheading 3.1 (*see* **Note 14**).
7. Grow initial transformants, which are usually heterokaryons, for several vegetative cycles in selective media to select for homokaryons, which are expected to harbor the mutation in all their nuclei (*see* **Note 10**).
8. Correct disruption of the gene has to be confirmed by PCR and/or Southern hybridization (*see* Subheading 3.4).

3.3 Stable Gene Overexpression

Mucor strains that overexpress a particular gene can be generated by genetic transformation using a self-replicative plasmid containing the coding region of the gene under either its native or a heterologous strong promoter [26]. However, these strains must be maintained under selective pressure to reduce plasmid loss. Thus, integration of a copy of the gene in *M. circinelloides* genome is recommended to provide a stable overexpression. This can be achieved by generating an RF containing both a selectable marker

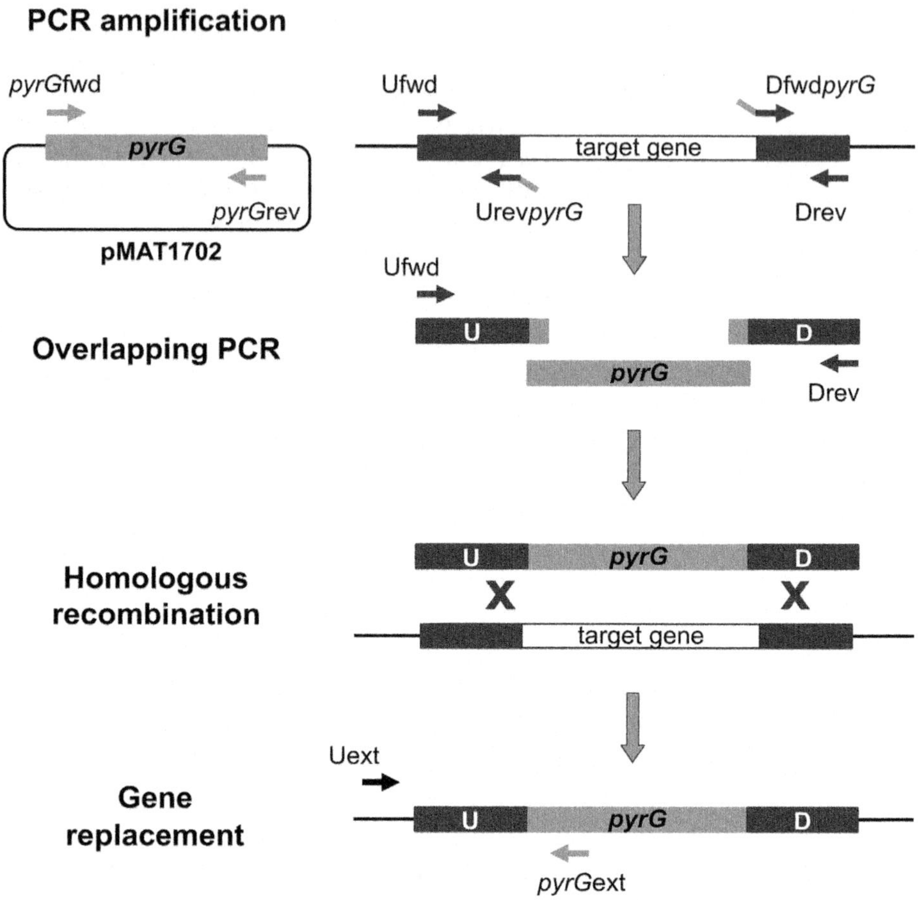

Fig. 2 Schematic representation of a replacement fragment obtained by overlapping PCR. Arrows represent primers for PCR amplification (*see* **Note 12**), U and D are upstream and downstream flanking regions of the target gene, and *pyrG* is the selectable marker. HR is checked by PCR amplification with primer Uext and *pyrG*ext. HR sites are marked with X

and a copy of the gene under the control of an appropriate high-expressing promoter, flanked by upstream and downstream sequences designed to target the whole construction to a particular genomic locus. Current *Mucor* vectors (*see* **Notes 4–6**) for overexpressing genes are designed to disrupt *carRP* locus because its disruption blocks colored-carotenoid production, rendering albino colonies that allow a fast screening for transformants with the construction in the *carRP* locus. In case of studies concerning carotenoid biosynthesis, new vectors that target a locus unrelated to carotenogenesis should be developed. The construction of an overexpression construction based in pMAT1477 (Fig. 3) and selection of transformants are as follows:

1. Set up a PCR reaction to amplify the coding region of the gene to be overexpressed using appropriate primers with added restriction sites for *Xho*I and/or *Sac*II at their 5′ ends to facilitate cloning.

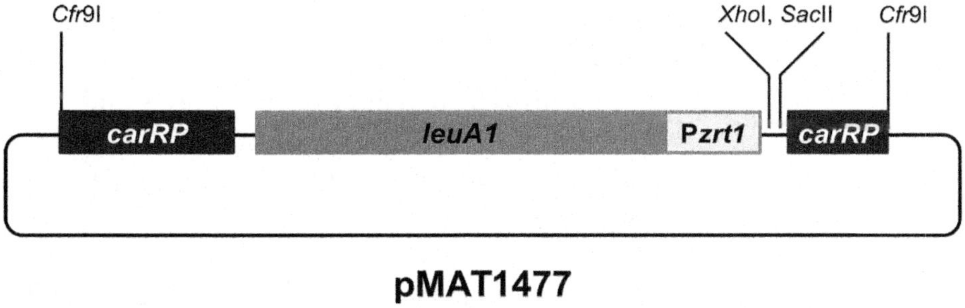

Fig. 3 Schematic representation of the overexpressing vector pMAT1477. *carRP* boxes represent flanking regions for integration in the *carRP* locus, *leuA1* is the selectable marker, and P*zrt1* is the promoter of gene *zrt1*. Restriction sites for designated enzymes are shown

2. The PCR product is digested with *Xho*I and/or *Sac*II and cloned in pMAT1477 previously cut with the same enzymes.
3. Transform *Escherichia coli* and select clones with recombinant plasmids.
4. Recombinant plasmids are cut with the restriction enzyme *Cfr*9I to release the whole construction, which is used to transform *Mucor* protoplasts following the procedure described in Subheading 3.1 (*see* **Note 15**).
5. Pick initial transformants, which are usually heterokaryons, in selective media pH 4.5 to favor the generation of albino patches, which reveal integration in *carRP* locus.
6. Harvest spores from the albino patches by scratching them using a micropipette tip or small laboratory spatula.
7. Dip the tip or spatula in 0.5 mL of sterile distilled water and shake it to release the spores.
8. Plate the spores in selective media pH 3.2 and incubate for 2–3 days at 26 °C. Collect spores from white colonies for subsequent plating. After repeating this process from 1 to 3 times, colonies that produce only albino descendants are obtained, indicating that they are homokaryons for the integration of the overexpressing construction.
9. Confirm the correct integration of the whole construction in the *carRP* locus by PCR and/or Southern hybridization (*see* Subheading 3.4).

3.4 Rapid Genomic DNA Purification and Analysis of Transformant DNA

Identification of knockout mutants with successful gene replacement or strains with an integrated overexpressing construction requires confirmation by PCR and/or Southern hybridization. These analyses need a straightforward and quick procedure to purify genomic DNA from selected transformants. The method described below, which is based on the silica binding method developed by Carter and Milton [27], works with high reliability in *Mucor*.

1. Weigh about 100 mg of mycelium and freeze it in liquid nitrogen.
2. Grind the frozen mycelium with a mortar and pestle until a fine powder is obtained.
3. Transfer the powder to a 2-mL microcentrifuge tube, avoiding tissue thawing.
4. Immediately add 500 μL of Lysis buffer. Vortex until the mycelium is completely resuspended, and centrifuge at 13,000 × *g* for 5 min.
5. Transfer the supernatant to a new 2-mL tube without disturbing the cell-debris pellet.
6. Add 1 mL of DNA binding solution. Ensure that the binding buffer is mixed completely before proceeding because silica particles precipitate quickly.
7. Mix the tube content by inversion during 5 min at room temperature to allow DNA binding to silica particles. Centrifuge for 1 min at 13,000 × *g*. Discard the supernatant with a micropipette.
8. Resuspend the pellet in 700 μL of washing buffer. Centrifuge for 1 min at 13,000 × *g*. Discard the supernatant with a micropipette. Repeat this step.
9. Add 700 μL of acetone. Centrifuge for 1 min at 13,000 × *g*. Discard the supernatant with a micropipette.
10. Repeat the centrifugation to remove any residual acetone from the pellet.
11. Dry the pellet by putting the tube open in an oven for 5 min at 37 °C.
12. Add 50 μL of elution buffer and incubate at 55 °C for 5 min. Centrifuge 1 min at 13,000 × *g*.
13. Collect the supernatant containing the genomic DNA. Transfer it to a new 2 mL tube and store the purified DNA at −20 °C. DNA concentration should be between 20 and 50 ng/μL (*see* **Note 16**).

Once genomic DNA is isolated from transformants and wild-type strains, gene replacement can be detected by PCR using discriminatory primers. For instance, a PCR reaction can be set up using a reverse primer of the selectable marker (*pyrG*ext) and forward primer (Uext) that anneals upstream to the flanking region used to generate the replacement fragment (Fig. 2); thus assuring that PCR amplification can only occur if gene replacement has taken place by HR. It is recommended to apply the following rapid PCR amplification procedure:

1. Set up PCR amplification reactions with 8% of DMSO in a final volume of 25 μL using 1 μL of gDNA and primers pyrGext and Uext. Use Herculase II Fusion DNA Polymerase and follow supplier's recommendations.
2. Run a PCR program with an initial denaturation step at 95 °C for 5 min followed by 30 cycles, each one consisting of 95 °C for 30 s, 55–60 °C (depending on primer length and sequence) for 30 s, and 72 °C for 30 s/kb. A final step of 5 min a 72 °C is included to finish all DNA chains.
3. Analyze the PCR results by agarose gel electrophoresis. Transformants showing positive amplification should be analyzed by Southern hybridization to confirm proper integration.

3.5 Silencing Vectors and Functional Genomic Screenings

RNAi is an alternative procedure to knock out genes in order to study gene function, particularly useful in the case of essential genes. The silencing vectors used in *Mucor* so far are self-replicative and harbor two main elements: a selectable marker (*leuA1* or *pyrG*) and a gene construction that triggers silencing. The triggering construction (transgene) can contain either inverted repeated sequences of the gene to be silenced cloned under a strong promoter or a gene fragment under the control of inverted strong promoters. The first type of plasmids is ideal to silence a specific gene because it transcribes hairpin RNA structures that produce the highest silencing frequencies (nearly 95% of the colonies) [28]. The second type is particularly useful when a large number of genes have to be analyzed, as in functional genomic screenings, because the construction of a hairpin-producing vector for each gene is not feasible. Detailed procedures to construct the first type of silencing vectors and analyze small RNAs were previously reported [5]. This section describes the procedure to generate a genomic library using the second type of silencing vectors, in particular vector pMAT1702 (*see* **Note 7**), and its use to perform functional genomic screenings in *Mucor* to look for genes involved in a particular process (Fig. 4). This strategy has been used to identify virulence determinants in mucormycosis [14], an opportunistic fungal infection of humans caused by Mucorales.

1. Digest vector pMAT1702 with *Xho*I and fill in with dCTP-dTTP to avoid self-ligation [29].
2. Partially digest *Mucor* gDNA with *Sau*3A and electrophorese it in an agarose gel.
3. Purify fragments of 0.5–4 kb from the agarose gel using a commercial kit such as GeneJET Gel Extraction Kit.
4. Fill in the purified fragments with dGTP-dATP to avoid self-ligation [29].

5. After ligation of *Mucor* DNA fragments and vector, transform *E. coli* to obtain a number of transformants higher than 73,680, which ensures that the probability that any particular fragment is found in the library is higher than 99% (*see* **Note 17**).
6. Collect all *E. coli* transformants using Luria-Bertani (LB) medium to create pools with around 10,000 colonies.
7. Purify plasmid DNA from each pool (*see* **Note 18**).
8. Plasmid DNA from each pool is used to transform *Mucor* (*see* Subheading 3.1) to obtain a number of transformants above of the confidence level of 99%. Phenotypes can be screened at this point, or spores from each pool can be harvested to perform later screenings.

3.6 Identification of the Genes Responsible for Phenotypes Selected in Functional Genomic Screenings

The self-replicative nature of pMAT1702 facilitates the identification of the sequences responsible for the phenotypes selected in screenings of silenced transformants. The procedure is as follows:

1. Purify gDNA from the selected transformant following the procedure described in Subheading 3.4.
2. Set PCR reactions using Herculase II Fusion DNA Polymerase amplification and primers *peuka-1* and *pzrt1* to amplify the DNA fragments cloned in pMAT1702 (Fig. 4a) (*see* **Note 19**).
3. Sequence the PCR fragments using primers *peuka-1* and *pzrt1*, and confront their sequences with the genome sequence of *Mucor* (http://genome.jgi.doe.gov/Mucci2/Mucci2.home.htmL) to identify the gene silenced in each transformant (*see* **Note 20**).

4 Notes

1. *M. circinelloides* R7B, the standard strain used as wild type, is a leucine auxotroph derived from *M. circinelloides* CBS277.49 [22].
2. *M. circinelloides* MU402, the standard strain used for transformation, is a leucine and uridine auxotroph derived from R7B strain [23].
3. For regular growth, adjust pH to 4.5 all described media with 1 M HCl or 1 N NaOH before autoclaving. When controlled colony growth is required (for viability counts, in transformation plates, etc.), the media are adjusted to pH 3.2 with 1 M HCl before autoclaving. YPG and YNB are, respectively, the complete and minimal media for *M. circinelloides*. MMC is a rich medium that allows the growth of leucine auxotrophs, but not uridine auxotroph, being frequently used in transformations selecting *pyrG* gene. To avoid hydrolysis of agar, double-

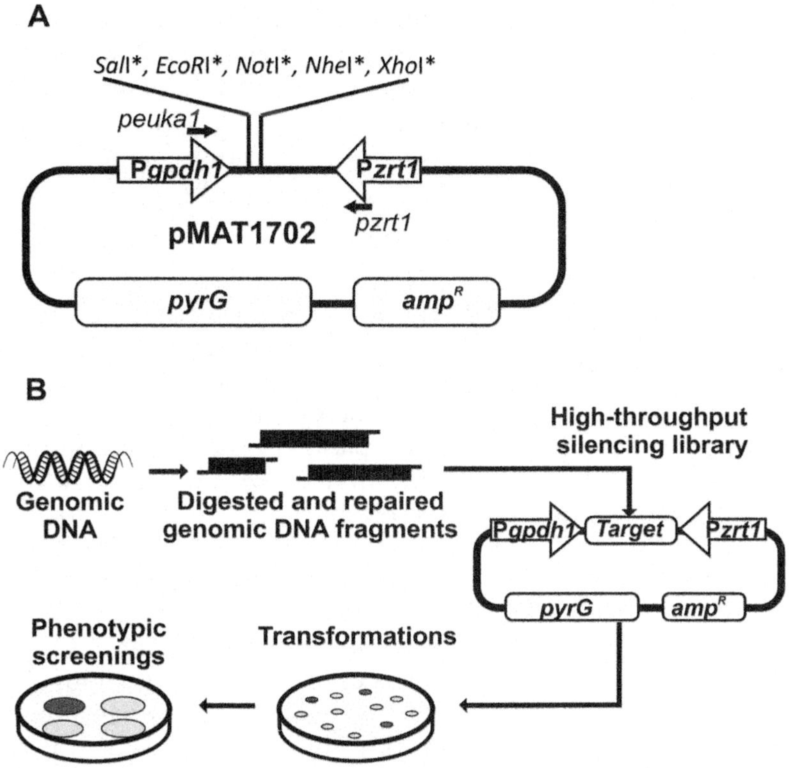

Fig. 4 Schematic diagram of the RNA silencing vector pMAT1702 (**a**) and the platform for functional genomics in *Mucor* (**b**). P*zrt1*, *Mucor zrt1* promoter; P*gpd1*, *Mucor gpd1* promoter; *pyrG*, the selectable marker; *amp*^{R}, ampicillin-resistant gene. Asterisks indicate unique restriction sites in the vector

strength solutions of agar and the other media components are autoclaved separately and mixed after cooling at 50 °C. When it is required, uridine or leucine is supplemented to the corresponding media at a final concentration of 200 and 20 μg/mL, respectively.

4. pMAT1477 (Fig. 3) contains *leuA1* as a selectable marker, followed by promoter of *Mucor zrt1* gene without a translation start codon and a cloning site for *Xho*I and *Sac*II restriction enzymes. This whole construction is flanked by upstream and downstream sequences of the *carRP* locus, to favor HR after transformation [6].
5. pMAT1476 comprises all pMAT1477 elements except for *zrt1* promoter and is designed to clone the target gene with its native promoter in *Nhe*I, *Xho*I, or *Sac*II sites [6].
6. pMAT1552 derives from pMAT1477 by replacing the *leuA* marker by *pyrG* gene. As a consequence of this replacement, this plasmid lacks *Sac*II as cloning site.

7. pMAT1702 is a high-throughput silencing vector that harbors a multiple cloning site (MCS) flanked by two convergent strong promoters (P*gpd1* and P*zrt1*) of *Mucor* (Fig. 4a).
8. Cell wall digestion and appearance of protoplasts are monitored with a light microscope (protoplasts are identified because cell wall refringence is lost). This usually takes about 90 min.
9. Usually, each transformation mixture is spread on three plates.
10. The self-replicative nature of the plasmids used for *Mucor* transformation causes that only a proportion of the spores produced by initial transformants contain plasmid. Moreover, these transformants have to be grown in selective medium to avoid plasmid loss. In the case of integrative transformation with linear DNA fragments, the presence of several nuclei in the protoplast results in heterokaryotic transformants, which have to be grown for several vegetative cycles in selective media to increase the proportion of transformed nuclei. Since the two main selectable markers used in *Mucor* are genes *leuA1* and *pyrG*, the proportion of the Leu+ or PyrG+ spores is used as an indicator of DNA integration. When PyrG+ transformants must be selected, mainly when the recipient strain is a double mutant *pyrG*$^-$ *leu*$^-$ (MU402), the selective medium has to be MMC, because otherwise the mycelial growth is very poor and spore production scarce. When Leu+ transformants are selected, the selective medium has to be YNB. Once transformants have grown, spores from each transformant are plated on minimal medium YNB pH 3.2 with and without leucine or MMC with and without uridine to determine the proportion of Leu$^+$ or PyrG$^+$ spores, respectively. For each cycle, spores from each transformant are harvested from colonies grown under selective pressure, that is, without leucine or uridine. Integration is denoted by the gradual increase in the proportion of Leu+ or PyrG+ spores throughout the cell cycles. Homokaryotic transformants showing 100% stable Leu$^+$ or Ura$^+$ spore proportion are usually obtained after 2–4 vegetative cycles.
11. RFs based on *leuA1* marker are constructed using the 4.4-kb fragment derived from plasmid pLEU4 [30], which contains the *leuA1* wild-type allele with its own promoter.
12. Gene-specific primers (Ufwd, Drev, Uext, Urev*pyrG*, and Dfwd*pyrG*) should have 18–25 nucleotides, 40–60% GC content, melting temperature of approximately 65 °C, and 1–3 G/C bases at the 3′ end. Primers Urev*pyrG* (5′-CAAGTACCAATGCTGAGGCA[N]-3′) and Dfwd*pyrG* (5′-CGATAGCATGGCCAGTGTAC[N]-3′) include *pyrG* sequences and 18–25 nucleotides of the target sequence in the 3′ end (Fig. 2).

13. The program for overlapping PCR (Table 1) has the feature of including intervals above and below annealing temperature that have to course at a slow pace of approximately 0.3 °C/s. After the 20th cycle, elongation time can be increased by 10 s per cycle to counteract the possible loss of polymerase activity.
14. RF can be further reamplified by PCR with primers Drev and Ufwd (Fig. 2) to obtain enough linear DNA for genetic transformation. Alternatively, the RF can be cloned in a standard cloning vector, such as pUC series vectors, and released in linear form by specific enzymatic restriction.
15. Alternatively, the whole construction can be amplified by PCR using primers *carRP*-F1 and *carRP*-R1, though it is not recommended due to the high probability that the polymerase introduces mutations in the gene.
16. Before using DNA, centrifuge briefly the DNA to avoid that remaining silica particles interfere with further reactions.
17. Confidence level is calculated according to the formula: $N = \ln(1 - P)/\ln(1 - f)$, where "$N$" is the necessary number of recombinants, "P" is the desired probability that any fragment of the genome is represented in the library at least one time, and "f" is the fractional proportion of the genome in a single recombinant. "f" can be further shown to be $f = i/g$, where "i" is the insert size and "g" is the genome size [31]. Thus, in the case of *Mucor*, which has a genome size of

Table 1
Overlapping PCR program

Step	Temperature	Duration
1. Denaturation	95 °C	3 min
2. Denaturation	95 °C	30 s
3. Drop	70 °C	1 s
3–4 Interval	Temperature rate: 0.3 °C/s	
4. Annealing	60 °C	30 s
4–5 Interval	Temperature rate: 0.3 °C/s	
5. Elongation	72 °C	30 s/kb (add 10 s per cycle after the 20th cycle)
Repeat steps 2–5 for 30 cycles		
6. Elongation	72 °C	5 min

3.6×10^7 bp, for fragments of 2.25 kb average, the number of transformants has to be higher than 73,680 for a $P = 0.99$.

18. Plasmid DNA aliquots of each pool can be stored at −20 °C.
19. An alternative procedure to identify the DNA cloned in pMAT1702 consists in transforming *E. coli* with gDNA purified from the selected *Mucor* transformants. Then, the plasmid is isolated from *E. coli* transformants and sequenced with primers *peuka-1* and *pzrt1*.
20. It is possible that more than one *Mucor* gene is found in a recombinant pMAT1702. In these cases, each gene must be amplified separately and cloned again in pMAT1702. The resulting plasmids are individually transformed into *M. circinelloides*, and the phenotype of silenced transformants will be analyzed to identify the gene responsible for the phenotype observed in initial transformants [14].

References

1. Spatafora JW et al (2016) A phylum-level phylogenetic classification of zygomycete fungi based on genome-scale data. Mycologia 108(5):1028–1046
2. Morin-Sardin S et al (2017) *Mucor*: a Janus-faced fungal genus with human health impact and industrial applications. Fungal Biol Rev 31(1):12–32
3. van Heeswijck R, Roncero M (1984) High frequency transformation of *Mucor* with recombinant plasmid DNA. Carlsb Res Commun 49:11
4. Gutiérrez A, López-García S, Garre V (2011) High reliability transformation of the basal fungus *Mucor circinelloides* by electroporation. J Microbiol Methods 84(3):442–446
5. Torres-Martínez S et al (2012) Molecular tools for carotenogenesis analysis in the zygomycete *Mucor circinelloides*. Methods Mol Biol 898:85–107
6. Rodríguez-Frómeta RA et al (2013) Malic enzyme activity is not the only bottleneck for lipid accumulation in the oleaginous fungus *Mucor circinelloides*. Appl Microbiol Biotechnol 97(7):3063–3072
7. Corrochano LM et al (2016) Expansion of signal transduction pathways in fungi by extensive genome duplication. Curr Biol 26(12):1577–1584
8. Silva F, Torres-Martínez S, Garre V (2006) Distinct *white collar-1* genes control specific light responses in *Mucor circinelloides*. Mol Microbiol 61(4):1023–1037
9. Silva F et al (2008) A RING-finger protein regulates carotenogenesis via proteolysis-independent ubiquitylation of a white collar-1-like activator. Mol Microbiol 70(4):1026–1036
10. Idnurm A et al (2008) Identification of the sex genes in an early diverged fungus. Nature 451(7175):193–196
11. Lee SC et al (2008) Microsporidia evolved from ancestral sexual fungi. Curr Biol 18(21):1675–1679
12. Garre V et al (2014) The RNAi machinery in Mucorales: the emerging role of endogenous small RNAs. In: Sesma A, von der Haar T (eds) Fungal RNA biology. Springer International Publishing, Cham, pp 291–314
13. Torres-Martínez S, Ruiz-Vázquez RM (2016) RNAi pathways in *Mucor*: a tale of proteins, small RNAs and functional diversity. Fungal Genet Biol 90:44–52
14. Trieu TA et al (2017) RNAi-based functional genomics identifies new virulence determinants in mucormycosis. PLoS Pathog 13(1):e1006150
15. Zhao L et al (2016) Role of malate transporter in lipid accumulation of oleaginous fungus *Mucor circinelloides*. Appl Microbiol Biotechnol 100(3):1297–1305
16. Navarro E et al (2001) A negative regulator of light-inducible carotenogenesis in *Mucor circinelloides*. Mol Gen Genomics 266(3):463–470
17. Zhang Y et al (2016) A new regulatory mechanism controlling carotenogenesis in the fungus *Mucor circinelloides* as a target to generate beta-carotene over-producing strains by genetic engineering. Microb Cell Factories 15:99

18. Ratledge C (2004) Fatty acid biosynthesis in microorganisms being used for single cell oil production. Biochimie 86(11):807–815
19. Roukas T (2016) The role of oxidative stress on carotene production by *Blakeslea trispora* in submerged fermentation. Crit Rev Biotechnol 36(3):424–433
20. Cerdá-Olmedo E (2001) *Phycomyces* and the biology of light and color. FEMS Microbiol Rev 25(5):503–512
21. Zhang Y et al (2017) Generation of lycopene-overproducing strains of the fungus *Mucor circinelloides* reveals important aspects of lycopene formation and accumulation. Biotechnol Lett 39(3):439–446
22. Roncero MI (1984) Enrichment method for the isolation of auxotrophic mutants of Mucor using the polyene antibiotic N-glycosyl-polifungin. Carlsb Res Commun 49(7):685–690
23. Nicolás FE et al (2007) Mutants defective in a *Mucor circinelloides dicer*-like gene are not compromised in siRNA silencing but display developmental defects. Fungal Genet Biol 44(6):504–516
24. Anaya N, Roncero MI (1991) Transformation of a methionine auxotrophic mutant of *Mucor circinelloides* by direct cloning of the corresponding wild type gene. Mol Gen Genet 230(3):449–455
25. Benito EP et al (1992) Cloning and sequence analysis of the *Mucor circinelloides pyrG* gene encoding orotidine-5′-monophosphate decarboxylase: use of *pyrG* for homologous transformation. Gene 116(1):59–67
26. Wolff AM, Arnau J (2002) Cloning of glyceraldehyde-3-phosphate dehydrogenase-encoding genes in *Mucor circinelloides* (Syn. *racemosus*) and use of the *gpd1* promoter for recombinant protein production. Fungal Genet Biol 35(1):21–29
27. Carter MJ, Milton ID (1993) An inexpensive and simple method for DNA purifications on silica particles. Nucleic Acids Res 21(4):1044
28. Calo S et al (2012) Two distinct RNA-dependent RNA polymerases are required for initiation and amplification of RNA silencing in the basal fungus *Mucor circinelloides*. Mol Microbiol 83(2):379–394
29. Werner E, Patel K, Holder AA (1997) Construction of a library for sequencing long regions of malarial genomic DNA. Biotechniques 23(1):20, 22, 24
30. Roncero MI et al (1989) Characterization of a *leuA* gene and an ARS element from *Mucor circinelloides*. Gene 84(2):335–343
31. Clarke L, Carbon J (1976) A colony bank containing synthetic col El hybrid plasmids representative of the entire *E. coli* genome. Cell 9(1):91–99

Chapter 14

Expression Vectors and Gene Fusions for the Directed Modification of the Carotenoid Biosynthesis Pathway in *Mucor circinelloides*

Enrique A. Iturriaga, María Isabel Alvarez, Arturo P. Eslava, and Tamás Papp

Abstract

Several fungal species, particularly some included in the *Mucoromycotina*, have been used to develop fermentation processes for the production of β-carotene. Oxygenated derivatives of β-carotene (xanthophylls) are desirable value-added products, and the preference by the market of carotenoids from biological sources has increased the research in different carotenoid-producing organisms. We currently use *Mucor circinelloides f. lusitanicus* as a model organism to develop strains with an increased content of new and more valuable carotenoids. The main carotenoid accumulated by *M. circinelloides* is β-carotene, although it has some hydroxylase activity and produces low amounts of zeaxanthin. On the other hand, in astaxanthin-producing organisms two enzymatic activities are required for the production of astaxanthin from β-carotene: a hydroxylase and a ketolase. In this chapter, we delineate part of our efforts to construct genetically modified strains that could advance in the improvement of carotenoid accumulation by this fungus and the diversification of its carotenoid content. Accordingly, we describe detailed and empirically tested protocols for the construction of functional expression vectors and gene fusions.

Key words Astaxanthin, Xanthophylls, *Mucor*, Expression vectors, Metabolic engineering, Gene fusions

1 Introduction

Carotenoids comprise a group of naturally occurring pigments with a broad range of biological functions such as a strong antioxidant activity [1], an enhancement of the immune response [2], or even they could be effective in preventing several types of cancer [3]. Consequently, its application is emerging in the pharmaceutical and personal care industries. Many of them are also used in the food and feed industries as a supplement in poultry and aquaculture industries [4]. Despite the availability of a variety of natural sources of carotenoids, synthetic ones are still cheaper and preferred by these industries, although this is changing due to some

Carlos Barreiro and José-Luis Barredo (eds.), *Microbial Carotenoids: Methods and Protocols*, Methods in Molecular Biology, vol. 1852, https://doi.org/10.1007/978-1-4939-8742-9_14, © Springer Science+Business Media, LLC, part of Springer Nature 2018

public concern over the safety of artificial food colorants. The availability of genes involved in carotenoid biosynthesis, isolated from many different organisms, together with the development of improved culture conditions, has led to an increase in carotenoid biotechnological research. Metabolic engineering of carotenoid production [5–7] has been carried out in naturally nonproducing or producing organisms such as some bacteria, *Saccharomyces cerevisiae* [8], and other fungi and the yeast *Xanthophyllomyces dendrorhous* and the algae *Haematococcus pluvialis*, the two major microorganisms able to synthesize natural astaxanthin [9, 10].

Carotenoid-producing fungi often accumulate a single main carotenoid. Among the *Mucoromycotina*, β-carotene is the key carotenoid accumulated by *Blakeslea trispora*, *M. circinelloides*, and *Phycomyces blakesleeanus*, while astaxanthin predominates in the basidiomycetous yeast *X. dendrorhous*. The basic technology for the industrial production of β-carotene by fungi is already set up. However, the best-known results for xanthophylls are still obtained at the laboratory and semi-industrial fermentation levels [9].

Several years ago, we started to use *M. circinelloides* as a model organism to study the biosynthesis of β-carotene and its regulation. The result was the isolation of many mutants and the structural genes involved in the pathway [11–14]. Later we took a preliminary applied research by improving strain and culture conditions for β-carotene production with this fungus. We obtained several β-carotene-overproducing mutants, a number of them able to grow as yeasts aerobically (*M. circinelloides* is a dimorphic fungus that shifts from hyphae to yeast in anaerobiosis) [15], but the genes affected in these mutants were not isolated and characterized.

To get an improvement in carotenoid content and diversification, we needed to ascertain several items: (1) Could heterologous genes for carotenogenesis be efficiently expressed in *M. circinelloides*? (2) Is there any bottleneck in the previous steps of the *M. circinelloides* carotenoid pathway so that increasing some of the gene products involved in the route would raise the total accumulation of β-carotene? (3) Could we fuse exogenous genes to the available ones in order to get the ketolase activity where β-carotene or zeaxanthin is being synthesized?

Blue light induces an increase in the expression of many carotenogenic genes [12–14], but even at their highest level, this rise is relatively low. Plasmids used to transform *M. circinelloides* tend to be autoreplicative, and they are sometimes present in several copies per nuclei in the transformants, but this fact is somehow random and not suitable to get satisfactory overproducing strains because they are usually lost during vegetative growth. Therefore, a preliminary question was clear: we needed to construct expression vectors to place the selected genes under the control of a strong promoter. We decided to use the *gpd1* promoter and terminator sequences to create vectors suitable for their use in transformation experiments [16] because the expression of this gene is highly regulated in response to carbon source, increasing with high levels of glucose in the medium [17, 18]. The whole aspects and tips of the strategy to construct these vectors are described below (*see* Subheading 3.3).

Once these vectors were available, we checked the possibility of *M. circinelloides* to produce astaxanthin by introducing in selected strains the *crtW* and *crtZ* genes from *Paracoccus* sp. N81106 (formerly *Agrobacterium aurantiacum*), which both together carry out the transformation of β-carotene into astaxanthin (Fig. 1c). Although we obtained *M. circinelloides* pink transformants, which accumulate astaxanthin, its amount with respect to the total carotenoid content was only about 10% [16]. This is not surprising, since the astaxanthin ratio with respect to the total carotenoid content in the native strain of *Paracoccus* sp. is about 19.5%. In general, it is assumed that in astaxanthin-synthesizing organisms, the rate-limiting step for astaxanthin production is the activity of the *crtW* enzymes (the ketolases), which show different activities and substrate specificities [19, 20]. The low amount of conversion of β-carotene to astaxanthin in the transformants could be due to other facts: first, it has been shown that there is a clear compartmentalization in the production of sterol, carotenoids, and other isoprenoids in *Zygomycetes* [21, 22]. Second, but related to the first, the product of the exogenous genes introduced in *M. circinelloides* could be not working where β-carotene is being synthesized.

We also introduced in the wild-type strain, under the control of the *gpd1* promoter, the genes *ipi*, *isoA*, and *carG*, responsible for the initial steps in the synthesis of β-carotene (IPP isomerase, FPP synthase, and GGPP synthase), getting an increase of up to four times in carotenoid content (Fig. 1a). The first bottleneck in the synthesis of β-carotene was found to be the step controlled by the *carG* gene, but overexpression of both the *ipi* and *isoA* genes together also showed an enhanced carotenoid accumulation probably due to a higher availability of precursors. Transformations with these isoprenoid genes in combination with the bacterial *crtW* gene resulted in *M. circinelloides* strains producing higher amounts of canthaxanthin and echinenone [23].

The third aspect to be considered was the possibility of constructing fusion genes in which the *crtW* product would be there where β-carotene or its derivatives are being synthesized. To do this, we fused the *crtW* of *Paracoccus* sp. to the *carRP* gene of *M. circinelloides*, responsible for the first and third steps of β-carotene biosynthesis in this fungus (*M. circinelloides* was shown to have some hydroxylase activity) [16] (Fig. 1B). The expected activity of this fusion gene would be the accumulation by *M. circinelloides* of canthaxanthin and probably some astaxanthin. A second construction

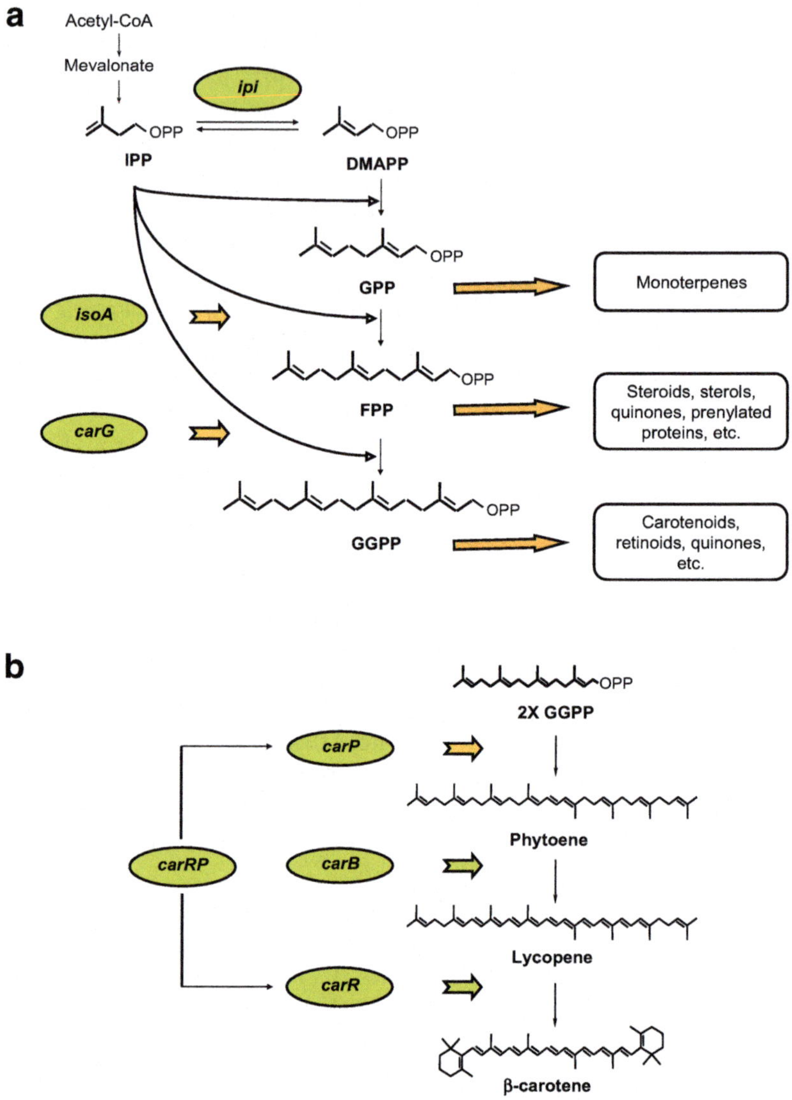

Fig. 1 The general mevalonic acid-isoprenoid biosynthetic pathway (**a**), the synthesis of β-carotene from GGPP (**b**), and the possible steps of β-carotene-astaxanthin conversion, as postulated by Misawa et al. [27]. (**a**) ***ipi***: isopentenyl pyrophosphate isomerase. ***isoA***: farnesyl pyrophosphate synthase. ***carG***: geranylgeranyl pyrophosphate synthase. (**b**) ***carRP***: a bifunctional gene with phytoene synthase (***carP***) and lycopene cyclase (***carR***) activities. ***carB***: phytoene dehydrogenase. (**c**) ***crtS***: ***X. dendrorhous crtS*** gene is able to produce astaxanthin in its own organism, but it only has hydroxylase activity in ***M. circinelloides***

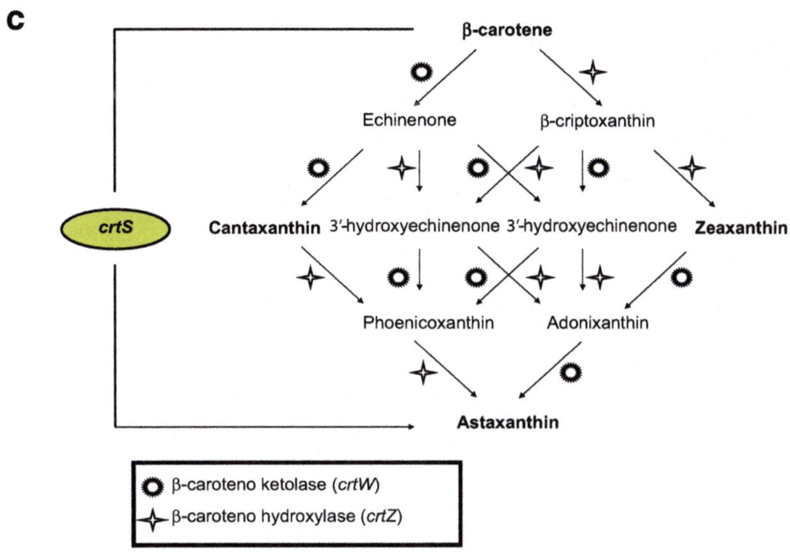

Fig. 1 (continued)

was the fusion of the *crtW* gene of *Paracoccus* sp. to the *crtS* gene of *X. dendrorhous*, responsible for the synthesis of astaxanthin from β-carotene in this fungus (Fig. 1c), but which was shown to have only hydroxylase activity in *M. circinelloides* [24]. The expected result in *M. circinelloides* transformants was the accumulation of astaxanthin.

Here we explain the rationale for the construction of expression vectors and describe an example and a detailed protocol for the construction of gene fusions of different genes for the conversion of β-carotene into astaxanthin in *M. circinelloides* that in our hands are currently working well and efficiently.

2 Materials

1. Suitable strains of *M. circinelloides f. lusitanicus*. Wild-type strain CBS 277.49 and mutant R7B (*leuA*$^-$), as well as plasmid pLeu4 [25], were obtained from Prof. D. von Wettstein (Carlsberg Laboratory, Copenhagen, Denmark). Mutant MS12 (*leuA*$^-$, *pyrG*$^-$) was obtained at our lab [26] (*see* **Note 1**).
2. Plasmids used are summarized in Table 1.
3. YPG liquid medium: 0.3% yeast extract, 1% peptone, and 2% glucose, pH = 4.5.
4. SSE buffer: 0.5 M sucrose, 0.1 M KCl, 10 mM EDTA, 1 mM spermine, 4 mM spermidine, 1 mM PMSF, and 14 mM β-mercaptoethanol.

Table 1
Plasmids used in the study

Plasmid	Description	Origin	Source or reference	Notes
pBluescript II SK+	General cloning vector		Stratagene	
pGEM-T Easy	General cloning vector		Promega	
pLeu4	*leuA* gene, arsM		[25]	(1)
pAVB107	*leuA* gene	USAL	[12]	
pEPM9	*pyrG* gene	USAL	[26]	
pAK96K	*crt* gene cluster of *A. aurantiacum*		[27]	(2)
pCA10	*ipi* gene	USZG	[23]	
pAVB160	*isoA* gene	USAL	[14]	
pAVB161	*carG* gene	USAL	[14]	
pALMC11	*crtS* cDNA of *X. dendrorhous*		[24]	(3)
pAVB103	*carRP* cDNA	USAL	[12]	
pPT36	*gpd1* PCR product	USAL/USZG	[16]	
pPT37	*gpd1* promoter (Apa-Xho)	USAL/USZG	[16]	
pPT38	*gpd1* terminator (Sac-Not)	USAL/USZG	[16]	
pPT39	*crtW* of *A. aurantiacum* (Xho-Not)	USAL/USZG	[16]	
pPT40	*crtZ* of *A. aurantiacum* (Xho-Not)	USAL/USZG	[16]	
pPT41	pPT37/ApaI, XhoI+pSK/ApaI, XhoI	USAL/USZG	[16]	
pPT43	pPT41/SacI, NotI+pPT38/SacI, NotI	USAL/USZG	[16]	(4)
pPT43leu	*leuA* gene in pPT43	USAL/USZG	[23]	
pPT43pyr	*pyrG* gene in pPT43	USAL/USZG	[23]	

Unless otherwise stated, all genes are from *M. circinelloides*. *USAL* University of Salamanca, Spain, collection, *USZG* University of Szeged, Hungary, collection

5. Lysis buffer A: 50 mM Tris–HCl, 20 mM EDTA pH = 8.0, and 1% sarkosyl.
6. Phenol-chloroform-isoamyl alcohol (25:24:1 v/v).
7. Ultracentrifuge and accessories.
8. Lysis buffer B: 4 M guanidine isothiocyanate, 30 mM sodium acetate pH = 5.2, and 1 M β-mercaptoethanol.
9. Separating solution: 5.7 M CsCl and 2 mM EDTA.
10. Chloroform-butanol (4:1 v/v).
11. DEPC (diethyl pyrocarbonate)-treated water.

12. Oligo-dT column.
13. SuperScript cDNA synthesis kit (Gibco BRL Spain, Invitrogen S.A., Barcelona, Spain).
14. Primer #1 (*crtSW01*): 5′-GGGTATCGATAATGTTCATCT-3′. Includes the *ClaI* recognition sequence for the directed cloning in pPT43 and the first nucleotides of the coding region of the *crtS* cDNA.
15. Primer #2 (*crtSW02*): 5′-CTCGCGGCCGCTCATGCGGTGTCCCCCTTG-3′. Includes the *NotI* recognition sequence for the directed cloning in pPT43 and the last nucleotides of the *crtW* gene.
16. Primer #3 (*crtSW03*): 5′-GGCATGTGCGCTCATTTCGACCGGCTTGACCTGC-3′.
17. Primer #4 (*crtSW04*): 5′-GCAGGTCAAGCCGGTCGAAATGAGCGCACATGCC-3′.
18. Pwo SuperYield DNA Polymerase (Roche Applied Science, Barcelona, Spain).
19. Expand Long Template PCR System (Roche Applied Science, Barcelona, Spain).
20. Restriction enzymes (Roche Applied Science, Barcelona, Spain).
21. Thermocycler.
22. Thin-walled 0.2 mL PCR tubes.
23. NucleoSpin Extract II (Macherey-Nagel, Düren, Germany).
24. pGEM-T Easy cloning vector (Promega Biotech Ibérica SL, Madrid, Spain).
25. Other main laboratory equipment: shaker incubators, a laminar flow hood, controlled temperature cabins, etc.

3 Methods

3.1 DNA Purification

High molecular weight pure DNA is always the best source to work with fungal genomic DNA. This simple procedure lets us obtain a high amount of ultrapure DNA from a single 1-week DNA extraction for years (*see* **Notes 5** and **6**).

1. Inoculate 200 mL of YPG liquid medium with 10^5 spores/mL of wild-type *M. circinelloides* CBS277.49 strain.
2. Grow the culture for 16–20 h at 28 °C and 180 rpm. Harvest the mycelium by paper filtration and wash it several times with distilled water. Dry the mycelium between several filter paper towels.

3. Put the dried mycelium in an ice-cooled mortar, and add 5–10 mL of SSE buffer, just to wet it. Freeze immediately adding liquid nitrogen.
4. Disrupt the mycelia (5–10 g) with a pestle in the mortar adding liquid nitrogen regularly until a fine powder is obtained.
5. Pass the powder to an appropriate baker that will maximize the interface in the phenol extraction (see below), and allow it to reach slowly room temperature. Then add 20 mL of lysis buffer A.
6. Incubate at 65 °C for 15 min and gently homogenate the mix using a sterile glass rod.
7. Add an equal volume of phenol-chloroform-isoamyl alcohol and shake at 100–120 rpm for 3–4 h at 4 °C in the baker.
8. Separate the two phases by spinning at 3500 × *g* for 15 min and 4 °C.
9. Collect the aqueous phases, and add CsCl to a 49% w/v and bisbenzimide to 0.5 mg/mL, final concentrations. Set up ultracentrifugation tubes and devices, and centrifuge at 20 °C, for 48 h at 100,000 × *g*.
10. After centrifugation, extract the DNA from the ultracentrifuge tubes with a syringe and needle and put it into dialysis tubing. Dialyze several times against distilled water and quantify. The final concentration of DNA should be about 200 ng/μL.

3.2 RNA Purification and cDNA Synthesis

Total RNA extraction from *M. circinelloides*, mRNA isolation, and cDNA synthesis were carried out following standard methods.

1. Get a liquid nitrogen-frozen fine powder (about 20 g) of *M. circinelloides* mycelium (*see* Subheading 3.1).
2. Add 50 mL of lysis buffer B. Vortex for 5–10 min at full speed.
3. Centrifuge at 28,500 × *g* for 30 min at 4 °C to eliminate cellular debris.
4. Overlay 5 mL of the supernatant onto previously prepared ultracentrifuge tubes filled with 20 mL of the separating solution, and centrifuge in a swinging bucket rotor at 25,000 × *g*, 20 °C, for 12–18 h. Eliminate the supernatants.
5. Resuspend the precipitates in 2–3 mL of sterile DEPC-treated water.
6. Extract 3–4 times with chloroform-butanol to eliminate salts (*see* **Note 7**).
7. Collect the aqueous phases, and precipitate the total RNA with sodium acetate pH = 5.2 (to 0.3 M) and ethanol (2.5 volumes).

8. Wash with the same volume of 70% ethanol and store the tubes at −80 °C until needed. When required, desiccate the tubes under vacuum and resuspend in DEPC-treated water. Once resuspended, make aliquots, and store at −80 °C.
9. To get the mRNA, pass the total RNA through an oligo-dT column following standard protocols.
10. To obtain single-stranded cDNA, use the SuperScript cDNA synthesis kit or equivalent. Use the population of single-stranded cDNA molecules as the DNA source for PCR amplification of the desired cDNA.

3.3 Construction of Expression Vectors

If we are considering the construction of an expression vector, we must study in depth all possible situations in order to develop a suitable strategy. We decided to use the *gpd1* promoter and terminator sequences because this gene has a strong level of expression that increases with the carbon source concentration. The idea is to create a vector with a polylinker between these two sequences. This can be done in two ways: one is by putting the two sequences together in a vector and later including the polylinker in the correct place. Alternatively, we can place the sequences at both sides of a pre-existing polylinker. There are different plans, and we must explore the whole steps needed in every situation to minimize the number of manipulations and maximize the number of restriction sites in the polylinker. Here we describe some tips about this task using the steps we did as an example. In our case it seemed easier the second option.

1. Consider first the restriction sites present in your sequences, which could interfere with the sites in the final polylinker and the target sequences that will be introduced in the oligonucleotides. Our sequences were relatively short, and we did not find problems with the target sites used (see next).
2. Design oligonucleotides to amplify the desired sequences. Our original idea was to amplify the promoter and terminator sequences with modified oligonucleotides that included in their sequences the restriction targets corresponding to two sites located at each of the ends of the pBluescript II SK polylinker. We used these oligonucleotides to amplify the promoter and terminator sequences from genomic DNA using the Pwo SuperYield DNA Polymerase (Roche) following the suppliers' recommendations (*see* **Note 8**). This enzyme has proofreading activity.

   Therefore, for the *gpd1* promoter, we designed two oligonucleotides:

   (gpd1P ApaI) 5′-GG<u>GGGCCC</u>GTGTTGGATGAACAAT TCAT-3′ and

(gpd1P XhoI) 5′-GGCTCGAGATTTATAAAATATATA GAGATA-3′.

And we designed two other ones to amplify the terminator sequence:

(gpd1T NotI) 5′-GGGCGGCCGCATCGTAAATCATTTC TAGTC-3′ and

(gpd1T SacI) 5′-GGGAGCTCGCGATGACAATAACGG CGAT-3′.

3. The gpd1P .(promoter) amplified sequence was digested with ApaI and XhoI and cloned at the left end of the polylinker. The gpd1T (terminator) amplified sequence was digested with SacI and NotI and cloned at the right end of the polylinker. The resulting vector has all the target sites in the pBluescript II SK available for subsequent cloning experiments between the XhoI and NotI sites. The resulting plasmid (pPT43) is thus a very convenient expression vector in which we can clone any open reading frame amplified with 5′-XhoI and NotI-3′ ends. In any case, it is very important to have in mind the direction of the sequences in the final vector when designing the oligonucleotides; sometimes this is the cause of time-consuming mistakes.
4. Plasmid expression vectors are the first step in expressing a gene, but in addition to their features (promoter and terminator sequences flanking a polylinker), we construct them in order to introduce selected genes in our organism. Therefore, we need to construct derivatives by including selectable markers for transformation experiments. In *M. circinelloides* the two most used selectable markers are the *leuA* and *pyrG* genes. The *leuA* gene is a long one (about 4.3 kb), and therefore it contains restriction sites for many enzymes. It is almost impossible to try to clone it in pPT43, and we found the solution doing the opposite: we cloned the pPT43 gpd1P-polylinker-gpd1T cassette in the available sites of pLeu4 giving rise to pPT43leu. Several restriction sites in the polylinker cannot be used in this vector because they are already present in pLeu4, but there are still enough restriction sites in the polylinker to clone any gene under the *gpd1* sequences.
5. The *pyrG* gene is smaller (about 0.8 kb) and was inserted directly in the ApaI site of pPT43 (outside the gpd cassette) giving rise to pPT43pyr.

3.4 Construction of Gene Fusions

A simple two-step PCR strategy was used to get a single gene from two different ones. The key point consists in the construction of two partially complementary molecules (PCR products A and B in Fig. 2B) that would be used as a template for a second amplification reaction (*see* **Note 8**). Primers #1 and #2 (Fig. 2) correspond to the 5′ end of the first gene and the 3′ end of the second gene,

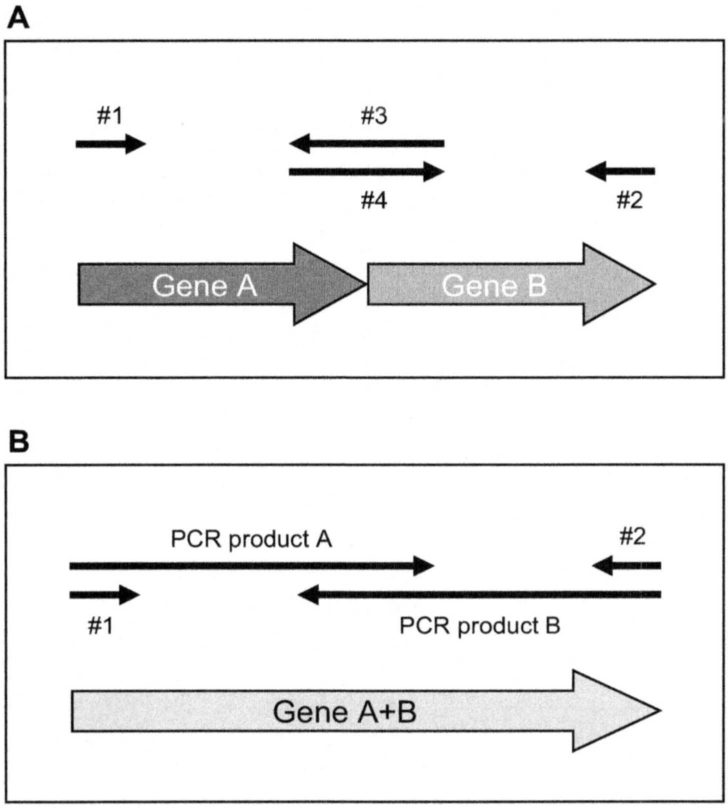

Fig. 2 PCR strategy for gene fusions. Four oligonucleotides (#1, #2, #3, and #4) are needed to create a partially complementary molecule. Oligonucleotides #3 and #4 are complementary and designed in an opposite orientation. They are used in the first step of the PCR amplification (**A**). When PCR products A and B are isolated and purified, they are used as DNA templates in the second step to create the fusion gene using oligonucleotides #1 and #2 in PCR amplification (**B**)

respectively, and are normally about 20 nt long (*see* **Note 9**). Primers #3 and #4 (Fig. 2A) are usually complementary (*see* **Note 10**) and include the 3′ end region of the first gene and the 5′ end region of the second gene. These primers must be longer (about 40 nt) and will be used in the first step to create the partially complementary molecules (*see* **Notes 11** and **12**). The system we have used (Expand Long Template PCR System) consists of a mix of thermostable DNA polymerases (*Taq* polymerase and *Tgo* polymerase, the latter with proofreading activity), with the availability of three different buffers to efficiently amplify DNA up to 20 kb (*see* **Note 13**). The optimal conditions (template DNA, incubation times, temperatures, Mg^{2+} concentration) depend on the sizes of the DNAs to be amplified and the thermal block cycler and must be determined empirically.

1. Thaw and equilibrate all buffers at 37–56 °C before use. If crystals are visible, incubate the 10× buffers until they are dissolved (*see* **Note 14**).
2. Pre-warm the PCR thermocycler.
3. To a sterile 0.2 mL thin-walled PCR tube, add (on ice) the following components in strict order: (a) Sterile distilled water (up to 50 μL). (b) dNTP nucleotide mix (final concentration 350–500 μM each, depending on the buffer used). (c) Primers (final concentration 300 nM). (d) 5 μL 10× PCR buffer (*see* **Note 15**). (e) Template DNA (*see* **Note 16**).
4. Mix well (do not vortex).
5. Spin down and add 0.75 μL of the enzyme mix (5 U/μL).
6. Start the PCR reactions immediately. Do not delay or store the reactions.
7. Standard PCR conditions (*see* **Note 17**): (a) Initial denaturation step: 92–94 °C for 2 min (*see* **Note 18**). (b) Ten cycles of denaturation (92–94 °C for 10 s), annealing (45–65 °C for 30 s), and elongation (68 °C for 45 s–30 min) (*see* **Note 19**). (c) 15–25 cycles of denaturation (92–94 °C for 10 s), annealing (45–65 °C for 30 s), and elongation (68 °C for 45 s–30 min + 20 s elongation time for each successive cycle) (*see* **Note 20**). (d) A final elongation step at 68 °C for 7 min.
8. Second set of PCR conditions (*see* **Note 17**): (a) Initial denaturation step: 94 °C for 2 min. (b) Ten cycles of denaturation (94 °C for 10 s), annealing (62 °C for 30 s), and elongation (68 °C for 6 min). (c) Ten cycles of denaturation (94 °C for 10 s), annealing (62 °C for 30 s), and elongation (68 °C for 6.5 min). (d) Ten cycles of denaturation (94 °C for 10 s), annealing (62 °C for 30 s), and elongation (68 °C for 7 min). (e) A final elongation step at 68 °C for 7 min.
9. After checking the PCR reactions by gel electrophoresis, isolate the first-step PCR products from preparative gels using a silica-based purification kit (*see* **Note 21**).
10. Use an appropriate amount (10–50 ng) of the two first-step PCR products to set up the second-step PCR (*see* **Note 22**).
11. Check and purify the second-step PCR product (Fig. 2) as in 1.
12. Digest pPT43 with ClaI and NotI in two consecutive steps. Gel-purify the DNA fragment (*see* **Note 23**).
13. Ligate the pPT43-digested vector (about 10–20 ng) to the purified second-step PCR product in a 1:10 ratio. Leave the ligation at room temperature for at least 3 h.
14. Transform any competent *E. coli* strain (we use DH5α), by chemical standard methods or electroporation.

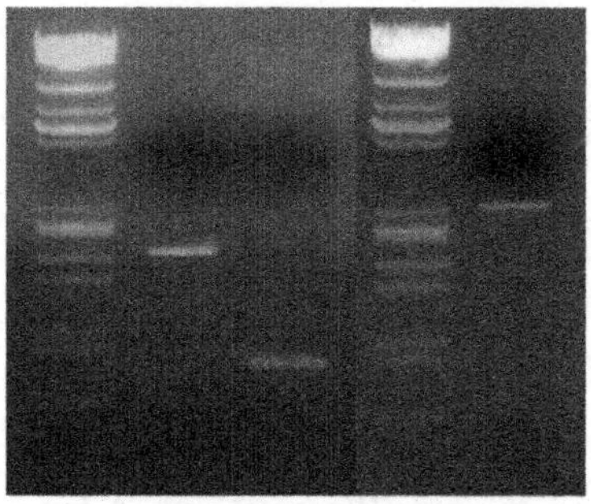

Fig. 3 First and second steps of the gene fusions. First step: amplification of ***crtS*** (lane 2) and ***crtW*** (lane 3). Second step: amplification of the fusion product ***crtSW*** (lane 5). Lanes 1 and 4 show a molecular weight marker prepared by the mix of λ DNA digested with Hind III and EcoRI+Hind III

15. Analyze several clones, and select one containing the fusion gene under the control of the *gpd1* gene promoter and terminator sequences. Check by PCR using oligos #1 and #2 and comparison to the PCR products in 3 (Fig. 3).

3.5 Transformation of M. Circinelloides

We used the protoplast-based PEG-mediated transformation method developed for *M. circinelloides*. This and other transformation methods have recently been reviewed [28] for this fungus.

Here we present the basic steps to construct gene expression vectors and a fast and efficient method to get gene fusions based on simple PCR strategies. Preliminary analyses of the transformants showed an increase of up to four times in the relative content of astaxanthin when using the *crtSW* fusion gene (with reference to transformants with the *crtS* and *crtW* in separate plasmids [16]).

4 Notes

1. *M. circinelloides* CBS277.49 (wild type) was used as the source of DNA and RNA. *M. circinelloides f. lusitanicus* strains CBS277.49 and R7B have been indistinctly used as standards of this organism. The source of genomic DNA and RNA has always been from one or another. During a long time, there has been some controversy with strains *M. circinelloides* CBS277.49 and *Mucor racemosus* ATCC 1216b, which were considered synonyms. Although Schipper [29] already detected this disagreement in 1976, the mistake continued for years. Finally, Díaz-Mínguez et al. [30] showed that the two strains are different and, in fact, belong to opposite mating types. The complete genome of *M. circinelloides f. lusitanicus* is available at http://genome.jgi.doe.gov/Mucci2/Mucci2.home.html.

M. circinelloides MS12 mutant was used as the recipient strain in transformation experiments.

2. Plasmid pAK96K containing the genes of *A. aurantiacum* (*crt* gene cluster) was kindly provided by Dr. N. Misawa, Central Laboratories for Key Technology, Kirin Brewery Co., Ltd., Kanagawa, Japan.
3. Plasmid pALMC11 including the *X. dendrorhous crtS* cDNA was kindly provided by Dr. J.L. Barredo (Antibióticos SA, León, Spain) [24].
4. Plasmid pPT43 was constructed by inserting the *M. circinelloides gpd1* promoter and terminator sequences flanking the pBluescript II SK+ polylinker and was used to clone all genes to be overexpressed [17, 23] (see text).
5. At the beginning of this work, we had plasmids with the *carRP* gene of *M. circinelloides*, its corresponding cDNA, and also with the cDNA of the *carS* gene of *X. dendrorhous*, but not the genomic DNAs of *gpd1* and *ipi* genes. Whenever possible, cDNA was preferred as the source of DNA to be PCR amplified because it maintains the length of the gene fusions as short as possible. It all depends on the sizes of the genes to be fused, the availability of cDNA copies, and the organism where we want them to be expressed. For instance, the *crtS* gene of *X. dendrorhous* contains 17 introns, and the final destination in our case is *M. circinelloides*. It could happen that this zygomycete does not process efficiently all of the introns, so the DNA source of election will be then the cDNA. We checked using total genomic DNA instead of the cloned cDNA of the *carRP* gene of *M. circinelloides* as a starting material for PCR amplifications, and it worked equally well in the transformation experiments.
6. When a plasmid (containing either the genomic sequences or the cDNAs) is used as a source of DNA amplification by PCR, the region to be amplified is always previously isolated from the bacterial sequences in the plasmid. This reduces the possible background during the amplifications.
7. Extractions with chloroform-butanol reduce the volume of the aqueous phase containing the RNA. Be aware of adding more DEPC-treated water to maintain the volume of the RNA solution. If too much water dissolves in the butanol solution, the RNA gets stuck to the walls of the tube, and it is almost impossible to resuspend it.
8. Since the final use of the overexpressed or fused genes will be its expression in *M. circinelloides* transformants, the use of thermostable DNA polymerases with proofreading activity is strongly recommended because it minimizes the risk of getting point mutations. Several commercial standard and long-range,

thermostable DNA polymerases with proofreading activity were used in preliminary experiments; we chose Pwo Polymerase for standard and the Expand Long Template PCR System to amplify long PCR products on the basis of its consistent results. In any case, it is advised to sequence the final PCR amplification product. We have found less than one nucleotide substitution per kilobase with these systems.

9. The length of the oligonucleotide primers must be enough to get a G+C content of at least 50%. It is opportune to introduce in these oligonucleotides recognition sequences for restriction enzymes that will facilitate the cloning of the gene fusion into the vectors used for the transformation experiments. Never put this recognition sequence at the 5′ end. Insert at least 3–4 nt between the 5′ end and the recognition sequence, since many enzymes are unable to digest 5′ ended targets.
10. In fact, they do not have to be necessarily fully complementary, but should maintain a complementarity of at least 30 nt. Since they are not used at the same time in first-step PCR reactions, there are no problems with possible primer-dimer reactions.
11. A common mistake when designing these primers is to maintain in their sequence the stop codon of the first gene. The stop codon must be released from the sequence. Be aware of it.
12. The spatial structure of the fused protein must be taken into account. It could happen that the tertiary and quaternary structures would be nonfunctional because of amino acid interactions at the fusion point. If problems arise, a good solution is to introduce 4–5 neutral amino acid codons between the last codon of the first gene and the initiation codon of the second.
13. Although the system is prepared for the amplification of genomic DNA fragments, it works efficiently using cloned genes.
14. Do not vortex, but mix by inversion (specially buffer 3). It contains detergents and bubbles will be formed.
15. Three buffers are available from the system. They differ in $MgCl_2$ concentration and buffer 3 also contains detergents. They are prepared to fit different ranges of amplification: from 0.5 to 9 kb (buffer 1), from 9 to 12 kb (buffer 2), and for >12 kb (buffer 3). The suppliers recommend increasing the concentration of Mg^{2+} when increasing the dNTP concentration. This Mg^{2+} increase is already included in the buffers, so only in particular circumstances should be taken into account. We used buffer 2. In our hands (amplifying DNA fragments from 0.7 to 6.0 kb), buffer 1 gave always high non-specific backgrounds, but buffers 2 and 3 gave us clear bands (*see* Fig. 2).

16. Keep the template DNA as minimum as possible. When using fungal total genomic DNA, never exceed 100 ng. One to 10 ng of DNA isolated from plasmids (*see* **Note 6**) gave the best, clear results. It is preferable to have a relatively low yield than getting a high background.
17. We used two sets of PCR conditions: the standard conditions suggested by the suppliers and a serendipity one that worked really well with the amplification of our DNA fragments (0.7–6.0 kb). The standard conditions gave us a high amplification yield but with some background. The second set of conditions gave us cleaner amplification reactions but to a lower yield.
18. The suppliers recommend keeping denaturation steps as short as possible and denaturation temperatures as low as possible. We checked these parameters and it is really true. Maintaining short and low-temperature denaturation steps minimizes background in the reactions.
19. The annealing temperature depends on the primers used. Start using a high annealing temperature. If there is no amplification, repeat the assay lowering it by 2 °C. The elongation times depend on the fragment length, say 2 min for up to 3 kb, 4 min for up to 6 kb, 8 min for up to 10 kb, etc.
20. For instance, cycle number 11 is 20 s longer than cycle number 10, cycle number 12 is 40 s longer than cycle number 10, and so on.
21. Isolating the PCR products from a preparative gel instead of purifying the whole reaction ensures the elimination of primers, template DNA, and background DNA that could affect negatively the second-step reaction.
22. In the first cycles of the second-step PCR reaction, we need that the two partially complementary molecules form a single molecular structure (i.e., that they complement each other). Half of these molecules will be good templates for the polymerases, because of their 3′OH ends. But it could happen, if we add the two end primers (#1 and #2) at the beginning, that they would be amplifying only each first-step PCR product. We checked running five cycles of PCR before adding these oligos, to obtain more complete molecules of the fusion product before the amplification. A higher yield was obtained but with a higher background.
23. ClaI and NotI sites are relatively close in the polylinker. It is always preferable to digest the vector first with one enzyme, check out the complete digestion, and add the second enzyme. A single double digestion could result in plasmids only digested by either one or another, which will religate easily in the ligation reaction, reducing the cloning efficiency. Purification of the plasmid vectors from preparative gels is also advised to eliminate the polylinker.

Acknowledgments

This work was supported by different grants of the Junta de Castilla y León (Spain) (GR64) and the Hungarian Scientific Research Fund and the National Office for Research and Technology (OTKA CK80188).

References

1. Nishikawa Y et al (2005) Effects of astaxanthin and vitamin C on the prevention of gastric ulcerations in stressed rats. J Nutr Sci Vitaminol 51:135–141
2. Chew BP, Park JS (2004) Carotenoid action on the immune response. J Nutr 134:257–261
3. Vainio H, Rautalahti M (1998) An international evaluation of the cancer preventive potential of carotenoids. Cancer Epidemiol Biomark Prev 7:725–728
4. Johnson EA, Schroeder WA (1995) Microbial carotenoids. In: Fiechter A (ed) Advances in biochemical engineering and biotechnology. Springer, Berlin
5. Misawa N, Shimada H (1997) Metabolic engineering for the production of carotenoids in non-carotenogenic bacteria and yeasts. J Biotechnol 59:169–181
6. Schmidt-Dannert C, Umeno D, Arnold FH (2000) Molecular breeding of carotenoid biosynthesis pathways. Nat Biotechnol 18:750–753
7. Umeno D, Tobias AV, Arnold FH (2005) Diversifying carotenoid biosynthetic pathways by directed evolution. Microbiol Mol Biol Rev 69:51–78
8. Ukibe K et al (2009) Metabolic engineering of *Saccharomyces cerevisiae* for astaxanthin production and oxidative stress tolerance. Appl Environ Microbiol 75:7205–7211
9. de la Fuente JL et al (2010) High-titer production of astaxanthin by the semi-industrial fermentation of *Xanthophyllomyces dendrorhous*. J Biotechnol 148:144–146
10. Domínguez-Bocanegra AR, Ponce-Noyola T, Torres-Muñoz JA (2007) Astaxanthin production by *Phaffia rhodozyma* and *Haematococcus pluvialis*: a comparative study. Appl Microbiol Biotechnol 75:783–791
11. Iturriaga EA, Velayos A, Eslava AP (2000) The structure and function of the genes involved in the biosynthesis of carotenoids in the Mucorales. Biotechnol Bioprocess Eng 5:263–274
12. Velayos A, Eslava AP, Iturriaga EA (2000) A bifunctional enzyme with lycopene cyclase and phytoene synthase activities is encoded by the *carRP* gene of *Mucor circinelloides*. Eur J Biochem 267:5509–5519
13. Iturriaga EA et al (2001) The genetics and molecular biology of carotenoid biosynthesis in *Mucor*. In: Pandalai SG (ed) Recent research developments in current genetics. Research Signpost, Trivandrum
14. Velayos A et al (2003) Expression of the *carG* gene, encoding geranylgeranyl pyrophosphate synthase, is up-regulated by blue light in *Mucor circinelloides*. Curr Genet 43: 112–120
15. Iturriaga EA et al (2005) Strain and culture conditions improvement for β-carotene production with *Mucor*. In: Barredo JL (ed) Microbial processes and products. Humana Press, Totowa, NJ
16. Papp T et al (2006) Heterologous expression of astaxanthin biosynthesis genes in *Mucor circinelloides*. Appl Microbiol Biotechnol 69:526–531
17. Wolff AM, Arnau J (2002) Cloning of glyceraldehyde-3-phosphate dehydrogenase-encoding genes in *Mucor circinelloides* (Syn. *racemosus*) and use of the gpd1 promoter for recombinant protein production. Fungal Genet Biol 35:21–29
18. Larsen GG et al (2004) Characterization of the *Mucor circinelloides* regulated promoter gpd1P. Curr Genet 45:225–234
19. Tao L et al (2006) Engineering a β-carotene ketolase for astaxanthin production. Metab Eng 8:523–531
20. Fraser PD, Miura Y, Misawa N (1997) *In vitro* characterization of astaxanthin biosynthetic enzymes. J Biol Chem 272:6128–6135
21. Domenech C et al (1996) Separate compartments for the production of sterols, carotenoids and gibberellins in *Gibberella fujikuroi*. Eur J Biochem 239:720–725
22. Kuzina V, Domenech C, Cerdá-Olmedo E (2006) Relationships among the biosynthesis of ubiquinone, carotene, sterols, and triacylglycerols in Zygomycetes. Arch Microbiol 186:485–493

23. Csernetics A et al (2011) Expression of three isoprenoid biosynthesis genes and their effects on the carotenoid production of the zygomycete *Mucor circinelloides*. Fungal Genet Biol 48:696–703
24. Álvarez V et al (2006) The *crtS* gene of *Xanthophyllomyces dendrorhous* encodes a novel cytochrome-P450 hydroxylase involved in the conversion of β-carotene into astaxanthin and other xanthophylls. Fungal Genet Biol 43:261–272
25. Roncero MIG et al (1989) Characterization of a *leuA* gene and an ARS element from *Mucor circinelloides*. Gene 84:335–343
26. Benito EP et al (1995) Isolation, characterization and transformation, by autonomous replication, of *Mucor circinelloides* OMPdecase-deficient mutants. Mol Gen Genet 248:126–135
27. Misawa N et al (1995) Structure and functional analysis of a marine bacterial carotenoid biosynthesis gene cluster and astaxanthin biosynthetic pathway proposed at the gene level. J Bacteriol 177:6575–6584
28. Garre V, Barredo JL, Iturriaga EA (2015) Transformation of *Mucor circinelloides f. lusitanicus* protoplasts. In: van den Berg MA, Maruthachalam K (eds) Genetic transformation systems in fungi, Fungal biology, vol 1. Springer International Publishing, Cham
29. Schipper MAA (1976) On *Mucor circinelloides*, *Mucor racemosus* and related species. Stud Mycol 12:1–40
30. Díaz-Mínguez JM, López-Matas MA, Eslava AP (1999) Complementary mating types of *Mucor circinelloides* show electrophoretic karyotype heterogeneity. Curr Genet 36:383–389

Chapter 15

Lycopene Production by Mated Fermentation of *Blakeslea trispora*

Sonia Martínez-Cámara, Sara Rubio, Hannah del Río, Marta Rodríguez-Sáiz, and José-Luis Barredo

Abstract

Lycopene is a carotenoid mainly present in red-colored fruits and vegetables. Its value in the pharmaceutical and food industry is linked to its benefits for the human health, including properties against cancer and cardiovascular diseases, and its use as a food colorant. Lycopene can be produced either by synthetic or natural means, but there is a preference for the second, since it is considered a more eco-friendly and less harmful process. Among natural methods for obtaining lycopene, microbial fermentation is a good alternative to extraction from plants that naturally contain lycopene, since it implies obtaining higher and more specific amounts of this carotenoid. This chapter describes lycopene production by fermentation of the fungus *Blakeslea trispora*, a naturally carotenoid producer, at 30 L scale. This procedure involves separated growth of the two sexual mating types of *B. trispora* during the vegetative stages and the use of a lycopene cyclase inhibitor to achieve lycopene accumulation during the production stage.

Key words Lycopene, Carotenoids, Pilot scale, *Blakeslea trispora*, Fermentation

1 Introduction

Carotenoids are lipophilic pigments with many applications, ranging from feed supplements, colorants, and nutraceuticals to medicine and cosmetics. They are synthesized by plants, bacteria, and fungi and are responsible for some of the vivid colors present in nature, such as the yellow, red, and orange colors observed in fruits, flowers, and vegetables. In addition, they have a large and varied number of biological activities such as provitamin A function, antioxidant activity, blue light filtering, intracellular communication, and immune function [1].

Some of the carotenoids with a greater demand in the market are lycopene, β-carotene, astaxanthin, lutein, canthaxanthin, and annatto. They are mostly commercialized as food pigments,

Carlos Barreiro and José-Luis Barredo (eds.), *Microbial Carotenoids: Methods and Protocols*, Methods in Molecular Biology, vol. 1852, https://doi.org/10.1007/978-1-4939-8742-9_15,

animal feeding additives, coloring agents in the pharmaceutical and cosmetic industries, and as dietary supplements [2].

Lycopene is an intermediate of the β-carotene biosynthetic pathway (Fig. 1) mainly present in red-colored fruits and vegetables such as tomato, papaya, peach, or watermelon. This carotenoid is a 40-atom open chain containing 11 conjugated and 2 non-conjugated double bonds. It lacks the terminal β-ionic ring, which means that it is not a vitamin A precursor. Given lycopene's large number of conjugated double bonds, it has the highest singlet oxygen quenching capacity among all carotenoids. This implies that lycopene is also a powerful antioxidant that could have many health-benefiting properties [3].

Lycopene has many uses as a pigment in the food industry. As other colorants, it can be used for several purposes, such as enhancing an already existing color, restoring it after processing, or just to give color to uncolored food. It is used, for example, in beverages, sauces, soups, cereals, bread, dairy products, dietary products, and dietary supplements [4, 5]. But the growing interest in lycopene is not only due to its role as an efficient pigment in the food industry but also in human health as a valuable compound that could prevent diseases such as cancer or cardiovascular diseases. Several studies suggest that it decreases cholesterol levels and can be used as a treatment against oxidative stress. It also inhibits cell growth in various human cancer cell lines, such as prostate cancer; it lowers the risk of cardiovascular diseases, including atherosclerosis, myocardial infarction, and stroke; and it has anti-inflammatory activity [6, 7]. According to BBC Research, lycopene market accounted for $60 million in 2010, expecting to reach $84 million in the next few years [8].

Commercial lycopene can be obtained by chemical synthesis, by extraction from natural sources, and by fermentation. Synthetic lycopene is established by the double Wittig condensation of a symmetrical dialdehyde with two identical phosphonium salts [9]. However, there is a preference in using natural ingredients rather than synthetic compounds in food, cosmetics, and pharmaceuticals, so natural methods to obtain lycopene are given more attention nowadays. To date, the most common method to obtain lycopene is extracting it from plants, mainly tomato, but this involves variable and relatively low extraction yields, as lycopene levels in the fruit are low and depend on variety, maturity stage, and agrochemical practices of cultivation. There are also some complications regarding the purification process due to the presence of other related carotenoids in plants [10, 11].

All these difficulties regarding lycopene production have triggered the extensive research into an effective way of producing lycopene. Several microorganisms have been studied for their ability to produce lycopene, such as mucoral fungi of the genera

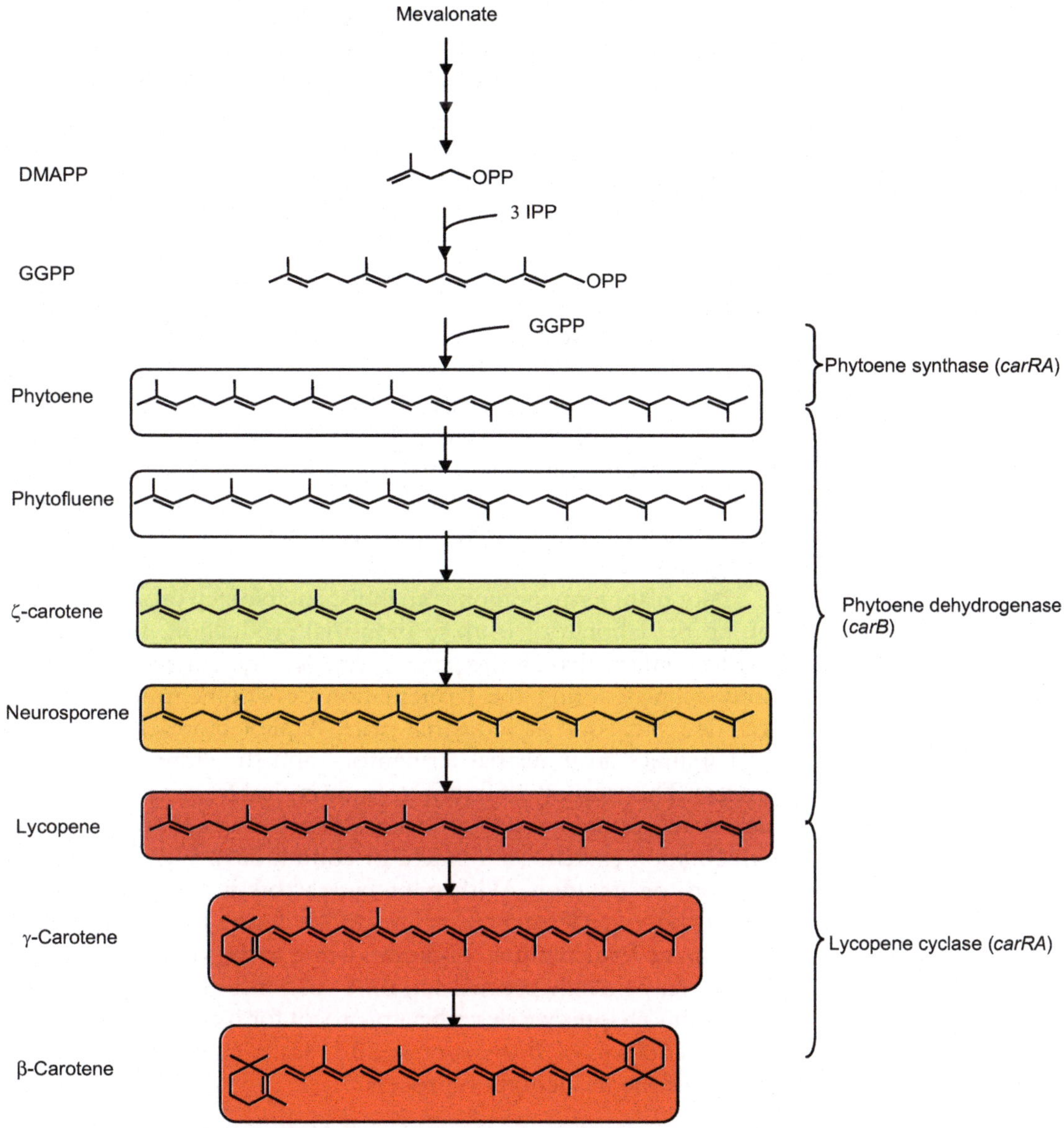

Fig. 1 β-Carotene biosynthetic pathway in *B. trispora*. One molecule of dimethylallyl pyrophosphate (DMAPP) and three molecules of isopentenyl pyrophosphate (IPP) are transformed into geranylgeranyl pyrophosphate (GGPP). Afterwards, phytoene synthase, encoded by the A domain of the *carRA* gene, catalyzes the junction of two molecules of GGPP to form phytoene. Subsequently, phytoene dehydrogenase, encoded by the *carB* gene, introduces four double bonds to yield sequentially phytofluene, ζ-carotene, neurosporene, and lycopene. Finally, lycopene cyclase, encoded by the R domain of the *carRA* gene, sequentially modifies each of the acyclic ends to form γ-carotene and finally β-carotene. The colored box framing each molecule indicates approximately their color

Blakeslea [12] and *Phycomyces* [13], *Flavobacterium* sp. [14], a mutant strain of *Streptomyces chrestomyceticus* [15], or genetically engineered strains of *Escherichia coli* [16, 17], *Pichia pastoris* [18], *Yarrowia lipolytica* [19], *Saccharomyces cerevisiae* [20, 21], and *Candida utilis* [22] among others.

To date, only lycopene produced by fermentation of *Blakeslea trispora* is accepted by the European Commission for commercial use [5]. This microorganism is a fungus with two mating types (+) and (−) that produces β-carotene by the cyclization of lycopene. Mated fermentation of both mating types (+) and (−) has proved to be successful by increasing carotenoid production [23, 24]. To produce lycopene using this microorganism, it is necessary to use a strain lacking lycopene cyclase or feed it with cyclase inhibitors such as imidazole or pyridine derivatives, in order to avoid formation of β-carotene and therefore accumulate lycopene [12, 25, 26]. Moreover, there are other factors, including strain improvement, culture media, and process optimization, that increase lycopene production [27, 28].

Pilot plant fermentation is an indispensable step previous to scaling up the laboratory work to industrial production. It is the best way to confirm that all the parameters and the fermentation conditions defined during the fermentation development at laboratory scale are successful at industrial scale. A pilot bioreactor of about 30 L mimics an industrial fermentor—and therefore an industrial process. This is a powerful tool to study critical fermentation parameters that cannot be tested at flask scale such as vessel design, type of fermentation (batch, semi-batch, or continuous), vegetative stages, aeration, or agitation. With pilot bioreactors it is also possible to identify potential fermentation limitations during the scaling-up and to optimize fermentation conditions being at the same time an environmentally friendly process and time- and cost-effective.

In this chapter we describe a protocol for lycopene production by fermentation of *B. trispora* at 30 L scale, emphasizing the key aspects of the fermentation process.

2 Materials

2.1 Vegetative Stages

2.1.1 First Vegetative Stage (Flask Scale)

1. *Blakeslea trispora* F-816 (+) and *Blakeslea trispora* F-744 (−) obtained from the Russian National Collection of Industrial Microorganisms (VKPM).
2. VM medium: Soybean meal 23 g/L, corn meal 47 g/L, KH_2PO_4 0.5 g/L, and thiamine hydrochloride 2 mg/L. Adjust to pH 6.3.
3. 500 mL Erlenmeyer flasks.
4. Flow laminar cabinet.
5. Orbital shaker.

2.1.2 Second Vegetative Stage (Bioreactor Scale)

1. Fermentor BIOSTAT® CPlus (Sartorius), or equivalent, with 10 L operating volume.
2. VM medium (*see* Subheading 2.1.1) with antifoam agent 1 g/L.
3. Inoculation flask: Sterile Erlenmeyer 500 mL flask with silicone tube and stainless steel needle.

2.2 Production Stage

1. Fermentor BIOSTAT® CPlus (Sartorius), or equivalent, with 30 L operating volume.
2. FM medium: Soybean meal 44 g/L, corn meal 19 g/L, KH_2PO_4 0.5 g/L, Na_2HPO_4 3.5 g/L, NaH_2PO_4 0.2 g/L, thiamine hydrochloride 2 mg/L, soybean oil 100 g/L, and antifoam agent 0.175 g/L. Adjust to pH 7.5.
3. Imidazole 20% (*see* **Note 1**).
4. Peristaltic pump.

2.3 Analytical Methods

2.3.1 Cell Growth

1. Graduate tubes.
2. Swinging-head centrifuge.

2.3.2 Viscosity

1. Viscometer Brookfield, model DV-II, or equivalent.
2. Spindle S-63.

2.3.3 Glucose Quantification

For glucose quantification, use the Megazyme D-Glucose (glucose oxidase/peroxidase, GOPOD) Assay Kit or equivalent.

1. Glucose determination reagent (Megazyme D-Glucose Assay Kit): Dilute the content of bottle I (GOPOD reagent buffer) to 1 L with distilled water. Dissolve bottle II (GOPOD reagent enzymes) in approximately 20 mL of the previous solution, and transfer this to the bottle containing the remainder solution I. Mix thoroughly and store at 4 °C.

2.3.4 Phosphate Quantification

1. Phosphate solution 50 ppm in water (prepared with $Na_2HPO_4 \cdot 12H_2O$).
2. Ammonium molybdate solution: $(NH_4)_6Mo_7O_{24} \cdot 4H_2O$ 25 g/L and H_2SO_4 28 mL/L, dissolved in water.
3. Tin chloride II solution ($SnCl_2$ 12.5 g/L dissolved in glycerol:H_2O 50:50).

2.3.5 Carotenoid Analysis

1. Laboratory vibration mill.
2. Glass beads (0.5 mm diameter).
3. Lycopene 85% purity (Sigma-Aldrich).
4. Carotene 95% purity (Sigma-Aldrich).
5. Methanol HPLC grade.

6. Dichloromethane.
7. Acetone.
8. Analytical balance.
9. HPLC equipment. Waters Alliance e2695 HPLC system with Waters 2998 Photodiode Array (PDA) Detector or equivalent.
10. Column: Hypersil ODS column (5 µm, 100 × 4.6 mm) or equivalent.
11. Mobile phase: Acetonitrile/methanol/chloroform (47/47/6).

2.3.6 General Equipment

1. Micropipette P200.
2. Micropipette P1000.
3. Desktop centrifuge.
4. Incubator 25–50 °C.
5. Spectrophotometer UV/visible.
6. pH meter.
7. Autoclave.
8. Microscope.
9. Magnetic stirrer.

3 Methods

3.1 Vegetative Stages

Vegetative stages are necessary to grow the microorganism and get enough biomass for the following steps of the fermentation process. Number and time of vegetative stages depend on the microorganism speed rate of growth and the volume of the production broth to be inoculated. In this case, the 30 L lycopene fermentation needs two vegetative stages to obtain enough biomass to inoculate the production tank.

Fermentation of *B. trispora* to produce lycopene requires separated growth during the vegetative stages for (+) and (−) strains to then inoculate the production media with a mixture of both strains.

3.1.1 First Vegetative Stage (Flask Scale)

1. Seed spores from (+) and (−) strains in separate vegetative flasks with 100 mL of VM culture medium (*see* **Notes 2** and **3**).
2. Incubate in an orbital shaker at 25 °C and 250 rpm for 48 h.

3.1.2 Second Vegetative Stage (Bioreactor Scale)

1. Set up the fermentors, calibrate the pH probes (*see* **Note 4**), and place pH and DO_2 probes in their corresponding inputs of the vessel.
2. Weigh the raw materials, and dissolve them sequentially in approximately 5 L of distilled water. Adjust to pH 6.3 using 20% NaOH.

3. Pour the previously prepared medium into the fermentors and add distilled water to get a final volume of 10 L.
4. Close all the ports of the fermentation vessels, and start the sterilization cycle according to the following sterilization parameters: temperature 121 °C, agitation 100 rpm, and sterilization time 30 min (*see* **Note 5**).
5. Cool down the fermentors, and set up the following fermentation conditions: 25 °C, 0.7–1.5 vol/vol/min airflow, 250 rpm, and 0.5 kg/cm^2 of back pressure.
6. Calibrate the DO_2 probe assigning a theoretical value of 100% before inoculating. This is the maximal oxygen supply at the initial aeration, pressure and agitation conditions (*see* **Note 6**).
7. Transfer each first vegetative flask independently (+ and – strains) to 10 L second vegetative tanks (0.1% seeding) (*see* **Note 3**) and incubate for 48 h.
8. Take a sample each 8 h to check lack of microbial contamination, cell growth, pH, viscosity, glucose and phosphate (*see* Subheading 3.3). At 48 h, microorganism should be well-grown and should have consumed most of the nutrients, so glucose and phosphate levels should be near zero (*see* **Note 7**).
9. After 48 h of incubation, for (−) strain pH should be around 5.5, PCV value around 30%, glucose levels less than 2 g/L and phosphate concentration in broth between 0 and 2 mg/L. For (+) strain, pH values should reach pH 8, PCV value around 35% approximately, glucose levels should be below 0.5 g/L and phosphate concentration in broth between 0 and 2 mg/L.

3.2 Production Stage

1. Set up the fermentor, calibrate the pH probe (*see* **Note 4**) and place pH and DO_2 probes in their corresponding inputs of the vessel.
2. Weigh the raw materials, except the oil, and dissolve them sequentially in approximately 5–10 L of distilled water. Adjust to pH 7.5 using 20% NaOH.
3. Pour the previously prepared medium into the fermentor, add the oil and distilled water to get a final volume of 25 L.
4. Close all the ports of the fermentation vessel and start the sterilization cycle according to the following sterilization parameters: temperature 121 °C, agitation 100 rpm, sterilization time 30 min (*see* **Note 5**).
5. Cool down the fermentor and set up the following fermentation conditions: 25 °C, airflow rate 1.0 vol/vol/min, agitation rate 250 rpm, and back pressure 0.5 kg/cm^2.
6. Calibrate the DO_2 probe assigning a theoretical value of 100% before inoculating. This is the maximal oxygen supply at the initial aeration, pressure, and agitation conditions (*see* **Note 6**).

7. Transfer 5 L of a mix of both second vegetative cultures (20%) to inoculate the production fermentor (*see* **Note 8**). Taking into account the poorer growth of the (−) strain, a proportion of 1 (+)/10 (−) ratio is used for the seeding (*see* **Note 9**).
8. During incubation, it is advisable to set a dissolved oxygen control over 30% to keep an optimal oxygen supply for the microbial growth (*see* **Note 10**). Keep agitation between 250 and 600 rpm and airflow rate between 1.0 and 1.2 vol/vol/min.
9. After 25–35 h of fermentation, add 120 mL of a sterile 20% imidazole solution using a peristaltic pump to get a final concentration of imidazole of 0.8 g/L in the fermentation broth (*see* **Note 11**).
10. Take a sample of the broth each 12–24 h to check lack of contamination, cell growth, pH, viscosity, and lycopene and β-γ-carotene production (*see* Subheading 3.3).
11. Lycopene production is mostly associated with secondary metabolism of *B. trispora*: maximum production rate is found during the second half of the fermentation when the metabolism of the microorganism decreases and nutrients of the broth are finishing. Fermentation lasts approximately 144 h, when lycopene production reaches about 1 g/L.

3.3 Analytical Methods

There are some critical parameters that need to be measured during the vegetative and/or productive stages of this fermentation process: cell growth, viscosity of broth, glucose and phosphate concentration in the fermentation broth, and carotenoid production. In this section, we have selected one among the various analytical methods available to measure each parameter, but an alternative method can be chosen.

3.3.1 Cell Growth (Packed Cell Volume, PCV)

The biomass content can be expressed through the packed cell volume (PCV).

1. Fill a graduated 10 mL glass tube with culture broth.
2. Centrifuge in a swinging-head centrifuge at 2000 × *g* for 10 min.
3. Express the biomass content as a percentage of the centrifuged broth.

3.3.2 Viscosity

1. Deep the spindle of the viscometer in a sample of 100 mL of fermentation broth, and set the run speed at 60 rpm for 20 s at room temperature.

3.3.3 Glucose Quantification

1. Mix 3 mL of the glucose determination reagent with 0.1 mL of the sample supernatant and incubate at 40–50 °C for 20 min.

2. Prepare as well a mix for the reagent blank and the D-glucose standard (*see* Subheading 2.3.3 for reagent preparation).
3. Read the absorbance at 510 nm. The absorbance results of this colorimetric assay are proportional to the concentration of D-glucose in the sample (*see* **Note 12**).

3.3.4 Phosphate Quantification

Standard Curve

1. Prepare samples of 1, 5, 10, and 20 ppm phosphate concentration from a 50 ppm phosphate stock solution.
2. Prepare a 2.2 mL microtube for each phosphate concentration with the following reagents: 840 μL distilled water, 50 μL ammonium molybdate solution, 10 μL tin chloride II solution, and 100 μL of each phosphate concentration (*see* Subheading 2.3.4 for reagent preparation).
3. Prepare a blank following the procedure described above but adding 100 μL of distilled water instead of phosphate solution.
4. Incubate at 25 °C for 15 min.
5. Read absorbance at 700 nm.
6. Represent the results of absorbance at 700 nm against phosphate concentration to obtain the regression equation (*see* **Note 13**).
7. Use this regression equation to calculate the phosphate concentration of the sample from the measured absorbance at 700 nm.

Phosphate Quantification Procedure

1. Add to a 2.2 mL microtube the following reagents: 840 μL distilled water, 50 μL ammonium molybdate solution, 10 μL tin chloride II solution, and 100 μL of the sample supernatant (*see* Subheading 2.3.4 for reagent preparation) (*see* **Note 14**).
2. Prepare a blank following the procedure described above but adding 100 μL of distilled water instead of sample.
3. Incubate at 25 °C for 15 min.
4. Read absorbance at 700 nm.

3.3.5 Carotenoid Analysis

Sample Extraction

1. Take 1–2 mL of the fermentation broth, and homogenize it with a laboratory vibration mill for 15 min using glass beads and 4 mL of methanol/dichloromethane (50/50).
2. Take 100 μL of the homogenized broth and mix with 900 μL of methanol/dichloromethane (50/50).
3. Stir at 50 °C for 20 min.
4. Centrifuge at 10,000 × *g* for 10 min.
5. Take 100 μL of supernatant, evaporate, and redissolve in 1 mL of acetone.

Carotenoid Quantification

1. Analyze 10 μL of the sample according to the following HPLC conditions: temperature 50 °C, flow 1 mL/min, and wavelength 470 nm (*see* **Note 15**).

4 Notes

1. Sterilize imidazole dissolved in water (20%) using a filter of nylon 0.2 μm.
2. Both strains, *B. trispora* F-816 (+) and *B. trispora* F-744 (−), can be maintained either in potato dextrose agar slants at room temperature for short-term maintenance or at −80 °C as a glycerol stock solution (40% v/v) for long-term maintenance.
3. Seed a TSA Petri plate and incubate at 31 °C for 24–48 h to detect potential microbial contaminations.
4. Use 7.02 and 4.02 buffer standard solutions for calibration.
5. It is advisable to do a pressure test before sterilization to check the tank is hermetically sealed. To do it, close the air output, open the air inlet until a pressure of 1 bar is reached inside the tank, and stop the airflow inlet. The 1 bar pressure should be maintained at least for 5 min. Before starting with the sterilization cycle, do not forget to open the air output to release the pressure.
6. A value of 0% can be obtained by briefly disconnecting the lead cable from the DO_2 electrode and setting the zero.
7. The (+) strain shows a higher growth rate than the (−) strain. This means that (+) strain may finish glucose before 48 h of incubation, PCV values of the (+) strain vegetative tank are higher than those for the (−) strain, and pH rises earlier due to the consumption of nitrogen sources from the culture medium for growth.
8. Higher seeding volume could lead to uncontrolled growth, with high viscosity and poor oxygenation, affecting to lycopene production.
9. Extreme proportions of both strains in the seeding can lead to a significant decrease in lycopene production, probably due to the inability of growth of one of the strains, thus avoiding mating stimulation for lycopene synthesis.
10. The most useful DO_2 control is the agitation cascade: when DO_2 decreases, the agitation increases to keep DO_2 level at about 30%. Another option is using a combined DO_2 control with agitation and aeration. Increasing aeration will ease DO_2 control without increasing too much agitation, which could affect negatively the microorganism growth during the first hours of growth.

11. Using lower imidazole concentrations, lycopene cyclase activity is not completely inhibited, leading to accumulation of β- and γ-carotene. On the other hand, concentrations of imidazole higher than 0.8 g/L affect biomass growth and reduce lycopene production.
12. The absorbance of the samples must be in the range of 0–1, so you may need to prepare a dilution of the sample.
13. The regression curve is a linear equation.
14. You may need to dilute the samples taking into account their expected phosphate content.
15. Prepare lycopene, β-carotene, and γ-carotene standards of the desired concentration in acetone to quantify lycopene and carotene production.

References

1. Britton G, Liaanen-Jensen S, Pfander H (2004) Carotenoids handbook. Birkhauser, Basel
2. Bauernfeind JC (1981) Carotenoids as colorants and vitamin A precursors: technical and nutritional applications. Academic Press, New York, NY
3. Kong KW et al (2010) Revealing the power of the natural red pigment lycopene. Molecules 15(2):959–987
4. Commission decision of 23 April 2009 authorising the placing on the market of lycopene as A novel food ingredient under regulation (EC) no. 258/97 of the European Parliament and the Council (notified under document number C(2009) 2975). Official Journal of the European Union 106, 55-9 (2009)
5. Commision decision of 23 October 2006 authorising the placing on the market of lycopene from *Blakeslea trispora* as a novel food ingredient under regulation (EC) no. 258/97 of the European Parliament and of the Council (notified under document number C(2006) 4973). Official Journal of the European Union (2006)
6. Müller L et al (2016) Lycopene and its antioxidant role in the prevention of cardiovascular diseases—a critical review. Crit Rev Food Sci Nutr 56(11):1868–1879
7. Aghajanpour M et al (2017) Functional foods and their role in cancer prevention and health promotion: a comprehensive review. Am J Cancer Res 7(4):740–769
8. BBC Research (2011). The global market for carotenoids [Online]. Available at: https://www.bccresearch.com/pressroom/fod/global-carotenoids-market-reach-$1.4-billion-2018. Accessed 29 May 2017
9. Ernst H (2002) Recent advances in industrial carotenoid synthesis. Pure Appl Chem 74:2213–2226
10. Ben-Yehuda M et al (1997) Industrial processing of tomatoes and lycopene extraction. WO Patent 97/48287
11. Mantzouridou F, Tsimidou MZ (2008) Lycopene formation in *Blakeslea trispora*, chemical aspects of a bioprocess. Trends Food Sci Technol 19(7):363–371
12. López-Nieto MJ et al (2004) Biotechnological lycopene production by mated fermentation of *Blakeslea trispora*. Appl Microbiol Biotechnol 66(2):153–159
13. Hsu WJ, Yokoyama H, DeBenedict C (1990) Chemical bioregulation of carotenogenesis in *Phycomyces blakesleeanus*. Photochemistry 29(8):2447–2451
14. McDermott JC, Britton G, Goodwin TW (1973) Carotenoid biosynthesis in a *Flavobacterium* sp.: stereochemistry of hydrogen elimination in the desaturation of phytoene to lycopene, rubixanthin and zeaxanthin. Biochem J 134(4):1115–1117
15. Bianchi ML et al (1969) Microbiological process for the production of lycopene. US Patent 3,467,579
16. Farmer WR, Liao JC (2000) Improving lycopene production in *Escherichia coli* by engineering metabolic control. Nat Biotechnol 18(5):533–537
17. Cho NR et al (2014) Method of producing lycopene using recombinant *Escherichia coli*. US Patent 8,828,697

18. Araya-Garay JM et al (2012) Construction of new *Pichia pastoris* X-33 strains for production of lycopene and β-carotene. Appl Microbiol Biotechnol 93(6):2483–2492
19. Matthäus F et al (2014) Production of lycopene in the non-carotenoid producing yeast *Yarrowia lipolytica*. Appl Environ Microbiol 80(5):1660–1669
20. Yamano S et al (1994) Metabolic engineering for production of beta-carotene and lycopene in *Sacharomyces cerevisiae*. Biosci Biotechnol Biochem 58(6):1112–1114
21. Chen Y et al (2016) Lycopene overproduction in *Saccaharomyces cerevisiae* through combining pathway engineering with host engineering. Microb Cell Factories 15(1):113
22. Miura Y et al (1998) Production of the carotenoids lycopene, beta-carotene, and astaxanthin in the food yeast *Candida utilis*. Appl Environ Microbiol 64(4):1226–1229
23. Lampila LE, Wallen SE, Bullerman LB (1985) A review of factors affecting biosynthesis of carotenoids by the order Mucorales. Mycopathologia 90(2):65–80
24. Wang JF et al (2012) Optimization of the mated fermentation process for the production of lycopene by *Blakeslea trispora* NRRL 2895 (+) and NRRL 2896 (−). Bioprocess Biosyst Eng 35(4):553–564
25. Mehta BJ, Obraztsova IN, Cerdá-Olmedo E (2003) Mutants and intersexual heterokaryons of *Blakeslea trispora* for production of beta-carotene and lycopene. Appl Environ Microbiol 69(7):4043–4048
26. Ninet L, Renaut JA, Tissier R (1968) Production of lycopene. US Patent 3(369):974
27. Wang Q et al (2013) Enhanced lycopene content in *Blakeslea trispora* by effective mutation-screening method. Appl Biochem Biotechnol 171(7):1692–1700
28. Sun Y, Yuan QP, Vriesekoop F (2007) Effect of two ergosterol biosynthesis inhibitors on lycopene production by *Blakeslea trispora*. Process Biochem 42(10):1460–1464

Chapter 16

HPLC Analysis of Carotenoids in Neurosporaxanthin-Producing Fungi

Dámaso Hornero-Méndez, M. Carmen Limón, and Javier Avalos

Abstract

The ascomycetous fungi *Fusarium fujikuroi* and *Neurospora crassa* are widely used as research models in the study of secondary metabolism and photobiology, respectively. Both fungi exhibit a similar carotenoid pathway, for which all the genes and enzymes have been identified. Under standard laboratory conditions, either *F. fujikuroi* or *N. crassa* accumulate a mixture of neurosporaxanthin, a carboxylic apocarotenoid acid, and several of its carotene precursors. We formerly described methods for the identification and quantification of neurosporaxanthin. However, the differences in polarity between this acidic xanthophyll and neutral carotenes make their global analysis cumbersome. Here we propose a simple HPLC methodology for the efficient separation of neurosporaxanthin and earlier pathway intermediates in a single HPLC run. This method should be useful to check the abundance of neurosporaxanthin under different experimental conditions and to evaluate the relative proportions of their different carotene precursors. To assess the validity of the method, we have compared the carotenoid profiles in samples of mycelia of *F. fujikuroi* and conidia of *N. crassa*, in both cases obtained from surface cultures of a wild strain of each species.

Key words RP-HPLC-DAD, Neurosporaxanthin, Xanthophyll, Acidic apocarotenoid, Neutral carotenoids, *Fusarium fujikuroi*, *Neurospora crassa*

1 Introduction

The fungi *Fusarium fujikuroi* and *Neurospora crassa* are well-known model systems for researching different biological problems, among which stand out the synthesis of secondary metabolites in the first case [1, 2] and the responses to light and circadian rhythmicity in the second [3, 4]. *F. fujikuroi* is a rice pathogen that produces gibberellins [5], growth-promoting plant hormones of the terpenoid family that share the first biosynthetic steps with carotenoids. In addition, this fungus produces bikaverin, fusarins, fusaric acid, fusarubin, among other secondary metabolites [6, 7]. On the other hand, the photobiology of *N. crassa* has received considerable attention because of the different photoresponses exhibited by this fungus [8] and the wide array of genetic tools for

Carlos Barreiro and José-Luis Barredo (eds.), *Microbial Carotenoids: Methods and Protocols*, Methods in Molecular Biology, vol. 1852, https://doi.org/10.1007/978-1-4939-8742-9_16,

its characterization, which allowed the identification of the relevant photoreceptor genes [3, 9]. Both fungal species have in common the induction by light of the synthesis of carotenoids, with neurosporaxanthin (NX) as a major end product [10, 11]. *F. fujikuroi* accumulates carotenoids mostly in the mycelium, but in the case of *N. crassa*, carotenogenesis is specially active in conidia [12], asexual spores whose formation is also stimulated by light.

NX was first identified as an acidic carotenoid in *N. crassa*, where it was found with neutral-colored carotenes, as lycopene, γ-carotene, and β-carotene [12]. The NX carotenoid pathway in this fungus was elucidated in the following years through the chemical identification of the accumulated carotenoids and the genetic and biochemical characterization of the genes and enzymes involved in the synthesis [11]. The chemical structure of NX was determined as a C35 apocarotenoid with a carboxylic acid end [13]. Earlier studies proposed the occurrence of a second xanthophyll, tentatively assigned as spirilloxanthin, formerly described in the bacterium *Rhodospirillum rubrum* [14]. The synthesis of spirilloxanthin by *N. crassa* was not corroborated in later reports, which, however, confirmed NX as the major product and added other neutral carotenoids, as ζ-carotene, 3,4-dehydrolycopene, and torulene, to those already found in this fungus [15, 16]. The NX precursors identified in *N. crassa* are consistent with the carotenoid pathway depicted in Fig. 1, in which four or five desaturations are achieved before the cyclization reaction. Chemical analysis of mutants in late steps of the pathway revealed altered patterns of the colored carotenoids, with the accumulation of different intermediates and the lack of the final acidic product [17–19]. Additionally, the order of the reaction steps was found to be altered when the fungus is illuminated at low temperature, leading to the accumulation of different carotene precursors of the NX biosynthetic pathway [20].

The first analyses of carotenoids in the wild type of *F. fujikuroi* were achieved together with those of a collection of mutants of the NX pathway [21]. Such analyses were consistent with previous studies in *Fusarium aquaeductuum* [22] and revealed the presence of similar carotenoid mixtures to those found in *N. crassa* with some exceptions, as the absence of 3,4-dehydrolycopene and the occurrence of β-zeacarotene in *F. fujikuroi*, that suggest differences in the order of the cyclization and the late desaturation reactions (Fig. 1).

We formerly described methods for NX analysis [23]. The chemical difference between this acidic xanthophyll and the neutral carotene precursors hinders the study in a single analysis of the carotenoid mixtures accumulated in these fungi. Here, we describe a reliable and simple HPLC procedure to separate neurosporaxanthin and most neutral precursors in a single chromatographic run. This method is useful to assess qualitative differences in the

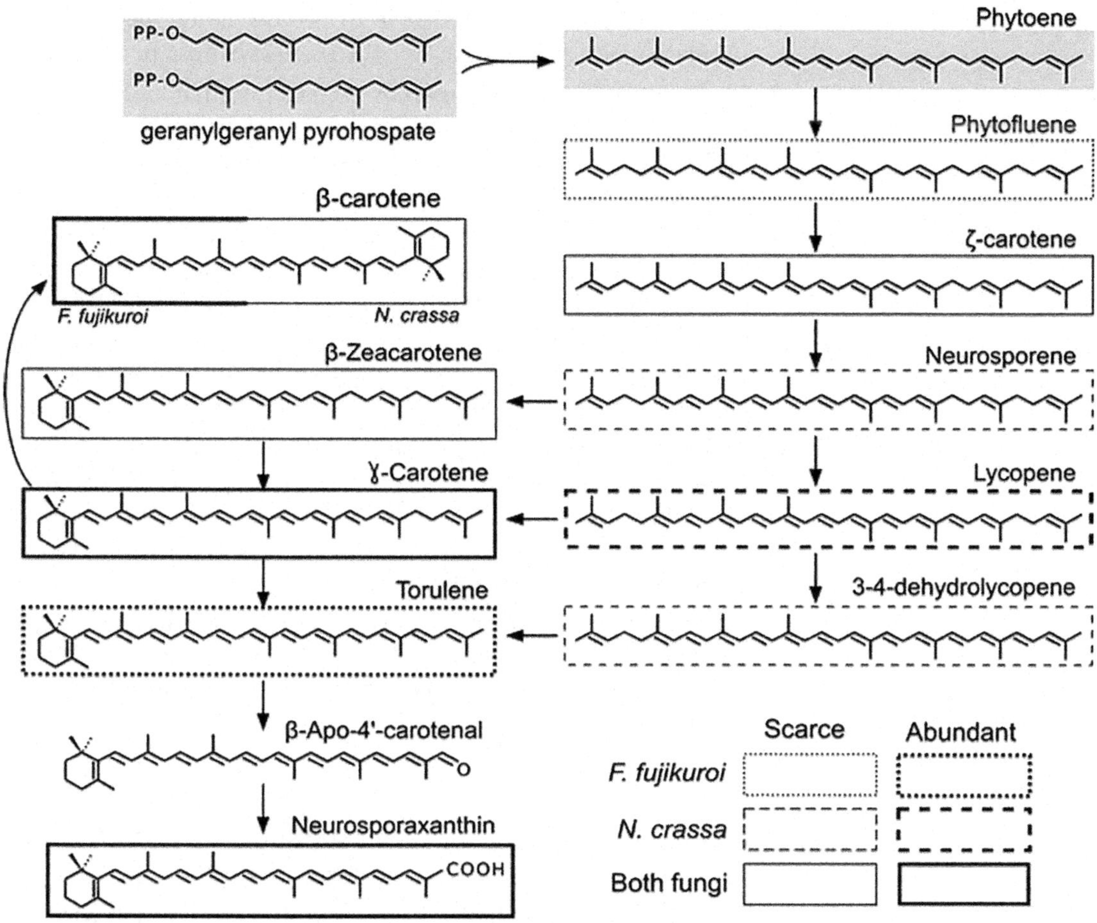

Fig. 1 Neurosporaxanthin biosynthetic steps in *F. fujikuroi* and *N. crassa*. Carotenoids found in either of these fungi in the analyses described in this chapter are boxed according to the indicated borders. Molecules not detected with the described HPLC method are shaded in gray

carotenoids mixtures between different *N. crassa* or *F. fujikuroi* strains or between different culture conditions. The method should be also applicable to any neurosporaxanthin-producing fungi, as other *Neurospora* [24] or *Fusarium* [10] species, *Verticillium* sp. [25, 26], or *Podospora anserina* [27]. Here we describe the procedures and the results obtained with carotenoid samples extracted from mycelia of *F. fujikuroi* and conidia of *N. crassa*, in both cases from wild strains grown under illumination.

2 Materials

2.1 Agar Cultures of *N. crassa* and *F. fujikuroi*

1. Wild strain IMI58289 of *Fusarium fujikuroi* (*Gibberella fujikuroi, mating population C*), obtained from the Imperial Mycological Institute, Kew, UK.

2. Wild strain ORSa (FGSC 2490) of *N. crassa*, obtained from the Fungal Genetics Stock Center (http://www.fgsc.net) [28].
3. Medium for *Neurospora crassa*: Vogel's minimal medium [29]. Add 15 g of sucrose, 15 g of agar, and 20 mL of Vogel salts to 965 mL distilled water. Sterilize by autoclaving for 20 min at 120 °C.
4. Vogel salts: dissolve 125 g of Na_3 citrate·$2H_2O$ in 775 mL of distilled water and dissolve afterward 250 g of KH_2PO_4, 100 g of NH_4NO_3, 10 g of $MgSO_4 \cdot 7H_2O$, 5 g of $CaCl_2 \cdot 2H_2O$ (formerly dissolved in 20 mL of H_2O to avoid precipitation), 2.5 mL of biotin solution (5 mg of biotin in 100 mL 50% EtOH), and 5 mL of microelement solution [29].
5. Medium for *Fusarium fujikuroi*: DG minimal medium [30]. Dissolve 30 g of glucose, 3 g of $NaNO_3$, 1 g of H_2KPO_4, 0.5 g of $MgSO_4 \cdot 7H_2O$, 0.5 g of KCl, 16 g of agar, and 2 mL of microelement solution in 970 mL of distilled water. Sterilize by autoclaving for 20 min at 120 °C.
6. Microelement solution: dissolve 0.5 mg of HBO_4, 5 mg of $CuSO_4$, 10 mg of $FeCl_3$, 1 mg of $MnCl_2$, 1 mg of $NaMoO_4$, and 100 mg of $ZnSO_4$ in 100 mL of distilled water [29].

2.2 Carotenoids Analyses

1. Lyophilizer.
2. Centrifuge and rotor for 50 mL Falcon tubes.
3. Centrifuge and rotor for microcentrifuge tubes (1.5–2 mL) with a minimum centrifugation force of 12,000 × *g*.
4. Cell disruptor: oscillating ball mill Retsch Model MM400 (Retsch, Haan, Germany).
5. HPLC system: Waters 2695 fitted with a Waters 2998 photodiode array detector and controlled with Empower2 software (Waters Cromatografía, S.A., Barcelona, Spain).
6. Reversed-phase C18 column Mediterranean SEA18, 3 mm, 20 × 0.46 cm (Teknokroma, Barcelona, Spain).
7. HPLC solvent system: (A) acetone (HPLC-grade acetone was supplied by BDH Prolabo, VWR International Eurolab, S.L., Barcelona, Spain), (B) deionized water (HPLC-grade ultrapure water produced with a Milli-Q Advantage A10 system, Merck Millipore, Madrid, Spain).

2.3 Equipment

1. Borosilicate filter: VitraPOR Filter-Crucible, ROBU® (Glasfilter-Geräte GmbH, Hattert, Germany).
2. New Brunswick Innova platform shaker (Eppendorf, Hamburg, Germany) or equivalent.
3. Vortex Heidolph REAX 2000 (Heidolph, Schwabach, Germany) or equivalent.

4. Lyophilizer Virtis SP Scientific Sentry 2.0 (SP Scientific, SP Industries, PA, USA).
5. Allegra™ X-22R centrifuge equipped with rotor for 50 mL Falcon tube tubes (Beckman Coulter, Brea, CA, USA).
6. Microcentrifuge Eppendorf 5424R (Eppendorf) or equivalent.
7. Oscillating ball mill Retsch Model MM400 (Retsch, Haan, Germany).
8. Waters 2695 Alliance chromatograph HPLC system fitted with a Waters 2998 photodiode array detector (Waters Cromatografía, S.A., Barcelona, Spain).

3 Methods

3.1 Preparation of a Mycelial Sample from a Culture of F. fujikuroi

1. Prepare Petri dishes of DG minimal medium (*see* **Note 1**) and inoculate with *F. fujikuroi*. In our case, inoculations are achieved with a sterile toothpick from a previously grown mycelial surface on seven symmetrically distributed positions on the agar (*see* **Note 2**).
2. Incubate at 30 °C for 1 week under light. In our case, the cultures were illuminated with four fluorescent tubes Philips TL-D 18W/840 at a distance of ca. 60 cm (*see* **Note 3**).
3. Separate mycelia from the agar with a clean scalpel blade. Remove excess of water of the mycelial sample by absorption on a filter paper, and lyophilize the sample.

3.2 Preparation of a Conidia Sample from a Culture of N. crassa

1. Prepare a 500 mL Erlenmeyer flask with 100 mL of Vogel's minimal agar. To facilitate conidiation, remove excess moisture on the flask walls letting it dry at 37 °C for 24 h (*see* **Note 4**).
2. Inoculate the flask with the wild strain of *N. crassa*. In our case, we transferred with a sterile inoculation loop a tiny mycelial sample from a slant culture to the center of the agar.
3. Incubate at 30 °C for 1 week under light. In our case, illumination was achieved as described for *F. fujikuroi* in the former section (*see* **Note 3**).
4. To obtain conidia, add 50 mL of 20% glycerol in water and agitate in an orbital shaker for 1 h at 200 rpm (*see* **Note 5**).
5. Remove mycelial debris by passing the conidia suspension through a borosilicate filter (*see* **Note 6**).
6. Centrifuge the filtered suspension at 3000 × *g* in 50 mL Falcon tubes at 4 °C for 5 min, remove supernatant, and add 20 mL of precooled water.
7. Repeat the former centrifugation step, remove supernatant, and lyophilize the conidia precipitate.

3.3 Carotenoid Extraction in Cell Disruptor

1. Transfer approximately 0.1 g of dry samples to a 2 mL microcentrifuge tube (preferably with a safe-lock click), and add two stainless-steel balls (5 mm Ø) and 1 mL acetone (*see* **Notes 7** and **8**) containing 0.1% BHT (*see* **Note 9**).
2. Mill in a cell disruptor equipment till bleaching of the sample. In our case, it was used an oscillating ball mill at 25 Hz for 1 min (*see* **Note 10**).
3. Centrifuge the extraction tubes at maximal speed (>12,000 × *g*) for 5 min at 4 °C, and transfer the acetone supernatant to a clean microcentrifuge tube.
4. If the pellet is still colored, repeat **steps 2–3** (*see* **Note 11**).
5. Evaporate the collected acetone phases under a gentle N_2 stream.
6. Dissolve the dry extract in 0.5 mL of HPLC grade acetone.
7. Centrifuge the sample at maximal speed (>12,000 × *g*) for 5 min at 4 °C (*see* **Note 12**).
8. Transfer the extract to a sample vial and analyze in the HPLC (*see* **Note 13**).

3.4 HPLC Separation of the Carotenoid Samples

1. Inject 5 μL of the sample in the HPLC system (*see* **Note 14**).
2. Run the sample with a binary gradient elution (A, acetone; B, deionized water) at a flow rate of 1 mL/min.
3. As a gradient elution system, use an initial composition of 75% A and 25% B, increase linearly to 95% A in 10 min, then raise to 100% A in 2 min, and maintain constant for 10 min. Before start, allow reaching initial conditions for 5 min. This elution system is based, with minor modifications, in the HPLC method of Mínguez-Mosquera and Hornero-Méndez [31], currently used for carotenoid profiling of different food matrices (especially for fruit and vegetables).
4. Keep column temperature at 25 °C (*see* **Note 15**).
5. Perform detection with a diode array detector, recording the signal at 450 nm and acquiring online spectra in the 350–600 nm wavelength range.
6. Analyze chromatogram at appropriate wavelength. In the case of the *F. fujikuroi* and *N. crassa samples*, 450 nm is chosen because of the high absorption of most carotenoids at this wavelength. Detection is performed at 364 nm for phytofluene.

3.5 Analysis of the Chromatograms of the F. fujikuroi and N. crassa Samples

1. Under the reversed-phase HPLC conditions used, the carotenoids with the higher polarity are eluted earlier than those with lower polarity (carotenes).

2. The use of a photodiode array detector allows retrieving the absorption spectra of all the peaks eluted from the column. The software of the HPLC system allows retrieving bidimensional (time × UV/visible absorption wavelength) or tridimensional (absorbance units × time × UV/visible absorption wavelength) representations of the elution profiles (Fig. 2). These representations are useful to choose an optimal wavelength for the chromatogram(s), as well as to select and compare the acquired UV/visible spectra of the eluted carotenoids.
3. Comparison of the UV/visible spectra with published data for the presumed NX precursors [32, 33] allows the identification of most of the peaks in the chromatograms.
4. The chromatogram of *F. fujikuroi* shows four major peaks (Fig. 3), corresponding to NX (1), torulene (2), γ-carotene (3), and β-carotene (6), representing together a majority of the carotenoids accumulated by this species under the investigated conditions. Minor peaks of ζ-carotene (4), β-zeacarotene (5), and phytofluene (7) are also detected, confirming former data on the capacity of the *F. fujikuroi* cyclase to recognize neurosporene as a substrate. A secondary peak (2) with the same UV/visible absorption spectrum of NX, therefore having a similar or identical chromophore but eluting later, may be interpreted as a NX isomer or a NX-derived ester. Identification of this carotenoid, and those for other unidentified peaks found in *N. crassa*, would require more refined chemical analysis.
5. The chromatogram of *N. crassa* conidia shows also four major peaks (1, 4, 6 and 10–11, Fig. 4), corresponding to NX (1), lycopene (4), γ-carotene (6), and an unidentified carotenoid (10), eluting together with a low amount of β-carotene (11). The chromatogram included also several minor peaks, which are identified as 3,4-dehydrolycopene (3), neurosporene (5), ζ-carotene (8), β-zeacarotene (9), and β-carotene (coeluting with 10). Four of the minor peaks (2, 7, 10, 12) could not be identified with the available information.
6. The comparison of the *N. crassa* and *F. fujikuroi* samples shows similar carotenoid mixtures, with NX as one of the predominant carotenoids (Fig. 2). NX has the same chemical nature in both fungi, as indicates its similar elution time and absorption spectra. The carotenoid mixture has a higher complexity in the case of *N. crassa*, with some carotenoids that could not be identified. In accordance with former analyses [21, 34], lycopene and 3,4-dehydrolycopene are not found in *F. fujikuroi*, confirming the preference for an earlier cyclization step in this fungus.

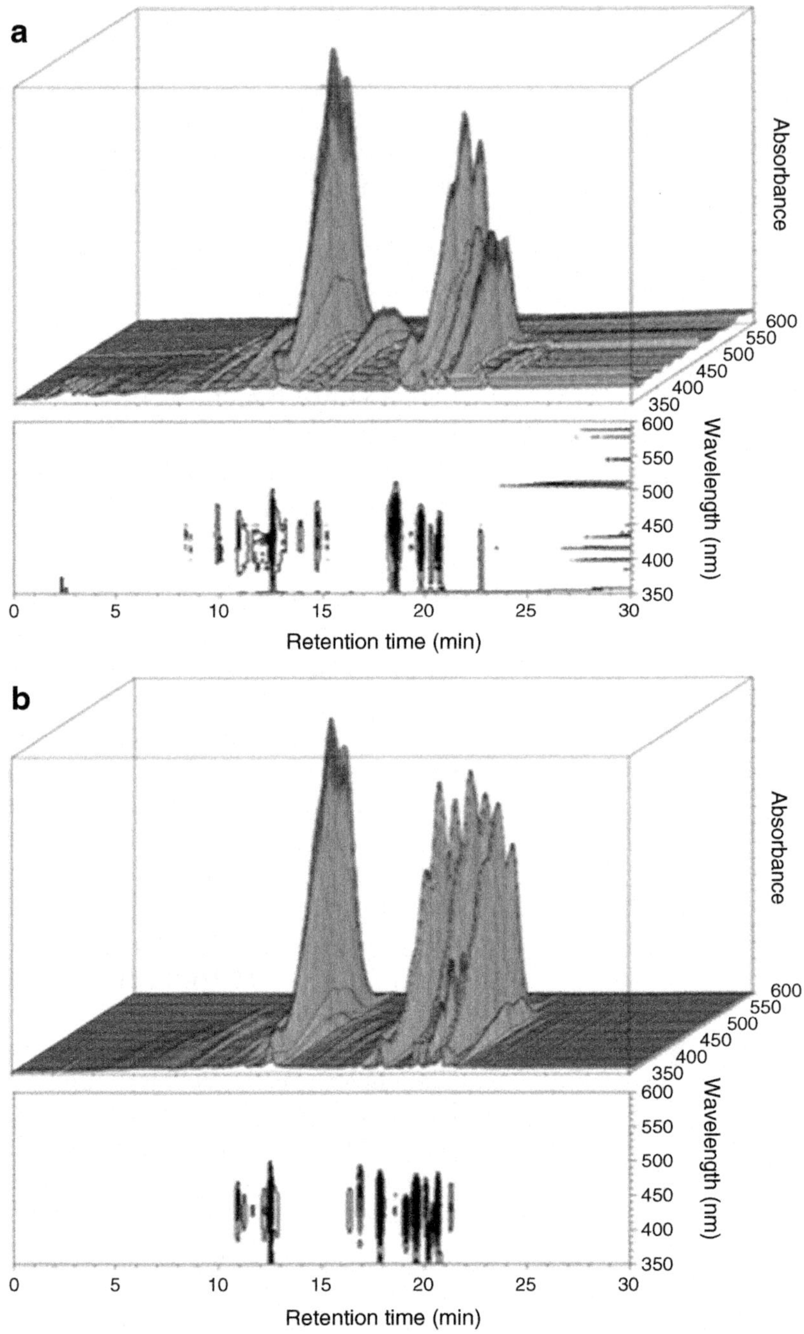

Fig. 2 Bidimensional (lower panels) and tridimensional (upper panels) representations of wavelength-absorption properties of the HPLC chromatograms displayed in Fig. 3 (**a**, *F. fujikuroi*) and Fig. 4 (**b**, *N. crassa*)

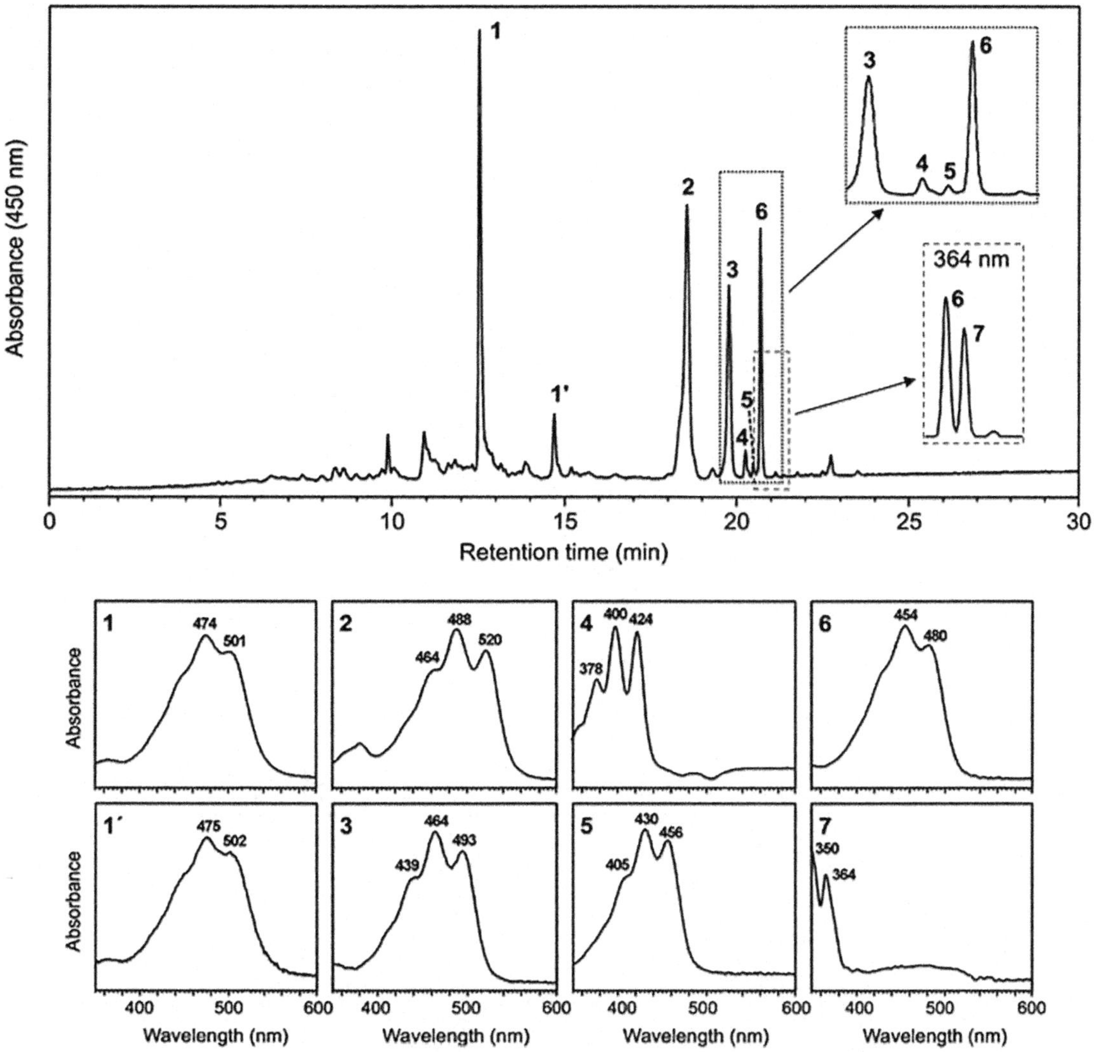

Fig. 3 Upper panel: chromatogram of the HPLC separation of the carotenoids accumulated by the wild strain of *F. fujikuroi* (detection wavelength at 450 nm). Lower graphs: UV/visible spectra of the peaks indicated on the chromatogram. Peaks assignation: 1, neurosporaxanthin; 1′, neurosporaxanthin isomer; 2, torulene; 3, γ-carotene; 4, ζ-carotene; 5, β-zeacarotene; 6, β-carotene; 7, phytofluene (detectable at 364 nm)

7. The late NX precursor β-apo-4′-carotenal is not detected in any of the analyzed chromatograms. This may be due to its rapid conversion to NX or to a high instability of this carotenoid, as previously suggested by the phenotype exhibited by the mutants blocked in this step of the pathway in *F. fujikuroi* [35] and *N. crassa* [19].

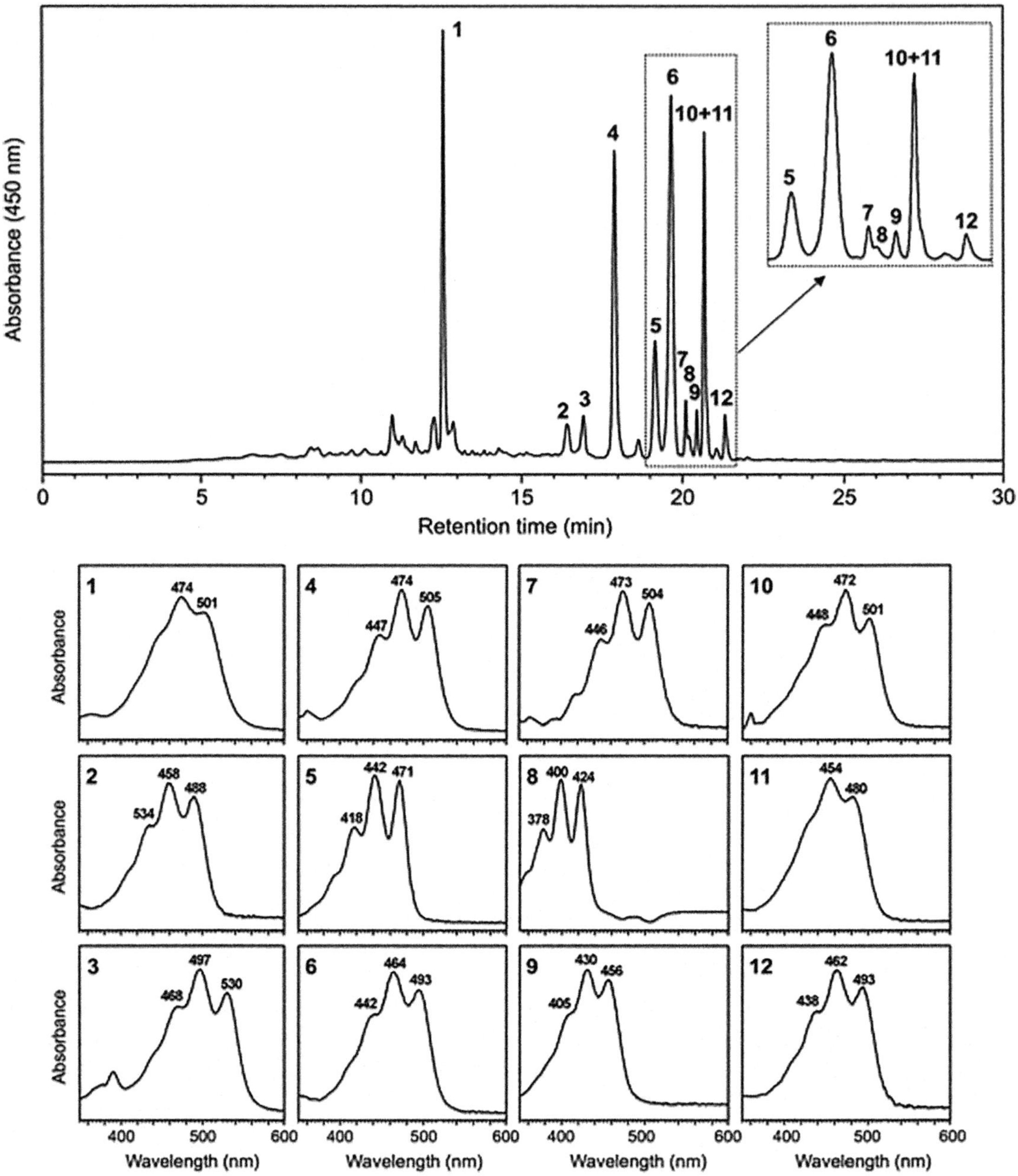

Fig. 4 Upper panel: chromatogram of the HPLC separation of the carotenoids accumulated in the conidia of the wild strain of *N. crassa* (detection wavelength at 450 nm). Lower graphs: UV/visible spectra of the peaks indicated on the chromatogram. Peaks assignation: 1, neurosporaxanthin; 2, not identified; 3, 3,4-dehydrolycopene; 4, lycopene; 5, neurosporene; 6, γ-carotene; 7, not identified; 8, ζ-carotene; 9, β-zeacarotene; 10, not identified (major component); 11, β-carotene (coelutes with 10 as minor component); 12, not identified

4 Notes

1. This has been the medium most frequently used in carotenoid studies in *F. fujikuroi*. In some works, a variant of this medium was used in which $NaNO_3$ was replaced by the same amount of asparagine (usually called DGasn medium).
2. The incubation of several colonies in the same Petri dish is recommended because of the slower radial growth of *F. fujikuroi* compared to other fungi, as *N. crassa*.
3. The fluorescent tubes should not be closer to avoid excessive heating of the cultures.
4. Moisture on the flask walls reduces the capacity of *N. crassa* to develop aerial mycelia on their surface.
5. Because of conidia hydrophobicity, the use of the glycerol solutions is recommended to facilitate suspension of the conidia.
6. If borosilicate filters are not available, the conidia suspension may be passed through a standard filter paper.
7. Acetone is an excellent solvent for both extraction and chromatography of carotenoids. However, acetone has a UV/visible cutoff wavelength of 350 nm which impaired the acquisition of the UV/visible at lower values. For this reason the phytofluene spectrum is uncompleted, and phytoene detection is not possible (270–300 nm).
8. All operations should be performed under dimmed light to prevent isomerization and photodegradation of carotenoids.
9. An antioxidant (BHT, butylhydroxytoluene) is added to the extraction solvent (acetone) at a concentration of 0.1% (w/v) in order to minimize the oxidative degradation of pigment during and after extraction.
10. Before and after cell disruption and pigment extraction, keep samples on ice to counteract the effect of heat during milling over the pigments.
11. Our extraction protocol is optimized for performing cell disruption and extraction in a single step. This is usually enough for most samples, but trials should be performed with unknown samples.
12. Prior to the chromatographic analysis, samples must be centrifuged at 13,000 × *g* for 5 min at 4 °C in order to clean the sample from any particles that might clog the column and the chromatographic system.
13. If carotenoids are not analyzed immediately after the extraction, samples should be stored at −30 °C (or lower) until chromatographic analysis (HPLC) in order to prevent degradation

of carotenoids. If possible, the air from the sample vial should be displaced with a stream of N_2.

14. If available, the injection should be achieved with an automatic injector.
15. If possible, the chromatographic system should be fitted with a cooling system for samples ($T < 15$ °C).

Acknowledgments

We thank the Spanish Government (projects BIO2012-39716, BIO2015-69613-R, AGL2014-53195R, and CaRed Network BIO2015-71703-REDT) and Junta de Andalucía (project CTS-6638) for financial support.

References

1. Wiemann P et al (2013) Deciphering the cryptic genome: genome-wide analyses of the rice pathogen *Fusarium fujikuroi* reveal complex regulation of secondary metabolism and novel metabolites. PLoS Pathog 9(6):e1003475
2. Avalos J et al (2007) Gibberellins and other metabolites of *Fusarium fujikuroi* and related fungi. Curr Org Chem 11(8):721–737
3. Chen CH, Dunlap JC, Loros JJ (2010) *Neurospora* illuminates fungal photoreception. Fungal Genet Biol 47(11):922–929
4. Baker CL, Loros JJ, Dunlap JC (2012) The circadian clock of *Neurospora crassa*. FEMS Microbiol Rev 36(1):95–110
5. Tudzynski B (2005) Gibberellin biosynthesis in fungi: genes, enzymes, evolution, and impact on biotechnology. Appl Microbiol Biotechnol 66(6):597–611
6. Studt L, Tudzynski B (2014) Gibberellins and the red pigments bikaverin and fusarubin. In: Martín J-F, García-Estrada C, Zeilinger S (eds) Biosynthesis and molecular genetics of fungal secondary metabolites. Springer New York, New York, NY, pp 209–238
7. Niehaus E-M et al (2014) Fusarins and fusaric acid in fusaria. In: Martín J-F, García-Estrada C, Zeilinger S (eds) Biosynthesis and molecular genetics of fungal secondary metabolites. Springer New York, New York, NY, pp 239–262
8. Corrochano LM, Avalos J (2010) Light sensing. In: Borkovich K, Ebbole D, Momany M (eds) Cellular and molecular biology of filamentous fungi. ASM Press, Washington, DC, pp 417–441
9. Chen C-H, Loros JJ (2009) *Neurospora* sees the light: light signaling components in a model system. Commun Integr Biol 2(5):448–451
10. Avalos J, Estrada AF (2010) Regulation by light in *Fusarium*. Fungal Genet Biol 47(11):930–938
11. Avalos J, Corrochano LM (2013) Carotenoid biosynthesis in *Neurospora*. In: Kasbekar DP, McCluskey K (eds) *Neurospora* genomics and molecular biology. Caister Academic Press, Norfolk, pp 227–241
12. Zalokar M (1954) Studies on biosynthesis of carotenoids in *Neurospora crassa*. Arch Biochem Biophys 50(1):71–80
13. Aasen AJ, Jensen SL (1965) Fungal carotenoids II. The structure of the carotenoid acid neurosporaxanthin. Acta Chem Scand 19(8):1843–1853
14. Goodwin TW, Osman HG (1953) Studies in carotenogenesis. 9. General cultural conditions controlling carotenoid (spirilloxanthin) synthesis in the photosynthetic bacterium *Rhodospirillum rubrum*. Biochem J 53(4):541–546
15. Harding RW, Mitchell HK (1968) The effect of cycloheximide on carotenoid biosynthesis in *Neurospora crassa*. Arch Biochem Biophys 128(3):814–818
16. Harding RW, Huang PC, Mitchell HK (1969) Photochemical studies of the carotenoid biosynthetic pathway in *Neurospora crassa*. Arch Biochem Biophys 129(2):696–707
17. Arrach N, Schmidhauser TJ, Avalos J (2002) Mutants of the carotene cyclase domain of *al-2*

from *Neurospora crassa*. Mol Genet Genomics 266(6):914–921

18. Saelices L et al (2007) Identification of the gene responsible for torulene cleavage in the *Neurospora* carotenoid pathway. Mol Genet Genomics 278:527–537
19. Estrada AF et al (2008) The *ylo-1* gene encodes an aldehyde dehydrogenase responsible for the last reaction in the *Neurospora* carotenoid pathway. Mol Microbiol 69(5):1207–1220
20. Estrada AF et al (2008) Novel apocarotenoid intermediates in *Neurospora crassa* mutants imply a new biosynthetic reaction sequence leading to neurosporaxanthin formation. Fungal Genet Biol 45(11):1497–1505
21. Avalos J, Cerdá-Olmedo E (1987) Carotenoid mutants of *Gibberella fujikuroi*. Curr Genet 11:505–511
22. Bindl E, Lang W, Rau W (1970) Untersuchungen über die lichtabhängige Carotinoidsynthese. VI. Zeitlicher Verlauf der Synthese der einzelnen Carotinoide bei *Fusarium aquaeductuum* unter verschiedenen Induktionsbedingungen. Planta 94(2):156–174
23. Avalos J, Prado-Cabrero A, Estrada AF (2012) Neurosporaxanthin production by *Neurospora* and *Fusarium*. Methods Mol Biol 898:263–274
24. Luque EM et al (2012) A relationship between carotenoid accumulation and the distribution of species of the fungus *Neurospora* in Spain. PLoS One 7(3):e33658
25. Valadon LR, Heale JB (1965) Carotenoids in mutants of *Verticillium albo-atrum*. J Gen Microbiol 41(3):341–348
26. Osman M, Valadon LR (1978) Studies on the near-UV effect on carotenogenesis in *Verticillium agaricinum*. Microbios 23(91): 53–64
27. Strobel I et al (2009) Carotenoids and carotenogenic genes in *Podospora anserina*: engineering of the carotenoid composition extends the life span of the mycelium. Curr Genet 55:175–184
28. McCluskey K (2003) The Fungal Genetics Stock Center: from molds to molecules. Adv Appl Microbiol 52:245–262
29. Davis RH, de Serres FJ (1970) Genetic and microbiological research techniques for *Neurospora crassa*. Methods Enzymol 17:79–143
30. Avalos J, Casadesús J, Cerdá-Olmedo E (1985) *Gibberella fujikuroi* mutants obtained with UV radiation and *N*-methyl-*N'*-nitro-*N*-nitrosoguanidine. Appl Environ Microbiol 49:187–191
31. Mínguez-Mosquera MI, Hornero-Méndez D (1993) Separation and quantification of the carotenoid pigments in red peppers (*Capsicum annuum* L.), paprika, and oleoresin by reversed-phase HPLC. J Agric Food Chem 41(10):1616–1620
32. Britton G, Liaaen-Jensen S, Pfander H (2004) Carotenoids: handbook. Birkhauser Verlag, Basel
33. Britton G (1995) UV/visible spectroscopy. In: Britton G, Liaaen-Jensen S, Pfander H (eds) Carotenoids. Birkhäuser Verlag, Basel, pp 13–62
34. Prado-Cabrero A et al (2009) Deviation of the neurosporaxanthin pathway towards β-carotene biosynthesis in *Fusarium fujikuroi* by a point mutation in the phytoene desaturase gene. FEBS J 276(16):4582–4597
35. Díaz-Sánchez V et al (2011) The gene *carD* encodes the aldehyde dehydrogenase responsible for neurosporaxanthin biosynthesis in *Fusarium fujikuroi*. FEBS J 278(17):3164–3176

Chapter 17

Extraction and Analysis of Carotenes and Xanthophylls Produced by *Xanthophyllomyces dendrorhous*

Carlos García-Estrada, Katarina Kosalková, and Isabel-Clara Sánchez-Orejas

Abstract

Xanthophyllomyces dendrorhous, a heterobasidiomycetous yeast that represents the teleomorphic state of *Phaffia rhodozyma*, is used as a natural source of several carotenoids, such as the xanthophyll astaxanthin. Here, we describe the culture procedure for the production of carotenoids in *X. dendrorhous* and a simple and rapid analytical method for the optimized extraction and HPLC determination of intracellular β-carotene, astaxanthin, canthaxanthin, and zeaxanthin.

Key words Carotenoids, Xanthophylls, β-Carotene, Astaxanthin, Zeaxanthin, Canthaxanthin, HPLC, *Xanthophyllomyces dendrorhous*

1 Introduction

Carotenoids are lipophilic and water insoluble secondary metabolites that belong to the family of terpenoid (the majority being C_{40} compounds) organic pigments. They are usually composed of eight isoprene units and have a long polyene central chain of conjugated double bonds that constitutes the chromophoric system, which is responsible for the characteristic color of these compounds.

There are more than 600 naturally occurring carotenoids, the majority being in the all-trans form [1], which are naturally produced by all photosynthetic species (including plants and algae), and by some classes of fungi and non-photosynthetic bacteria [2–5]. In general, animals are unable to synthesize carotenoids, which must be incorporated through diet. According to the oxygenation degree, these molecules can be classified into non-oxygenated carotenoids known as carotenes (e.g., beta-carotene, alpha-carotene, or lycopene) and oxygenated carotenoids with a hydroxy, epoxy, and/or oxo group known as xanthophylls (e.g., lutein,

Carlos Barreiro and José-Luis Barredo (eds.), *Microbial Carotenoids: Methods and Protocols*, Methods in Molecular Biology, vol. 1852, https://doi.org/10.1007/978-1-4939-8742-9_17,

zeaxanthin, canthaxanthin, or astaxanthin) [6]. Carotenoids are responsible for the bright colors (from yellow orange to reddish) of plants and animals and play different roles, as a source of retinoids (vitamin A precursors), as coloring and antioxidant agents, as a part of light harvesting systems, and as photoprotectors of photosynthetic machinery in photosynthetic organisms [7, 8].

Due to the increased demand for these natural products, carotenoids market in 2019 is supposed to reach $1.8 billion with a compound annual growth rate of 3.9% [9]. Biotechnological production of natural carotenoids for their use in cosmetics, food and feed, and chemical and pharmaceutical industries is achieved by means of algae [10] and yeasts [11]. One of these yeast species is *Xanthophyllomyces dendrorhous*, a red/pink-pigmented heterobasidiomycete that was isolated in the late 1960s from tree-exudates in Japan and Alaska [12]. This microorganism represents the teleomorphic state of *Phaffia rhodozyma* [13] and is mainly used for the production of the xanthophyll astaxanthin, whose global market in 2014 was valued at US$447 million and is expected to reach a value of US$1.1 billion by 2020 [14].

X. dendrorhous is able to produce several carotenoids, including beta-carotene, canthaxanthin, zeaxanthin, and astaxanthin (Fig. 1). One of the most widespread compounds among the carotenes is beta-carotene. This molecule, which has a strong red-orange color, possesses beta-rings at both ends. Beta-carotene also serves as an intermediary molecule for the biosynthesis of the xanthophyll astaxanthin [15, 16] and appears to be the most important vitamin A precursor for vertebrate animal species [17]. Canthaxanthin (β,β-carotene-4,4′-dione) is a xanthophyll of orange-red color with antioxidant properties that together with astaxanthin is widely used in aquaculture (to provide the characteristic salmonid flesh color) and poultry farming (as coloring agent of chicken skin and egg yolk) [18, 19]. Zeaxanthin (β,β-carotene-3,3′-diol) is a xanthophyll alcohol of yellow color that is present in high concentrations, together with lutein and meso-zeaxanthin, within the retina (especially in the macula) of the eye [20, 21]. Many studies have suggested that zeaxanthin and lutein play a protective role against various eye diseases, especially late forms of age-related macular degeneration [22]. The red-orange lipid-soluble pigment astaxanthin [3,3′-dihydroxy-β,β-carotene-4,4′-dione] has strong antioxidant properties and, as indicated above, is used as a feed supplement in aquaculture and egg production [18, 23]. Several studies have reported on the beneficial effects of astaxanthin in cardiovascular, immune, inflammatory, carcinogenic, and neurodegenerative diseases [24–28].

In *X. dendrorhous*, carotenoids are synthesized and stored into the cells. Therefore, recovery of these compounds requires extraction processes. In addition, carotenoids are highly unstable and can isomerize and oxidize during the analysis because of the presence

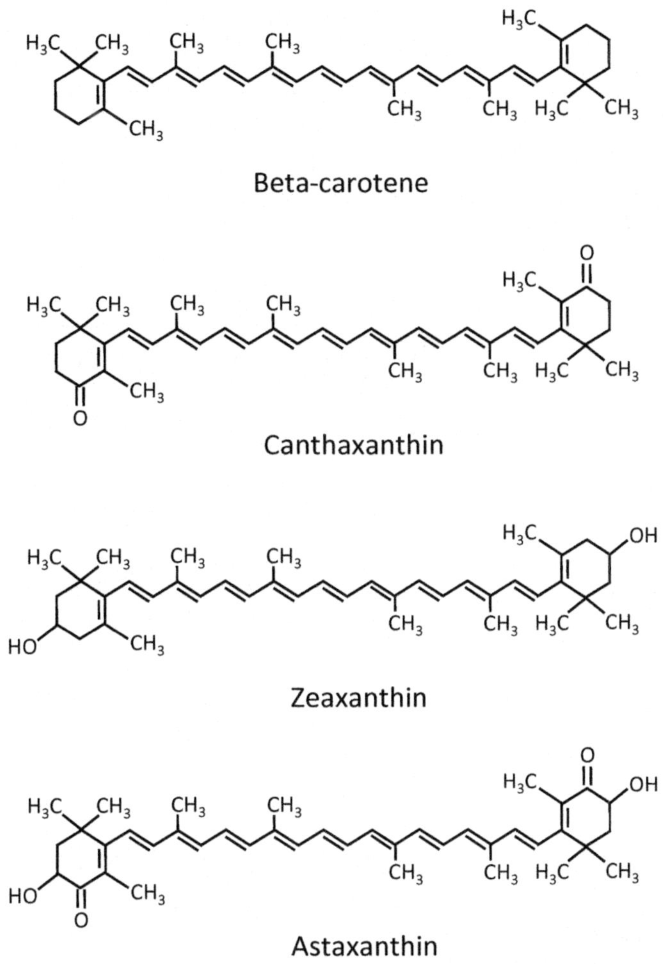

Fig. 1 Chemical structure of different carotenoids

of long-chain conjugated double bonds [29]. This processes can be minimized by means of making a rapid analysis, using antioxidants, protecting the compounds of light, and avoiding high temperatures and contact with acids [30–32]. The traditional methods used for extraction of carotenoids, which are carried out with solvents such as chloroform and benzene, may create a safety problem. In order to prevent it, chilled acetone can be used to homogenize the sample [33–35]. DMSO for cell breaking [33, 34] and *n*-hexane [36] for the final extraction are commonly used.

HPLC is the method of choice for separation and quantification of carotenes and xanthophylls [35, 37]. Both normal-phase [36, 38–40] and reversed-phase [41–43] chromatographic conditions have been widely used. The diode-array detector (DAD) is by far the most common detector used in carotenoids analysis as it makes use of the absorption properties of these compounds in the visible region. Several studies report the use of a Nucleosil 100

NH2 column [38] with a mobile phase consisting of ethyl acetate/*n*-hexane (80:20) [40]. Spectra of carotenoids cannot be used as conclusive evidence of structure, since many carotenoids present very similar absorption profiles [43].

2 Materials

All solutions must be prepared with ultrapure water (prepared by purifying deionized water, to attain a sensitivity of 18 MΩ cm at 25 °C) and analytical grade (peroxide-free) reagents.

Exposure to light, especially direct sunlight or UV light, induces trans-cis photoisomerization and photodestruction of carotenoids. Thus, carotenoid work must be done with amber glassware or wrapping the tubes containing carotenoids with aluminum foil and under subdued or red light [30–32].

2.1 Microbiology Materials

1. *X. dendrorhous* CECT 11028 (*X. dendrorhous* ATCC 24203), the wild-type astaxanthin producing strain, was employed in this study.
2. YM medium: 10 g glucose, 3 g malt extract, 3 g yeast extract, 5 g peptone, and 1 L water. pH was adjusted to 6.5, and the medium was sterilized in autoclave. For the preparation of YM solid medium, agar (20 g) was added prior sterilization.

2.2 Extraction Reagents and Equipment

1. 20% (w/v) NaCl (must be prepared using ultrapure water).
2. Analytical Balance.
3. Centrifuge.
4. Vortex.
5. Water bath at 60 °C.
6. Centrifugal evaporator.
7. Filtration system.
8. 400–600 μm diameter glass beads.
9. 0.22 μm syringe filters.
10. 10-mL amber glass tube with screw cap.
11. Acetone (must be chilled at 4 °C).
12. Dimethyl sulfoxide (DMSO) preheated at 60 °C.
13. *n*-Hexane.
14. Ethyl acetate/*n*-hexane (80:20).

2.3 Chromatrography

1. Mobile phase: ethyl acetate/*n*-hexane (80:20). Mobile phase is used to dissolve standards and samples after evaporation. Ethyl acetate and *n*-hexane should be of HPLC quality (*see* **Note 1**).

Measure 800 mL ethyl acetate with a 1 L graduate cylinder. Add into a 1 L volumetric flask and add *n*-hexane to complete 1 L volume (*see* **Note 2**). This mixture was filtered through a 0.45 μm filter before use in a filtration system. This removes any particulate matter that may cause blockages. After filtration, the solvents should be stored in a covered reservoir to prevent contamination with dust, etc. (*see* **Note 3**). The freshly prepared mobile phase should be thoroughly degassed by sonication for 15 min before it is pumped around the HPLC system in order to remove all dissolved gasses. Before attempting any solvent change, ensure that the solvent is already in the system and column is compatible with the new solvent. If the miscibility or physical properties of the two solvents are unknown, then it is better to mix the solvents in a beaker to see the reaction than to go ahead and pump the second solvent into the first on the HPLC instrument mixing problems that are easier to rectify before HPLC (*see* **Note 4**).

2. Standards: dissolve 10 mg of each zeaxanthin, β-carotene, canthaxanthin, and astaxanthin standard in 10 mL of ethyl acetate/*n*-hexane (80:20) in order to prepare 1 mg/mL stock solutions. Aliquots can be kept at 4 °C for current use and stored at −20 °C for up to 1 month. Standards were prepared at 0.2 mg/mL in ethyl acetate/*n*-hexane (80:20). Zeaxanthin, β-carotene, canthaxanthin, and astaxanthin standards with a purity of 98% were obtained from Sigma-Aldrich.
3. HPLC apparatus: Agilent system equipped with pump (G1312A), cooled auto sampler (G13130B), UV-visible photo-diode-array detector (G1315B-DAD), Peltier (G1330B), thermostatic column oven (G1316A TCC), Eluent Degasser (G1379A), and Chem Station LC 3D software (Rev. A.10.02) from Agilent, which is used for instrument control, data acquisition, and data processing.
4. Column 250 × 4.6 mm Nucleosil 100 NH2 5 μm (Supelco, Sigma-Aldrich).
5. Guard column 10 × 4.6 mm Nucleosil 100 NH_2 5 μm (Supelco, Sigma-Aldrich).

3 Methods

Caution should be taken to avoid losses of carotenes and xanthophylls. In addition to protect samples from direct light exposure, the analysis should be carried out in the shortest possible timeframe, avoiding high temperature and contact with acids and using high purity solvents free of peroxides (*see* **Note 5**).

3.1 Carotenoids Production in Submerged Cultures

1. For lab-scale fermentation experiments, inoculate one individual colony of *X. dendrorhous* grown on YM solid medium into 500-mL flasks containing 100 mL of YM liquid medium.
2. Shake the flasks at 220 rpm for 96 h at 21 °C.
3. Take 2-mL samples and proceed with the extraction protocol.

3.2 Extraction of Carotenoids

1. Transfer 2-mL samples obtained as indicated above to a 10-mL amber glass tube with screw cap and centrifuge it at 2500 × *g* for 5 min at 4 °C.
2. Add 1 mL of chilled acetone and vortex 1 min.
3. Centrifuge again at 2500 × *g* for 5 min at 4 °C. Remove the acetone with the extracellular carotenoids.
4. Add 1 mL of milli Q water and vortex for 1 min.
5. Centrifuge at 2500 × *g* for 5 min at 4 °C and remove the supernatant.
6. Disrupt the cells by adding 1 mL of glass beads (diameter 400–600 μm) and 1 mL DMSO (preheated at 60 °C), and vortex. The mixture is incubated 10 min at 60 °C.
7. Stir the mixture, and add 1 mL of chilled acetone, 1 mL of *n*-hexane, and 1 mL of 20% (w/v) NaCl. Mix gently (*see* **Note 6**).
8. Vortex 1 min and centrifuge for 10 min at 2500 × *g* at 4 °C.
9. Collect the upper phase (hexane) in a new microcentrifuge tube.
10. Add 1 mL of *n*-hexane, and repeat the procedure from **step 7** until the mixture becomes colorless (*see* **Note 7**).
11. Evaporate the hexane fraction mix to dryness in a centrifugal evaporator.
12. Keep the samples dry, in darkness, at −20 °C for further usage (less than a month) (*see* **Note 8**).
13. Resuspend the pellet in 1 mL ethyl acetate/*n*-hexane (80:20), filter with a 0.22 μm syringe filter, and transfer to a vial for chromatographic analysis.

3.3 Chromatographic Method

The method indicated below is based on that described by [44].

1. The mobile phase consisted in ethyl acetate/*n*-hexane (80:20).
2. Use isocratic elution (*see* **Note 9**) with the flow rate at 0.8 mL/min.
3. Set the run time at 20 min.
4. Wavelength detection is set up at 474 nm (*see* **Note 10**). Absorption spectra of each standard was measured between

200 and 500 nm and stored in a spectra library in order to support the identification.

5. During analysis the tray compartment containing sample vials is cooled at 4 °C.
6. After each working period (approximately 50 samples), it is necessary to rinse the column with *n*-hexane and 2% ethyl acetate at a flow rate of 0.8 mL/min for 20 min in order to eliminate highly hydrophobic compounds and prevent the loss of column efficiency (*see* **Note 11**).
7. The purge valve frit of the pump needs to be replaced every 200 samples approximately, to avoid excessive pressure (*see* **Note 12**).

3.4 Calibration Curve

1. Since samples are evaporated to dryness in speed back, the mobile phase represents the blank sample.
2. Each standard solution is injected alone to identify their retention time and spectra as parameters characterizing each molecule (Fig. 2).
3. A nine-point calibration curve is built from astaxanthin standard solution (Fig. 3).

3.5 Chromatographic Analysis

1. Inject a total of 20 μL of each sample for HPLC analysis.
2. For qualitative identification, do direct comparison between samples and the corresponding standard (Fig. 4).
3. Calculate concentrations of carotenoids in the samples using a linear regression $ax + b$ (concentration vs. area) using nine-point external standard curves (*see* **Note 13**).
4. Take into consideration recovery yield when calculating the concentration. Estimate recovery by comparison between the concentration provided by the standard and the concentration of the standard processed in the same way as samples. Recovery with this protocol was within the range of 50–55% (Fig. 5).
5. For quantitative analysis, prepare quality controls duplicates for 5% out of the total number of samples. Provide results as the mean value whenever the differences between individual results are below 10%. If differences are above this percentage, the analysis should be repeated.
6. The positive control consists of an astaxanthin standard at a concentration coincident with that obtained at the mid-point of the calibration curve. Inject controls every 10 samples. Consider the analysis valid when the values of the areas undergo a deviation below 5% of average values.

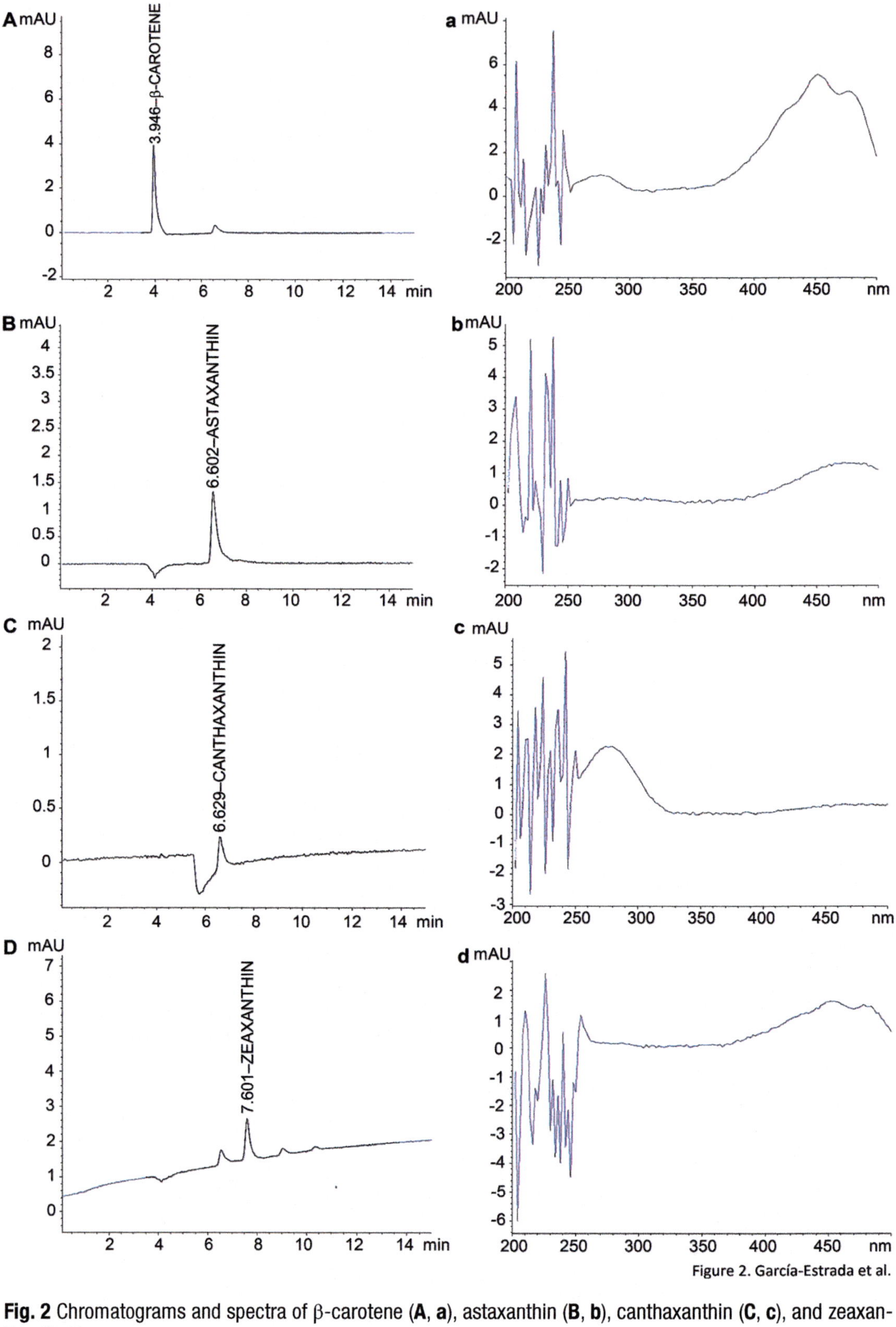

Fig. 2 Chromatograms and spectra of β-carotene (**A**, **a**), astaxanthin (**B**, **b**), canthaxanthin (**C**, **c**), and zeaxanthin (**D**, **d**)

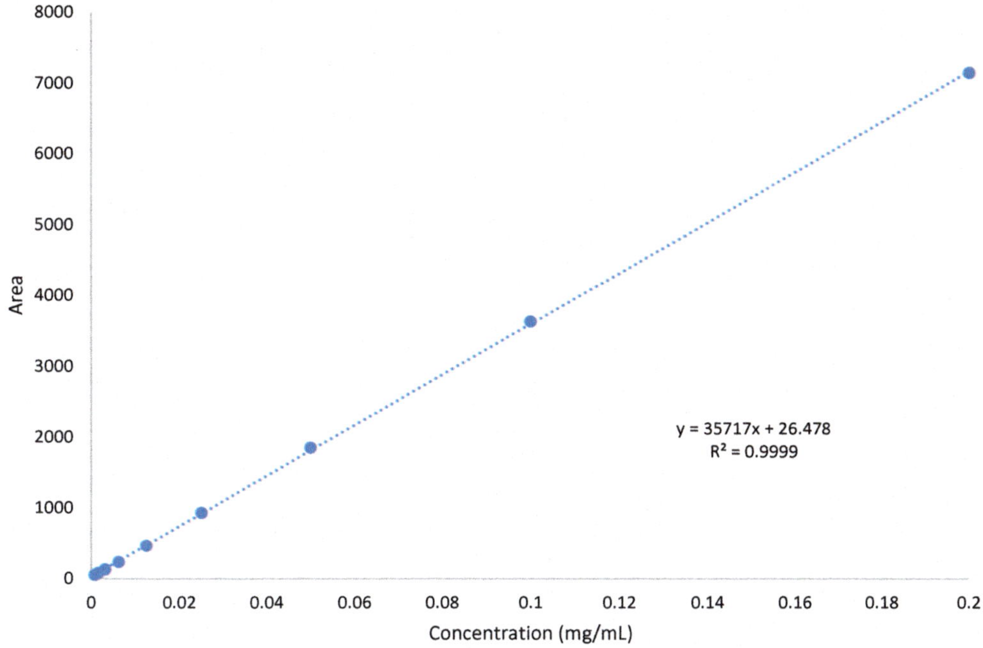

Fig. 3 Calibration curve of astaxanthin. Samples ranged between 7.8 and 200 μg/mL

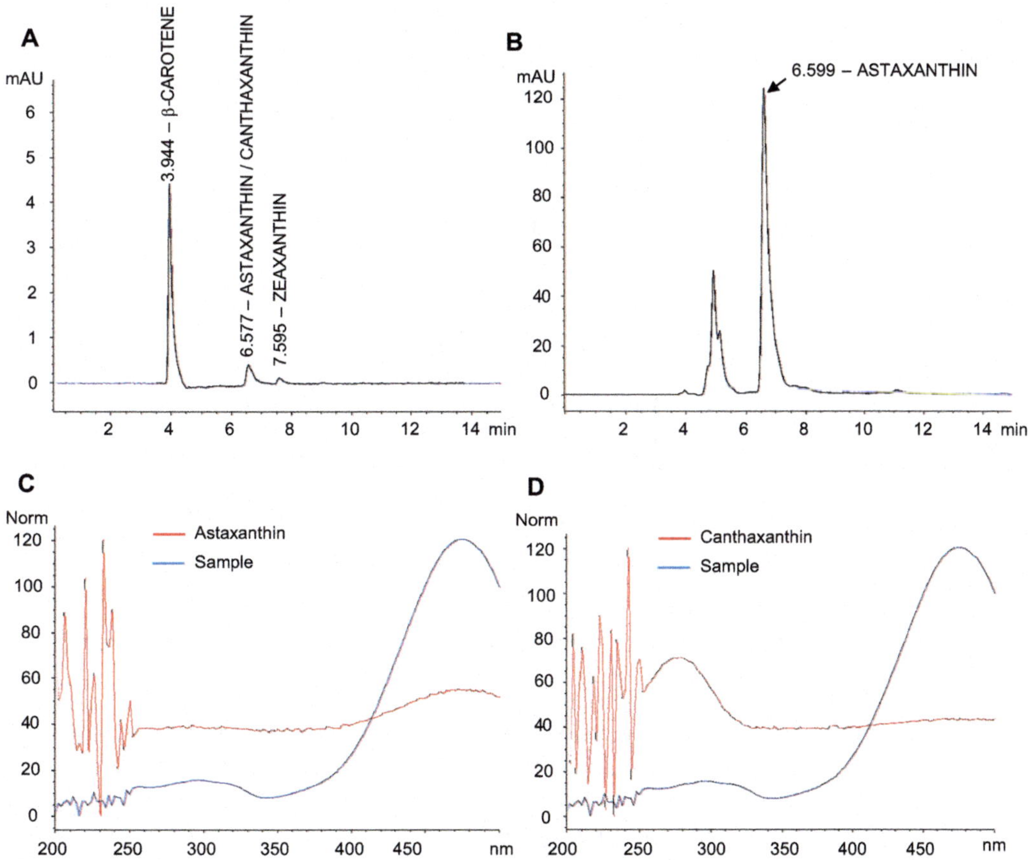

Fig. 4 Standard of β-carotene, canthaxanthin, astaxanthin, and zeaxanthin (**a**) and example of sample (**b**). Comparison between the sample spectrum at 6.6 min with astaxanthin spectrum (**c**) and canthaxanthin spectrum (**d**)

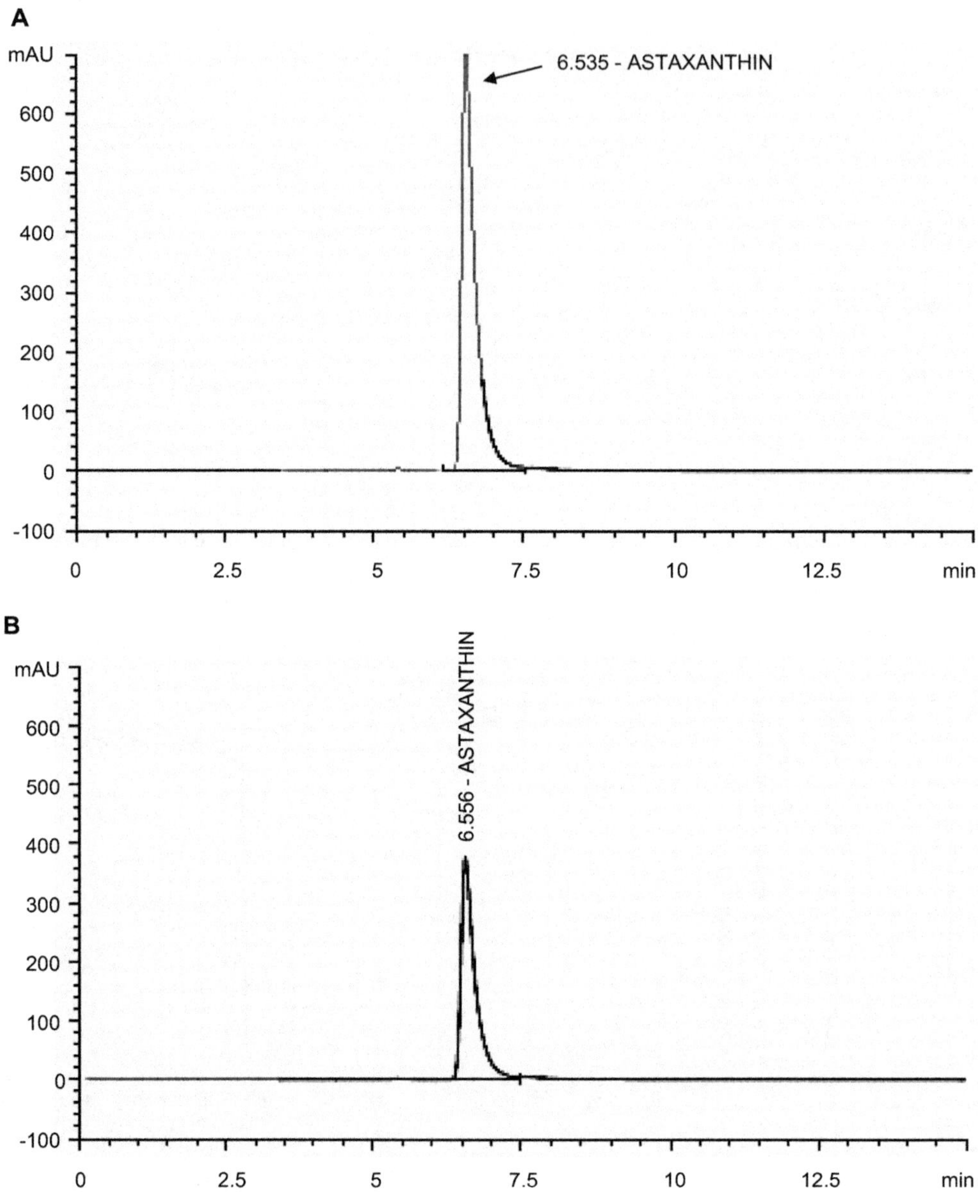

Fig. 5 Standards of astaxanthin (**a**) and astaxanthin processed as a sample (**b**)

4 Notes

1. HPLC-grade reagents may cost slightly more than lower-grade reagents, but the difference in purity is remarkable. HPLC grade reagents contain no impurities to produce spurious peaks in a chromatogram baseline.
2. If the method of mixing the solvent is not the same as that described above, differences can occur in the mobile phase that may affect the chromatography.

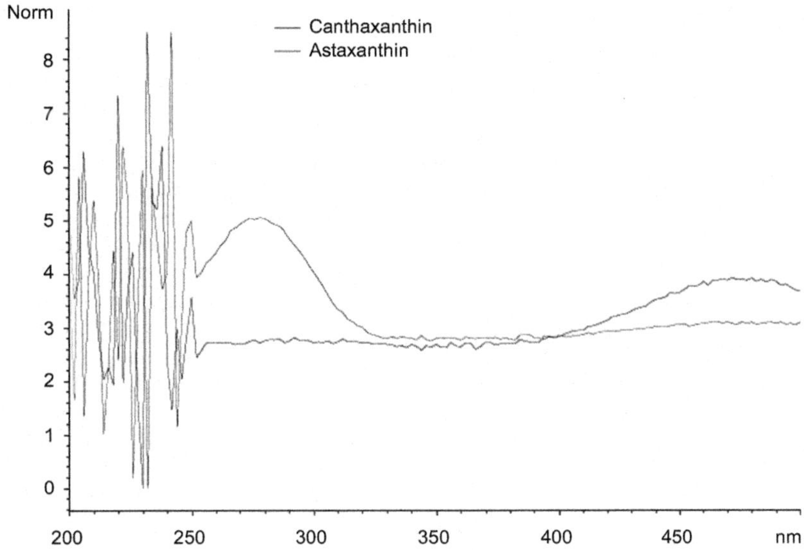

Fig. 6 Spectra of astaxanthin and canthaxanthin

3. Use amber (for aqueous solutions) and transparent (for organic solutions) glass bottles.
4. To convert a reversed-phase system/column to a normal-phase system/column, flush with a solvent that is miscible with both the current reversed-phase solvents and the proposed normal-phase solvents. If the reversed-phase solvents include a buffer, then it is advisable to move from a 5% aqueous methanol flush to a 100% methanol flush, then isopropanol/ethyl acetate (50:50), and finally normal-phase ethyl acetate/hexane.
5. You can use antioxidants like 0.01% butylated hydroxytoluene (BHT) if the reagents are not peroxides free [41, 42].
6. Open the tube gently to eliminate the gas that may have formed.
7. Two extractions should be enough.
8. For longer storage periods, samples should be lyophilized.
9. Although some methods include gradient elution, this is time-consuming because of the long-equilibration period required by the analytical column between each run [41].
10. Canthaxanthin maximum absorbance is achieved at 278 nm.
11. Regenerate the column with *n*-hexane and 2% ethyl acetate. Use mobile phase at 0.8 mL/min to wash the column. Use *n*-hexane and 2% ethyl acetate to store the column.
12. When the samples are filtered through a 0.22 μm filter, the purge valve frit is long-lasting.
13. Canthaxanthin and astaxanthin chromatographic peaks are difficult to resolve. However, there are big differences between their spectra, since canthaxanthin is detected at 278 nm and astaxanthin at 474 nm (Fig. 6).

References

1. Britton G, Liaaen-Jensen S, Pfander H (1995) Carotenoids today and challenges for the future. In: Britton G, Liaaen-Jensen S, Pfander H (eds) Carotenoids. Birkhäuser, Basel, pp 13–26
2. Sandmann G, Misawa N (2002) Fungal carotenoids. In: Osiewacz HD (ed) The Mycota X industrial applications. Springer Verlag, Berlin Heidelberg, pp 247–262
3. Sieiro C, Poza M, de Miguel T, Villa TG (2003) Genetic basis of microbial carotenogenesis. Int Microbiol 6:11–16
4. Fraser PD, BramLey PM (2004) The biosynthesis and nutritional uses of carotenoids. Prog Lipid Res 43:228–265
5. Britton G, Liaaen-Jensen S, Pfander H (2004) Carotenoids: handbook. Birkhauser, Boston
6. Bhosale P, Bernstein PS (2005) Microbial xanthophylls. Appl Microbiol Biotechnol 68:445–455
7. Blomhoff R, Blomhoff HK (2006) Overview of retinoid metabolism and function. J Neurobiol 66:606–630
8. Domonkos I, Kis M, Gombos Z, Ughy B (2013) Carotenoids, versatile components of oxygenic photosynthesis. Prog Lipid Res 52:539–561
9. The global market for carotenoids (2011) http://www.bccresearch.com/market-research/food-and-beverage/carotenoids-global-market-report-fod025e.htmL. Accessed 11 Jan 2017
10. Del Campo JA, García-González M, Guerrero MG (2007) Outdoor cultivation of microalgae for carotenoid production: current state and perspectives. Appl Microbiol Biotechnol 74:1163–1174
11. Mata-Gómez LC, Montañez JC, Méndez-Zavala A, Aguilar CN (2014) Biotechnological production of carotenoids by yeasts: an overview. Microb Cell Factories 13:12
12. Phaff HJ, Miller MW, Yoneyama M, Soneda M (1972) A comparative study of the yeast florae associated with trees on the Japanese Islands and on the west coast of North America. In: Terui G (ed) Fermentation technology today. Society of Fermentation Technology, Osaka, pp 759–774
13. Golubev WI (1995) Perfect state of *Rhodomyces dendrorhous* (*Phaffia rhodozyma*). Yeast 11:101–110
14. Global astaxanthin market—sources, technologies and applications, Report (2015) http://www.reportlinker.com/p02788641-summary/Global-Astaxanthin-Market-Sources-Technologies-and-Applications.htmL. Accessed 11 Jan 2017
15. Alvarez V, Rodríguez-Sáiz M, de la Fuente JL, Gudiña EJ, Godio RP, Martín JF et al (2006) The crtS gene of *Xanthophyllomyces dendrorhous* encodes a novel cytochrome-P450 hydroxylase involved in the conversion of beta-carotene into astaxanthin and other xanthophylls. Fungal Genet Biol 43:261–272
16. Ojima K, Breitenbach J, Visser H, Setoguchi Y, Tabata K, Hoshino T et al (2006) Cloning of the astaxanthin synthase gene from *Xanthophyllomyces dendrorhous* (*Phaffia rhodozyma*) and its assignment as a β-carotene 3-hydroxylase/4-ketolase. Mol Gen Genomics 275:148–158
17. Green AS, Fascetti AJ (2016) Meeting the vitamin A requirement: the efficacy and importance of β-carotene in animal species. ScientificWorldJournal 2016:7393620
18. Breithaupt DR (2008) Xanthophylls in poultry feeding. In: Britton G, Pfander H (eds) Carotenoids. Birkhäuser, Basel, pp 255–264
19. Zhang W, Zhang KY, Ding XM, Bai SP, Hernandez JM, Yao B et al (2011) Influence of canthaxanthin on broiler breeder reproduction, chick quality, and performance. Poult Sci 90:1516–1522
20. Bone RA, Landrum JT, Friedes LM, Gomez CM, Kilburn MD, Menendez E et al (1997) Distribution of lutein and zeaxanthin stereoisomers in the human retina. Exp Eye Res 64:211–218
21. Landrum JT, Bone RA (2001) Lutein, zeaxanthin, and the macular pigment. Arch Biochem Biophys 385:28–40
22. Lima VC, Rosen RB, Farah M (2016) Macular pigment in retinal health and disease. Int J Retina Vitreous 2:19
23. Higuera-Ciapara I, Felix-Valenzuela L, Goycoolea FM (2006) Astaxanthin: a review of its chemistry and applications. Crit Rev Food Sci Nutr 46:185–196
24. Wang X, Willen R, Wadstrom T (2000) Astaxanthin-rich algal meal and vitamin C inhibit *Helicobacter pylori* infection in BALB/cA mice. Antimicrob Agents Chemother 44:2452–2457
25. Park JS, Chyun JH, Kim YK, Line LL, Chew BP (2010) Astaxanthin decreased oxidative stress and inflammation and enhanced immune response in humans. Nutr Metab 7:18

26. Yasui Y, Hosokawa M, Mikami N, Miyashita K, Tanaka T (2011) Dietary astaxanthin inhibits colitis and colitis-associated colon carcinogenesis in mice via modulation of the inflammatory cytokines. Chem Biol Interact 193:79–87
27. Fassett RG, Coombes JS (2012) Astaxanthin in cardiovascular health and disease. Molecules 17:2030–2048
28. Fassett RG, Coombes JS (2009) Astaxanthin, oxidative stress, inflammation and cardiovascular disease. Futur Cardiol (4):333–342
29. Lin W, Chien J, Chen B (2005) Determination of carotenoids in spear shrimp shells (*Parapenaeopsis hardwickii*) by liquid chromatography. J Agric Food Chem 53:5144–5149
30. Rodríguez Amaya DB (2001) A guide for carotenoids analysis in foods. ILSI Press, Washington, DC
31. Rodriguez-Amaya DB, Kimura M (2004) Harvest plus handbook for carotenoid analysis. ILSI Press, Washington, DC
32. García S, Pérez RM (2012) Aspectos analíticos sobre la determinación de compuestos carotenoides en microalgas mediante cromatografía de líquidos con detector de diodos. Informes técnicos Ciemat 1248
33. Sedmak JJ, Weerashinge DK, Jolly SO (1990) Extraction and quantitation of *astaxanthin from Phaffia rhodozyma*. Biotechnol Tech 4:107–112
34. An GH, Cho MH, Johnson EA (1999) Monocyclic carotenoid biosynthetic pathway in the yeast *Phaffia rhodozyma* (*Xanthophyllomyces dendrorhous*). J Biosci Bioeng 88:189–193
35. Biswas AK, Sahoo J, Chatli MK (2011) A simple UV-Vis spectrophotometric method for determination of β-carotene content in raw carrot, sweet potato and supplemented chicken meat nuggets. LWT Food Sci Technol 44:1809–1813
36. Miller KW, Yang CS (1985) An isocratic high-performance liquid chromatography methods for the simultaneous analysis of plasma retinol, alpha-tocopherol and various carotenoids. Anal Biochem 145:21–26
37. Chauveau-Duriot B, Doreau M, Nozière P, Graulet B (2010) Simultaneous quantification of carotenoids, retinol and tocopherols in forages, bovine plasma and milk: validation of a novel UPLC method. Anal Bioanal Chem 397:777–790
38. Ohmacht R, Toth G, Voigt G (1987) Separation of serum carotenoids and vitamin A on chromsil-amino and -cyano phases by a bi-directional gradient elution technique. J Chromatogr 395:609–612
39. Panfili G, Manzi P, Pizzoferrato L (1994) High-performance liquid chromatographic method for the simultaneous determination of tocopherols, carotenes, and retinol and its geometric isomers in Italian cheeses. Analyst 119:1161–1165
40. McGeachin RB, Bailey CA (1995) Determination of carotenoid pigments, retinol, and alpha-tocopherol in feeds, tissues, and blood serum by normal phase high performance liquid chromatography. Poult Sci 74:407–411
41. Gueguen S, Herbeth B, Siest G, Leroy P (2002) An isocratic liquid chromatographic method with diode-array detection for the simultaneous determination of *m*-tocopherol, retinol and five carotenoids in human serum. J Chromatogr Sci 40:69–76
42. Abd El-Baky HH, El Baz FK, El-Baroty GS (2007) Production of carotenoids from marine microalgae and evaluation as safe food colorant and lowering cholesterol agents. J Agric Sci 2:792–800
43. Guaratini T, Cardozo HM, Pinto E, Colepicolo P (2009) Comparison of diode array and electrochemical detection in the C30 reverse phase HPLC analysis of algae carotenoids. J Braz Chem Soc 20:1609–1616
44. García-Guerra A (2012) Estudio de la producción de astaxantina por *Xanthophyllomyces dendrorhous*. PhD Thesis, University of León, Spain

Chapter 18

Isolation and Selection of New Astaxanthin-Producing Strains of *Phaffia rhodozyma*

Diego Libkind, Martín Moliné, and Fernando Colabella

Abstract

Astaxanthin is a xanthophyll pigment of high economic value for its use as a feeding component in aquaculture. *Phaffia rhodozyma* (*Xanthophyllomyces dendrorhous*) is a basidiomycetous fungi able to synthesize astaxanthin as its major carotenoid, the only known yeast species bearing the capability to produce this type of carotenoid and the only tremellomycetes with biotechnological application. Recently, the habitat and intraspecific variability of this species have been found to be wider than previously expected, encouraging the search for new wild strains with potential biotechnological applications. Here we describe effective procedures for isolation of *P. rhodozyma* from environmental samples, accurate identification of the strains, analysis of their astaxanthin content, and proper conservation of the isolates.

Key words Astaxanthin, *Phaffia*, *Xanthophyllomyces*, ITS, Selective culture media, Natural environments, Mycosporines, Sequencing

1 Introduction

The ability of the basidiomycetous yeast *Phaffia rhodozyma* (*Xanthophyllomyces dendrorhous*) to accumulate the antioxidant astaxanthin as its major carotenoid is responsible for the industrial use of this yeast as a microbial source of pigments for aquaculture [1]. Astaxanthin is economically important because it is the most expensive aquaculture feed component [2]; it is used for the artificial pigmentation of fish and crustaceans [3]. Astaxanthin is, in fact, one of the most expensive components of salmon farming, accounting for about 15% of total production costs [4].

Original isolations of *P. rhodozyma*, which have served for most of the studies involving this species, were carried out in the 1960s by Phaff et al. [5]. These isolates were obtained from slime exudates of various broad-leafed trees in different mountainous regions in the northern hemisphere, such as Japan and Canada. More recently, isolations of this species have been carried out in Italy [6], Germany [7], Argentina [8], the USA [9], and Antarctica [10].

Carlos Barreiro and José-Luis Barredo (eds.), *Microbial Carotenoids: Methods and Protocols*, Methods in Molecular Biology, vol. 1852, https://doi.org/10.1007/978-1-4939-8742-9_18,

The isolations carried out in Argentina (NW Patagonia) were the first to be reported in the southern hemisphere, and unlike all previous isolations, they were not from sap flow but rather from ascostroma of *Cyttaria hariotii* [8] parasiting the tree *Nothofagus dombeyi*. A few isolates were also obtained from a mountain lake, and all Patagonian strains of *P. rhodozyma* bear genetic differences with collection *P. rhodozyma* strains. Even more recently, a study of association of *Phaffia* with *Nothofagus-Cyttaria* in Australasia [11], the other region of the world where *Nothofagus* are endemic, revealed an ever higher *Phaffia* diversity, including two endemic and markedly divergent lineages which represent putative new species, both with the capability to produce astaxanthin and mycosporine-glutaminol-glucoside (MGG). The genetic diversity that was described in *P. rhodozyma* correlates with the observed association with the tree genera (*Betula*, *Cornus*, and *Nothofagus*), rather than with geography, and constitutes seven different lineages. Later, an even more divergent strain was found in soil samples [12].

Thus, it becomes clear that the natural geographic distribution of *Phaffia* is wider than expected and that its habitat is not restricted to slime fluxes but rather includes sugar-rich fungal stromata, leaves, soil, and freshwater. Furthermore, the genetic variability of this astaxanthin-producing yeast is also more diverse than what was previously thought, and recent genetic evidence indicates that there is more than one species in the genus [11, 12]. A draft genome sequence of the type strain of *P. rhodozyma* became recently available, and comparison with that of the Patagonian population suggested the latter deserves to be assigned to a distinct variety [13]. In this work, the genetic basis of several photoprotective and antioxidant strategies is described, indicating that *P. rhodozyma* is one of the fungi most well equipped to cope with environmental oxidative stress, a factor that has probably contributed to shaping its genome. The biotechnological potential of the Patagonian isolates has only been preliminary assessed [14].

In the present work, we describe the strategies currently used in our laboratory for the effective isolation of *Phaffia* spp. from environmental samples, the rapid and accurate identification of the strains, the analysis of their astaxanthin content, and the proper conservation of the isolates.

2 Materials

Prepare all solutions using distilled sterile water and analytical grade reagents. Prepare and store all reagents at room temperature (unless otherwise indicated). Follow all waste disposal regulations when disposing waste materials.

2.1 Isolation of P. rhodozyma from Environmental Samples

1. YM agar: 3 g/L yeast extract, 3 g/L malt extract, 5 g/L peptone, 10 g/L dextrose, 20 g/L agar. Adjust pH 4.5–5.0 and supplement with 200 mg/L chloramphenicol.
2. YPD agar: 10 g/L yeast extract, 20 g/L peptone, 20 g/L glucose, 20 g/L agar. Adjust pH 4.5–5.0 and supplement with 200 mg/L chloramphenicol.
3. YNB + Tre agar (selective isolation medium): 6.7 g/L YNB (yeast nitrogen base, DIFCO), 5 g/L trehalose, 20 g/L agar, and supplement with 200 mg/L chloramphenicol (*see* **Note 1**).
4. 0.9% NaCl.
5. Filter membranes (0.45 μm) (Millipore Corporation, Bedford, Massachusetts, USA).
6. Sterile filtering apparatus.

2.2 Identification of Wild P. rhodozyma Isolates Using Phenotypic Characteristics

1. YM agar (*see* Subheading 2.1).
2. Lugol's iodine: 0.33% w/v iodine (I_2) and 0.66% w/v potassium iodide (KI) in distilled water, to make a brown solution with a total iodine concentration of 150 mg/mL.
3. Fermentation base medium: 4.5 g/L yeast extract, 7.5 g/L peptone, 20 g/L glucose, and enough bromothymol blue to obtain a sufficiently dark green color.
4. Durham tubes.
5. Illumination apparatus emitting visible light (photosynthetically active radiation) with or without UVA (*see* **Note 2**).
6. 25% v/v methanol solution.
7. UV-visible spectrophotometer, with quartz cuvettes.
8. Thermostatic bath.
9. Incubation chamber with controlled temperature (20 °C).
10. Ribitol agar: 5 g/L ribitol and 20 g/L agar.

2.3 Molecular Identification of Wild P. rhodozyma Isolates

1. YM agar (*see* Subheading 2.1).
2. ITS5 forward primer: 5′-GGA AGT AAA AGT CGT AAC AAG G-3′.
3. LR6 reverse primer: 5′-CGC CAG TTC TGC TTA CC-3′.
4. NL1 forward primer 5′-GCA TAT CAA TAA GCG GAG GAA AAG-3′.
5. NL4 reverse primer 5′-GGT CCG TGT TTC AAG ACG G-3′.
6. ITS1 forward primer 5′-TCC GTA GGT GAA CCT GCG G-3′.
7. ITS3 forward primer 5′-GCA TCG ATG AAG AAC GCA GC-3′.

8. ITS4 reverse primer 5′-TCC TCC GCT TAT TGA TAT GC-3′.
9. PhR reverse primer 5′- GAC TTG TAC ACA GGC CGG CA-3′.
10. PCR reaction mix kit.
11. Thermocycler.
12. Genomic DNA extraction kit.
13. Lysing buffer: 50 mM Tris–HCl, 250 mM NaCl, 50 mM EDTA, and 0.3% w/v SDS, pH 8.
14. PCR product purification kit.

2.4 Extraction, Detection, and Quantification of Astaxanthin

1. C18 HPLC column (5 μm, 250 by 4.6 mm).
2. Acetonitrile/methanol/isopropanol: 40 mL acetonitrile, 50 mL methanol, 10 mL isopropanol.
3. Astaxanthin standard (Sigma, St. Louis, MO, USA).
4. Carotenoid extract from a reference *P. rhodozyma* collection strain (e.g., CBS 7918^{T}) [15].

2.5 Preservation of P. rhodozyma Wild Strains

1. YM agar (*see* Subheading 2.1).
2. Ultrafreezer (−80 °C).
3. 20% glycerol sterile solution.
4. Sterile 425–600 μm Ø glass beads.

3 Methods

Carry out all procedures at room temperature unless otherwise specified.

3.1 Isolation of P. rhodozyma from Natural Environments

To date, *P. rhodozyma* has been found in association to different types of natural substrates, including slime exudates, fungal stromata, leaves and, less frequently, in water samples. Slime exudates are typically found during the spring on birch or beech stumps, and when colonized by *P. rhodozyma*, have an orange-characteristic color. Photos of fluxes colonized by *P. rhodozyma* can be seen in Weber et al. [7]. Such fluxes have so far only been found in the northern hemisphere in tree species of the genera *Betula*, *Cornus*, *Carpinus*, and *Fagus*. In contrast, there are no reports of the presence of this yeast in exudates of trees from the Southern Hemisphere. However, *P. rhodozyma* has been found in the stromata of the *Nothofagus* spp. parasitic fungus *Cyttaria* and less frequently in leaves (*Nothofagus* and *Eucalyptus*) and freshwater samples [8, 11, 16]. In most environmental samples, it has been observed that isolation frequency is very low; thus a partially selective culture medium with trehalose as sole carbon source has been developed (here referred as YNB + Tre agar) [17]. This selective isolation medium quite effectively inhibits growth of most of the white co-occurring yeasts which typically appear with higher abundance and growth rates than *P. rhodozyma* (*see* **Note 3**). In this Subheading we provide hints for an effective isolation of this yeast species from the different substrates, each of which require a specific procedure to free yeast cells.

3.1.1 Isolation of P. rhodozyma from Exudate Samples

1. Collect exudate samples aseptically and streak out directly onto agar plates.
2. Alternatively, plate out 0.1 mL aliquots of a dilution series obtained by shaking 1 g (fresh weight) of exudate in 10 mL 0.9% sterile aqueous NaCl or distilled water and stepwise dilution of the resulting suspension in 0.9% NaCl or distilled water.

3.1.2 Isolation of P. rhodozyma from Stromata Samples

1. Collect aseptically stromata of *Cyttaria hariotii*, an ascomycetous fungus parasite of *Nothofagus dombeyi* ("Coihue").
2. Place them in sterile plastic bags with a known volume of distilled sterile water (5–50 mL).
3. Crush manually *Cyttaria* fruit bodies through the plastic bag, and agitate in an orbital shaker for 30 min at 300 rpm.
4. Dilute the extract conveniently and spread aliquots on agar plates.

3.1.3 Isolation of P. rhodozyma from Exudate Samples Leaves Samples

1. Collect aseptically whole leaves (as fresh as possible).
2. Place leaves (1–15 depending on size) in a 10–15 mL Falcon tube with 8–10 mL distilled water (just enough to cover leaves).
3. Vortex vertically at maximum speed for 5 min.
4. Shake samples overnight at 150 rpm in an orbital shaker.
5. Dilute conveniently the extract.
6. Spread aliquots on agar plates.

3.1.4 Isolation of P. rhodozyma from Water Samples

1. Collect water samples in sterile flasks.
2. Filter them preferably in situ (<15 in. of mercury) through filter membranes (0.45 μm) with a sterile filtering apparatus.
3. Place the filters directly on agar plates.

3.1.5 Isolation of P. rhodozyma by Enrichment

This is an additional isolation method for yeasts that may be applied to any of the previous substrates:

1. Incubate small samples in shaken flasks (100 mL liquid medium with antibiotics) in an orbital shaker (120 rpm) for 48 h.
2. Filter through sterile glass wool.

3. Do dilution series in fresh medium.
4. Plate out on agar plates.

Recommended temperature range for plate incubation is 15–20 °C, with a 17 °C optimum. Normally, 5 days of incubation are enough for colony emergence (*see* **Note 4**). Transfer orange colonies to YM or YPD fresh medium and purify.

3.2 Identification of Wild P. rhodozyma Using Phenotypic Characteristics

P. rhodozyma possess such unique morphological and physiological characteristics that makes identification based on phenotypic traits a viable alternative. Even if not used as only mean of identification, the tests described in this subheading are useful for reducing the number of putative *P. rhodozyma* isolates to the minimum. The distinctive combination of phenotypic traits includes the ability to produce amyloid compounds, glucose fermentation capabilities [18], mycosporine-glutaminol-glucoside (MGG) production (Fig. 1) [19], development of unique sexual structures (Fig. 2) [15], and synthesis of astaxanthin as a main carotenoid pigment [20]. Any one of these features can be tested to eliminate non *P. rhodozyma* isolates, and the combination of all of these features will provide an accurate identification. In particular, unlike any other yeast species, *P. rhodozyma* simultaneously produces MGG and astaxanthin, both of which can be extracted in a single step [17]. The characteristic spectrum of extracts consists of a narrow peak, with a maximum absorbance at 309–310 nm, and a broad spectral band with a maximum at 477 nm, corresponding to MGG and astaxanthin, respectively (Figs. 1 and 3). This method is particularly useful for rapid

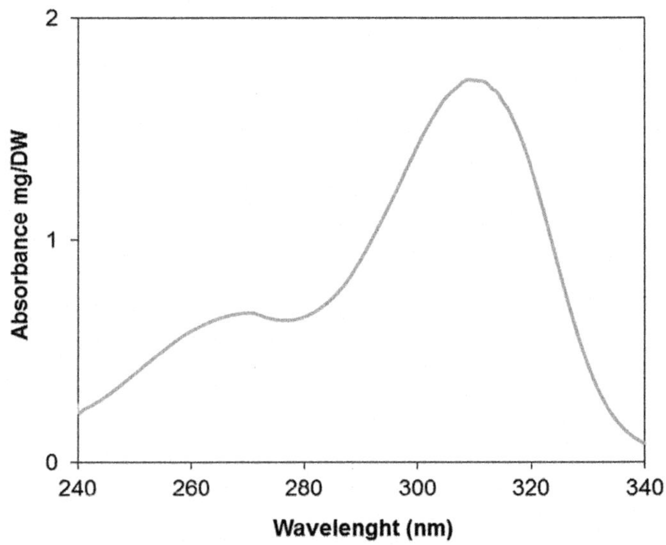

Fig. 1 UV-visible spectrum of mycosporine-glutaminol-glucoside in *P. rhodozyma* extract. *DW* dry weight

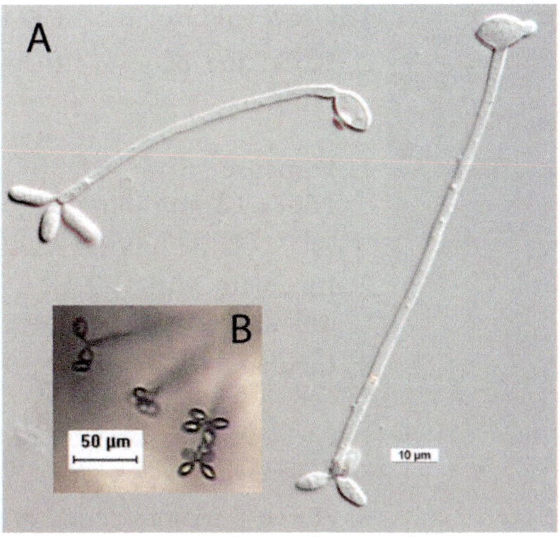

Fig. 2 Sexual structures of *P. rhodozyma*. (**a**) Daughter cell-mother cell conjugation (pedogamy) and holobasidium with terminal basidiospores. (**b**) Terminal basidiospores

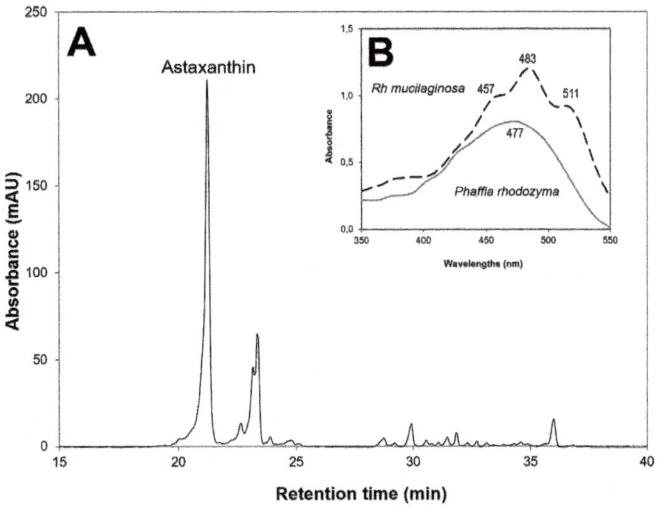

Fig. 3 Spectrophotometric and HPLC analysis of *P. rhodozyma* extracts. (**a**) Typical HPLC chromatogram of *P. rhodozyma*. (**b**) Comparison of carotenoid extracts UV-visible spectra with *Rhodotorula mucilaginosa* (ubiquitous pigmented yeast). *AU* absorbance units

identification to distinguish *P. rhodozyma* from other pigmented yeasts that share its habitat, e.g., *Cystofilobasidium* cannot release MGG nor carotenoids into extracts, while yeasts able to produce MGG have a different carotenoid spectrum which in most cases are absent (*A. pullulans* and *Rhodotorula* spp.) or consisted of three peaks around 455, 484, and 520 nm (*Dioszegia* spp.).

3.2.1 Production of Starch Compounds

1. Grow wild isolates on YM agar for 1 week at 20 °C.
2. Flood the plates with five times diluted Lugol's iodine (*see* **Note 5**).

3.2.2 Glucose Fermentation

1. Dispense 5–6 mL of base fermentation medium into 150 × 12 mm plugged test tubes containing a small inverted tube (Durham tube) approximately 50 mm high and 6 mm Ø.
2. Inoculate with 100 μL of a dense cell suspension of the yeast isolate.
3. Incubate for 1 week at 20 °C (*see* **Note 6**).

3.2.3 Mycosporine Production

1. Grow yeast isolates on YM agar plates under illumination conditions for 3 days at 20 °C (*see* **Note 2**).
2. Harvest biomass and transfer to Eppendorf tube containing 1 mL of 80% (v/v) ethanol.
3. Agitate vigorously with a vortex mixer and incubate in 85 °C bath for 2 h.
4. Clear extracts by centrifugation at 10,000 × *g* for 10 min.
5. Obtain full spectrum within range 190–700 nm in an UV-visible spectrophotometer using quartz cuvettes.
6. Compare with reference strain or with Fig. 1.

This method simultaneously releases astaxanthin together with MGG, allowing a rapid identification screen for pigmented yeasts, as it was mentioned above.

3.2.4 Sexual Structure Formation

1. Inoculate young cells (48 h in YM agar) in Ribitol agar.
2. Incubate plates for 1 week at 17 °C.
3. Compare with reference strain or with Fig. 2.

3.2.5 Astaxanthin Biosynthesis

For astaxanthin biosynthesis, *see* Subheading 3.4.

3.3 Molecular Identification of Wild *P. rhodozyma* Isolates

As described above, *P. rhodozyma* has a unique combination of phenotypic characteristics that in some cases may be sufficient for an appropriate identification. However, nowadays, accurate identification requires the use of molecular techniques.

DNA sequencing allows phylogenetic analysis using additional sequences from public gene databases (e.g., GenBank). Useful regions for *P. rhodozyma* identification through DNA sequencing are the gene encoding the D1/D2 domains of the large ribosomal subunit (26S) and the region comprising the Internal Transcribed Spacers (ITS) 1 and 2, and the 5.8S ribosomal subunit (hereinafter D1/D2 and ITS regions, respectively). The D1/D2 region is highly conserved in *P. rhodozyma*, and it is a good choice for accurate identification. The ITS region not only allows correct species assignation but also, given to a high intraspecific variability in *P. rhodozyma*, may provide information for distinguishing natural populations [8].

On the other hand, a quick and accurate PCR-based detection method was developed to detect isolates belonging to all known lineages of the genus *Phaffia* [21]. For this purpose, a primer of 20 nucleotides (called PhR) was designed to be used in combination with universal primers ITS3 and NL4 in a multiplex amplification. The reverse PhR primer was designed to anneal in the conserved D1/D2 domains of the 26S rRNA gene and to generate a 643 bp amplicon when it is used together with the universal forward primer ITS3. The universal reverse primer NL4 was included for amplification control since it will produce a single fragment around 1155 bp in any eukaryotic microorganism. The use of three primers in a multiplex reaction has the advantage of still obtaining an amplicon in a negative sample; therefore, the results are unambiguous and cannot be attributable to a failure in the amplification reaction.

1. Prepare a fresh culture of the *P. rhodozyma* strain on YM agar.
2. Extract the genomic DNA using appropriate commercial kits or alternative procedures (*see* **Note 7**).
3. Polymerase chain reaction (PCR).
    (a) PCR-based detection method: Amplify using primers ITS3, PhR, and NL4. Use the following PCR program: (1) initial denaturing step at 95 °C for 5 min. (2) Thirty-five cycles of 30 s at 95 °C, 60 s at 55 °C, and 60 s at 72 °C. (3) A final extension step of 7 min at 72 °C (*see* **Note 8**).
    (b) DNA sequencing: Amplify a specific region of rRNA genes or using primers ITS5 and LR6. Use the following PCR program: (1) initial denaturing step at 95 °C for 5 min. (2) Forty cycles of 45 s at 93 °C, 60 s at 50 °C, and 60 s at 72 °C. (3) A final extension step of 6 min at 72 °C. Expect a fragment size of approximately 1400 bp.
4. For DNA sequencing, purify the corresponding PCR product using an appropriate extraction kit, and sequence it using the pair of primers NL1/NL4 for D1/D2 region (ca. 660 bp) and/or ITS1/ITS4 (ca. 715 bp) for ITS region (*see* **Note 9**).

3.4 Extraction, Detection, and Quantification of Astaxanthin

P. rhodozyma strains typically produce astaxanthin as their major carotenoid pigment (approximately 80% of total carotenoids); consequently, this yeast possesses significant biotechnological relevance as a natural source of astaxanthin pigment for aquaculture feed. Other minor carotenoids produced by *P. rhodozyma* are β-carotene, echinenone, 3-hydroxyechinenone, phoenicoxanthin, and canthaxanthin [20]. This pool of pigments substantially differs from that of most pigmented yeasts, such as those of the genus *Rhodotorula*. Astaxanthin is a xanthophyll and is highly polar; hence, in HPLC chromatograms it normally appears first (Fig. 3a). *P. rhodozyma* extracts UV-vis spectra has a typical broad spectrum band with no evident absorption peaks and a maximum at 477 nm (Fig. 3b).

1. Harvest the cells of 20 mL of culture by centrifugation at 16,000 × *g*.
2. Wash twice with distilled water.
3. Suspend the pellet in 4.5 mL of distilled water and divide in four aliquots: two tubes with 1 mL each for carotenoid extraction and two previously tared tubes with 1 mL each for dry weight.
4. For carotenoid extraction, add 1.5 mL of glass beads (425–600 μm Ø), and incubate on ice for 2 min.
5. Mix in vortex for 2 min.
6. Add 2 mL of acetone, and incubate on ice for 2 min.
7. Mix in vortex for 2 more min.
8. Centrifuge at 1500 × *g* and collect the supernatant in clean tubes. Store at −20 °C.
9. Repeat the process until a white pellet is obtained.
10. Add 2 mL of petroleum ether (35–60 °C) at 5 °C to the tube containing the supernatants, and mix in vortex for 2 min.
11. Centrifuge the tubes at 10,000 × *g* for 10 min at 5 °C, and collect the supernatant, which contains the pigments, in a new tube.
12. Evaporate the petroleum ether under a continuous nitrogen flux.
13. Add 1 mL of *n*-hexane or petroleum ether for spectrophotometric quantification.
14. Measure the absorbance of the sample at 474 nm.
15. Estimate total carotenoid concentration using the following simplified formula.

$$\text{Carotenoids}_{\text{t}}\left(\mu\text{g}/\text{g}\right)=\frac{A_{474}\times 1\times 10^{4}\left(\mu\text{g/mL}\right)}{2100\times \text{dw}\left(\text{g/mL}\right)}$$

A_{474}: average absorbance (474 nm) of the two tubes (*see* **step 11**).
dw: average dry weight of the two tubes (*see* **step 2**).
2100: extinction coefficient of astaxanthin.

16. For HPLC detection of astaxanthin, suspend extracts in 100 μL acetone.

17. Analyze the carotenoid content using a C18 column, an isocratic mobile phase consisting of 85% acetonitrile/10% methanol/5% isopropanol, and a flow rate of 1 mL/min.
18. Monitor the eluted fractions with a photodiode array detector, scanning from 270 to 600 nm every 2 s.
19. Identify carotenoids by their retention times and by comparison of the spectral features with those of pure compounds or with those of a reference strain.

3.5 Preservation of *P. rhodozyma* Wild Strains

Long-term preservation of fungal strains is essential for their in-depth study; however, both the viability and the stability of living cells should be ensured during the preservation period. Wild isolates of *P. rhodozyma* normally do not survive long on regular culture media and need to be subjected to cryopreservation for prolonged preservation. Different methods have been reported for yeast preservation [22, 23], although the method described in this Subheading has proven to be effective for preserving *P. rhodozyma* wild strains (*see* **Note 10**).

1. Prepare a fresh culture of *P. rhodozyma* strain on a small YM agar plate.
2. Spread 1 mL of a sterile solution of 20% v/v glycerol over the colony.
3. Add 10–15 glass beads (2–3 mm diameter) onto the plate.
4. Use a sterile small spoon to mix beads, glycerol, and yeast biomass, and transfer yeast-soaked beads into a sterile cryotube.
5. Store tubes at −80 °C (*see* **Note 11**).
6. To obtain a fresh culture, remove one of the beads using a sterile pair of tweezers, rub bead on YM agar plate, and incubate at 17 °C.

4 Notes

1. Colony size on YNB + Tre agar is about 50% smaller than conventional medium. This problem can be solved by increasing trehalose concentration to 8–10 g/L, which increase the size of *P. rhodozyma* colonies obtained to comparable levels to that of the conventional medium, thus facilitating colony detection and picking.
2. This species normally constitutively produces detectable levels of MYCs, though photostimulation is recommended. Any illuminating device with a significant light intensity will stimulate mycosporine synthesis in *P. rhodozyma*. Temperature control is important when incubating under light due to the heat emitted by the lamps.

3. The fact that samples positive for *P. rhodozyma* in YNB + Tre agar medium are usually negative in conventional medium (YM agar) and vice versa, it is suggested that the best strategy to increase the recovery rate of *P. rhodozyma* from environmental samples is the use of both media.
4. An additional 2–3 days at 5 °C helps to intensify colony colors and thus helps detection *P. rhodozyma*-like colonies, which are orange or salmon orange.
5. *P. rhodozyma* colonies turn to a characteristic intense blue to violet color when flooded with diluted lugol.
6. A positive result is obtained when gas is accumulated inside the inverted Durham tube.
7. An alternative DNA extraction procedure is described by Libkind et al. [24]. Briefly, two loopfuls of YM agar-grown cultures are suspended in Eppendorf tubes containing 500 μL of lysing buffer and a volume of 200 μL of 425–600 μm of glass beads. After vortexing for 3 min, tubes are incubated for 1 h at 65 °C. After vortexing again for 3 min, the suspensions are centrifuged for 15 min at 4 °C and 20,000 × *g*. Finally, the collected supernatant is diluted 1:750, and 5 μL are used directly for PCR studies.
8. Expect a fragment size of 643 bp if the sample belong to the genus *Phaffia* with a faint band of approximately 1155 bp. Negative samples show a single 1155 bp band. This method proved efficient for all lineages reported by David-Palma et al. [11], though it did not work for the divergent strain TSN-67 reported by Yurkov et al. [12].
9. Type strain of *P. rhodozyma* is CBS 7918, and the rDNA sequences for comparison are AF075496 (D1/D2) and AF139628 (ITS).
10. An alternative preservation method for *P. rhodozyma* mutants, based on dehydrated gelatin drops, was described by Baeza et al. [25].
11. In our experience, survival is improved if tubes are initially stored at −20 °C (2–3 days) and then transferred to −80 °C.

Acknowledgments

This work was supported by UNComahue-CRUB grant B199, ANPCYT-FONCYT grant PICT 2542, and PIP 0392 to D.L.

References

1. Rodríguez-Sáiz M, de la Fuente JL, Barredo JL (2010) *Xanthophyllomyces dendrorhous* for the industrial production of astaxanthin. Appl Microbiol Biotechnol 88:645–658
2. Johnson EA (2003) *Phaffia rhodozyma*: colorful odyssey. Int Microbiol 6:169–174
3. Johnson EA, Conklin DE, Lewis MJ (1977) The yeast *Phaffia rhodozyma* as a dietary pigment source for salmonids and crustaceans. J Fish Res Board Can 34:2417–2421
4. Mann V, Harker M, Pecker I, Hirschberg J (2000) Metabolic engineering of astaxanthin production in tobacco flowers. Nat Biotechnol 18:888–892
5. Phaff HJ, Miller MW, Yoneyama M, Soneda M (1972) A comparative study of the yeast florae associated with trees on the Japanese Islands and on the west coasts of North America. In: Terui G (ed) Proceedings of the 4th IFS: Fermentation Technology today meeting. Society of Fermentation Technology, Osaka, pp 759–774
6. Weber RWS, Davoli P (2005) *Xanthophyllomyces* and other red yeasts in microbial consortia on spring sap-flow in the Modena province (Northern Italy). Atti Soc Nat Mat Modena 136:127–135
7. Weber RWS, Davoli P, Anke H (2006) A microbial consortium involving the astaxanthin producer *Xanthophyllomyces dendrorhous* on freshly cut birch stumps in Germany. Mycologist 20:57–61
8. Libkind D, Ruffini A, van Broock M, Alves L, Sampaio JP (2007) Biogeography, host specificity, and molecular phylogeny of the basidiomycetous yeast *Phaffia rhodozyma* and its sexual form, *Xanthophyllomyces dendrorhous*. Appl Environ Microbiol 73:1120–1125
9. Fell JW, Scorzetti G, Statzell-Tallman A, Boundy-Mills K (2007) Molecular diversity and intragenomic variability in the yeast genus *Xanthophyllomyces*: the origin of *Phaffia rhodozyma*? FEMS Yeast Res 7:1399–1408
10. Contreras G, Barahona S, Sepúlveda D, Baeza M, Cifuentes V, Alcaíno J (2015) Identification and analysis of metabolite production with biotechnological potential in *Xanthophyllomyces dendrorhous* isolates. World J Microbiol Biotechnol 31:517–526
11. David-Palma M, Libkind D, Sampaio JP (2014) Global distribution, diversity hot spots and niche transitions of an astaxanthin-producing eukaryotic microbe. Mol Ecol 23:921–932
12. Yurkov A, Wehde T, Kahl T, Begerow D (2012) Aboveground deadwood deposition supports development of soil yeasts. Diversity 4:453–474
13. Bellora N, Moliné M, David-Palma M, Coelho MA, Hittinger CT, Sampaio JP, Goncalvez P, Libkind D (2016) Comparative genomics provides new insights into the diversity, physiology, and sexuality of the only industrially exploited tremellomycete: *Phaffia rhodozyma*. BMC Genomics 17(1):901
14. Libkind D, Moliné M, de García V, Fontenla S, van Broock M (2008) Characterization of a novel South American population of the astaxanthin producing yeast *Xanthophyllomyces dendrorhous* (*Phaffia rhodozyma*). J Ind Microbiol Biotechnol 35:151–158
15. Golubev WI (1995) Perfect state of *Rhodomyces dendrorhous* (*Phaffia rhodozyma*). Yeast 11:101–110
16. Weber RWS, Becerra J, Silva MJ, Davoli P (2008) An unusual *Xanthophyllomyces* strain from leaves of *Eucalyptus globulus* in Chile. Mycol Res 112:861–867
17. Tognetti C, Moliné M, van Broock M, Libkind D (2013) Favored isolation and rapid identification of the astaxanthin-producing yeast *Xanthophyllomyces dendrorhous* (*Phaffia rhodozyma*) from environmental samples. J Basic Microbiol 53:766–772
18. Miller MW, Yoneyama M, Soneda M (1976) *Phaffia*, a new yeast genus in the Deuteromycotina (Blastomycetes). Int J Syst Bacteriol 26:286–291
19. Libkind D, Moliné M, van Broock MR (2010) Production of the UVB absorbing compound Mycosporine-glutaminol-glucoside by *Xanthophyllomyces dendrorhous* (*Phaffia rhodozyma*). FEMS Yeast Res 11(1):52–59
20. Andrewes AG, Phaff HJ, Starr MP (1976) Carotenoids of *Phaffia rhodozyma*, a red-pigmented fermenting yeast. Phytochemistry 15:1003–1007
21. Colabella F, Libkind D (2016) PCR-based method for the rapid identification of astaxanthin-accumulating yeasts (*Phaffia* spp.). Rev Argent Microbiol 48(1):15–20
22. Malik KA, Hoffmann P (1993) Long-term preservation of yeast cultures by liquid-drying. World J Microbiol Biotechnol 9:372–376

23. Bond CJ (2007) Cryopreservation of yeast cultures. In: Day JG, Pennington MW (eds) Cryopreservation and freeze-drying protocols, Methods in molecular biology, vol 38. Humana Press, Totowa, NJ, pp 39–47
24. Libkind D, Brizzio S, Ruffini A, Gadanho M, van Broock MR, Sampaio JP (2003) Molecular characterization of carotenogenic yeasts from aquatic environments in Patagonia, Argentina. Antonie van Leeuwenhoek 84:313–322
25. Baeza M, Retamales P, Sepúlveda D, Lodato P, Jiménez A, Cifuentes V (2009) Isolation, characterization and long term preservation of mutant strains of *Xanthophyllomyces dendrorhous*. J Basic Microbiol 49:135–141

Chapter 19

Engineering *Pichia pastoris* for the Production of Carotenoids

Patricia Veiga-Crespo, José Miguel Araya-Garay, and Tomás G. Villa

Abstract

Carotenoids are one of the most diverse and widely distributed classes of pigments in the biosphere and exhibit a variety of functions in the nature. Their importance and biotechnological applications are higher and higher, but their sources are not increasing in the same exponential way. Here we describe the process of bioengineering the yeast *Pichia pastoris* by sequential transformation to get an astaxanthin producer.

Key words Astaxanthin, *Pichia pastoris*, Yeast engineering, Carotenoids, Production

1 Introduction

Carotenoids are responsible for many of the plant colors (involved as light harvesters in photosynthesis) or act as precursors of plants hormones [1]. They can be synthetized by all the photosynthetic organisms but also by many non-photosynthetic, bacteria, plants, or fungi [2, 3]. This wide distribution represents also a variety of possible pathways of synthesis, including some common steps. The synthesis of carotenoids involves the condensation of two molecules of geranyl-geranyl diphosphate (GGDP, C_{20} precursor) to yield phytoene (C_{40}), and then a series of sequential reactions occurs to produce lycopene [4]. The most frequent carotenoids accumulated in plants are cyclic, carrying two β-ionone rings or one β-ionone ring and one ε-ionone ring, coming from the cyclation of lycopene [5].

The number of biotechnological applications of carotenoids, such as β-carotene, canthaxanthin, lutein, or astaxanthin, is increasing in the last years in many diverse fields such as biomedicine, food complements, poultry, and other industrial applications [2, 6–12]. Indeed, one could think that the high variety of non-microbial natural sources could supply carotenoids for the different fields, but they are not enough to account for all the industrial

Carlos Barreiro and José-Luis Barredo (eds.), *Microbial Carotenoids: Methods and Protocols*, Methods in Molecular Biology, vol. 1852, https://doi.org/10.1007/978-1-4939-8742-9_19,

needs, this encouraging the search of microbial sources of carotenoids.

The use of microalgae and seaweeds as an alternative as carotenoids producers has been also considered for years, and they could be an alternative and sustainable source of various bioactive natural carotenoids such as lutein, β-carotene, canthaxanthin, zeaxanthin, violaxanthin, or fucoxanthin [13, 14]. However, these natural sources are hampered by issues regarding the low efficiency (extraction), selectivity, the solvents amount needed for extraction, and so on [15]. However, some good experiments are carried out these days for making them a real alternative by optimizing the cultivation procedure [16, 17] or getting more efficient extraction techniques [15, 18], but yet the carotenoidic amount obtained is not enough to cover the market demands.

Astaxanthin is a keto-carotenoid widespread used in aquaculture, especially as a feed supplement in farmed trout, salmon, and prawns to obtain the desired red-pink coloration. The color in these species is not only an esthetic factor but also has an important commercial added value [19] since costumers prefer pigmented individuals over nonpigmented ones. As we mentioned before, astaxanthin also have a great interest in the biomedicine field as a powerful antiaging, anticancer, or against the onset of the degenerative disease [20, 21]. The genes involved in the astaxanthin biosynthesis have been characterized in different sources such as *Phaffia rhodozyma*, *Xanthophyllomyces dendrorhous*, *Haematococcus pluvialis*, *Brevibacterium linens*, *Paracoccus haeundaensis*, or *Agrobacterium aurantiacum* [22–26].

The synthetic sources of pigments have been an alternative for years, but today they seem to have an increasing social rejection. Microorganisms are an actual alternative for carotenoid production. Carotenoid biosynthesis has been achieved in engineered microorganism such as *Escherichia coli* [5, 27–30]. The use of engineered yeast as expression tools instead of bacteria has some advantage such as safety since many yeast species are considered as safe organisms and have a GRAS (generally recognized as safe) status by international organisms such as the US Food and Drug Administration (FDA) or their ability to exhibit an efficient isoprenoid metabolism. A major disadvantage is the lower number of tools for yeast engineering than for bacteria. Some good and promising examples include *Saccharomyces cerevisiae*, *Candida utilis*, *Rhodotorula toruloides*, or the fungus *Mucor circinelloides* [31–36].

In this chapter, we show the method for engineering *Pichia pastoris* X-33 strain in order to produce different carotenoids by a sequential introduction of the genes *crtE* (encoding geranylgeranyl diphosphatase synthase), *crtB* (phytoene synthase), *crtI* (phytoene desaturase), *crtL* (lycopene β cyclase), *crtW* (β-carotene ketolase), and *crtZ* (β-carotene hydroxylase).

2 Materials

*2.1 Construction of pGAPZαA**

1. Shuttle plasmid *E. coli–P. pastoris* pGAPZαA (Invitrogen, Carlsbad, CA, USA) (Table 1).
2. Site-directed mutagenesis (SDM) kit including PfuTurbo® DNA polymerase (2.5 U/μL), 10× reaction buffer, DpnI restriction enzyme (10 U/μL), and dNTP mix (QuikChange, Stratagene, La Jolla, CA, USA).
3. Pair of primers SDM-pGAPZαA* forward (F)/SDM-pGAPZαA* reverse (R) (Table 2).
4. Thermocycler.
5. Incubator shaker.

Table 1
List of plasmids used in this work

Plasmid	Description
pGAPZαA	*Pichia* integrative plasmid, Zeocin resistant
pGAPZαA*	pGAPZαA devoid of AvrII site
pAC*CRT*-EIB	*E. coli* lycopene expression plasmid [37]
pAK96K	*E. coli* pBlueScript II vector carrying *crtW* and *crtZ* genes from *Agrobacterium aurantiacum* [26]
pET21-Lycβ	*E. coli* pET21 carrying the gene *crtL* from *Ficus carica* [5]
pGAPZA-E	*crtE* in pGAPZA without alpha factor
pGAPZA*-B	*crtB* in pGAPZA* without alpha factor
pGAPZA*-I	*crtI* in pGAPZA* without alpha factor
pGAPZA*-I*	*crtI** in pGAPZA* without alpha factor; *crtI** is *crtI* without BamHI site
pGAPZA*-L	*crtL* in pGAPZA* without alpha factor
pGAPZA*-L*	*crtL** in pGAPZA* without alpha factor; *crtL** is *crtL* without BglII site
pGAPZA-EB	*crtE* and *crtB* in pGAPZA without alpha factor
pGAPZA-EBI*	*crtE*, *crtB* and *crtI* in pGAPZA without alpha factor for lycopene production
pGAPZA-EBI*L*	*crtE*, *crtB*, *crtI**, and *crtL** in pGAPZA without alpha factor for β-carotene production
pGAPZA-W	*crtW* in pGAPZA without alpha factor
pGAPZA*-Z	*crtZ* in pGAPZA* without alpha factor
pGAPZA-WZ	*crtW* and *crtZ* in pGAPZA without alpha factor

Table 2
List of primers used in this work

Technique	Primer	Description	Sequence
SDM	SDM-pGAPZαA* F	Removal of AvrII site in pGAPZαA	5′CACCGCCGTTACCGTCGCTAGGAAATTTTACTC3′
SDM	SDM-pGAPZαA* R	Removal of AvrII site in pGAPZαA	5′GAGTAAAATTTCCTAGCGACGGTAACGGCGGTG3′
SDM	SDM-*crt*I* F	Removal of BamHI site in *crtI*	5′CGTCTACAAGCTGCGGGTATCCCGTCTTACTGC3′
SDM	SDM-*crt*I* R	Removal of BamHI site in *crtI*	5′GCAGTAAGACGGGATACCCGCAGCTTGTAGACG3′
SDM	SDM-*crt*L* F	Removal of BglII site in *crtL*	5′CCAAAAACCCATCAAGGTCTTCAAGAGCAAGCA3′
PCR	Eu-*crt*E F	*crtE* with a Kozak sequence and SfuI restriction site	5′AACTATTTCGAAACGATGGCAGTCTGCGCAAAA3′
PCR	Eu-*crt*E R	*crtE* and EcoRI restriction site	5′GGAATTCTTAACTGACGGCAGCG3′
PCR	Eu-*crt*B F	*crtB* with a Kozak sequence and SfuI restriction site	5′AACTATTTCGAAACGATGGCAGTTGGCTCG3′
PCR	Eu-*crt*B R	*crtB* and EcoRI restriction site	5′GGAATTCCTAGAGCGGGCGCTGC3′
PCR	Eu-*crt*I F	*crtI* with a Kozak sequence and SfuI restriction site	5′AACTATTTCGAAACGATGGCACCAACTACGG3′
PCR	Eu-*crt*I R	*crtI* and EcoRI restriction site	5′GGAATTCTCAAATCAGATCCTCCAG3′
PCR	Fc-*crt*L F	*crtL* with a Kozak sequence and SfuI restriction site	5′AACTATTTCGAAACGATGGGGTACTCTTCCATATCC3′
PCR	Fc-*crt*L R	*crtL*and XbaI restriction site	5′GCTCTAGACTAAATGGTTTCAAGTGCCA3′
PCR	Aa-*crt*W F	*crtW* with a Kozak sequence and SfuI restriction site	5′AACTATTTCGAAACGATGGCACATGCCCTGCC3′
PCR	Aa-*crt*W R	*crtW* and EcoRI restriction site	5′GGAATTCTCAGCGGTGCCCC3′
PCR	Aa-*crtZ* F	*crtZ* with a Kozak sequence and SfuI restriction site	5′AACTATTTCGAAACGATGGCAAATTTCCTGGATCG3′
PCR	Aa-*crtZ* R	*crtZ* and EcoRI restriction site	5′GGAATTCTCACGTGCGCTCCTGC3′

2.2 Transformation of E. coli XL1-Blue Supercompetent Cells

1. Falcon® 2059 polypropylene tubes (15 mL).
2. *E. coli* XL1-Blue Supercompetent cells (*recA1*, *endA1*, *gyrA96*, *thi*, *hsdR17*, *supE44*, *relA1*, *lac*, F′, *proAB*, *lacIqZΔM15*, *Tn10*(tetr)).
3. Isopropyl-1-thio-β-D-galactopyranoside (IPTG).
4. 5-Bromo-4-chloro-3-indolyl-β-D-galactopyranoside (X-gal).
5. Luria Bertani (LB) low salt medium: Tryptone 10g/L, yeast extract 5 g/L, NaCl 5 g/L. Adjust to pH 7.4. Sterilize at 121 °C and 1.1 atm for 20 min.
6. LB-agar: LB and 2% agar.
7. LB-agar-Zeocin: LB-agar and 25 mg/mL Zeocin (Invitrogen) for the colony selection.
8. Orbital shaker.

2.3 Amplification and Cloning of the Genes Coding for the Ketocaroteno-genesis Synthesis in E. coli

1. Oligonucleotides pairs for the genes *crtB*, *crtE*, and *crtI* from *E. uredovora*, *crtL* from *Ficus carica*, and *crtW*, *crtZ* from *A. aurantiacum* (Table 2, *see* **Note 1**).
2. PCR reaction mix: PfuTurbo® DNA polymerase (2.5 U/μL); 10× reaction buffer and dNTP mix.
3. DNA template: Plasmid pAC*CRT*-EIB as template for *crtB*, *crtE*, and *crtI*, pET21-Lycβ as template for *crtL*, and plasmid pAK96K as template for *crtW* and *crtZ* (Table 1).
4. Agarose gels 1% (type II; medium EEO, Sigma-Aldrich) in TAE buffer
5. TAE buffer: Tris-acetate (Sigma Aldrich) 0.04 M and EDTA (Sigma Aldrich) 0.002 M.
6. Molecular weight marker: 1 kb Ladder (Promega)
7. T4 DNA ligase (Promega).
8. Illustra™ GFX™ PCR DNA and Gel Band Purification Kit (GE Healthcare).
9. *E. coli* TOP10 One Shot (Invitrogen) (F-mcrA, Δ(mrr-hsdRMS-mrcBC) φ80lacZΔM15ΔlacX74 nupG recA1 araD139 Δ(ara-leu) galE15 galK16 rpsL (Str^R) endA1 λ^{-1}).
10. T4 DNA ligase.
11. Restriction enzymes: BamHI, BglII, EcoRI, SfuI.
12. LB-agar-Zeocin: LB-agar and 25 mg/mL Zeocin (Invitrogen) for the colony selection.
13. Plasmid Maxi Kit for plasmid recovery from *E. coli* (Qiagen, Hilden, Germany).
14. Loading buffer: EDTA 25 mM (Sigma Aldrich), glycerol 50% (Sigma Aldrich), and bromophenol blue 0.25% (Sigma Aldrich).

15. Thermocycler.
16. Orbital shaker.
17. Electrophoresis cuvettes.

2.4 Cloning and Expression of the Genes in P. pastoris

1. YPD medium: Dissolve 1 g yeast extract, 2 g peptone, and 2 g glucose in 100 mL of water. Sterilize at 121 °C and 1.1 atm. for 20 min.
2. Electrocompetent cells *Pichia pastoris* X-33 (Invitrogen, CA, USA).
3. Sorbitol 1M.
4. YPDS medium: YPD medium + sorbitol 1 M.
5. YPDS agar-Zeocin medium: YPDS medium + 2% agar + Zeocin (100 mg/mL).
6. Rotary Shaker.
7. Centrifuge.
8. Double deionized water.
9. AvrII restriction enzyme.
10. Bio-Rad MicroPulser (Bio-Rad Laboratories, Inc., CA, USA).

2.5 Carotenoids Extraction and HPLC Analysis from P. pastoris

1. *Pichia pastoris* strains: *Pichia pastoris* Pp-EBI (lycopene producer), *Pichia pastoris* Pp-EBIL (β-carotene producer), and *Pichia pastoris* Pp-EBILWZ (astaxanthin producer).
2. YPD medium + Zeocin 200 μg/mL.
3. Lyophilizer.
4. DMSO.
5. Acetone (*see* **Note 2**).
6. Hexane (*see* **Note 2**).
7. Chloroform (*see* **Note 2**).
8. Phosphate buffer 0.1 M pH 7.4.
9. Polycarbonate 0.22 μm membranes (Millipore) (*see* **Note 3**).
10. HPLC equipment (Waters) equipped with a C30 carotenoids column (250 mm, 4.6 mm, 5 μm) (YMC Europe GMBH), accomplished to the acquisition software Empower2 (Waters), registering the absorbance from $\lambda = 250$ nm to $\lambda = 600$ nm with the photodiode detector accomplished to the HPLC.
11. Linear elution gradient: methanol/water/methyl-*tert*-butyl-ether (MTBE) (*see* **Note 2**)
12. Carotenoids standards: lycopene, β-carotene, and astaxanthin (Sigma-Aldrich, Madrid, Spain) (*see* **Note 4**).

3 Methods

*3.1 Construction of pGAPZαA**

1. Prepare the PCR mix reactions: 50 ng plasmid pGAPZαA as DNA template, 2.5 U/μL of *PfuTurbo* DNA polymerase, 5 μL of 10× reaction buffer, 125 ng of each oligonucleotide primer SDM-pGAPZαA* F and SDM-pGAPZαA* R, 1 μL of dNTP mix, and ddH_2O to a final volume of 50 μL (*see* **Note 5**).
2. Do the PCR according to these parameters: 1 cycle 95 °C, 30 s; 20 cycles, 95 °C 30s, 55 °C 1 min, and 68 °C 2 min/kb of plasmid.
3. Add 10 U/μL of DnpI restriction enzyme for the digestion of the parental supercoiled dsDNA. Incubate at 37 °C for 1 h.

*3.2 Transformation of E. coli XL1-Blue Supercompetent Cells with the Plasmid pGAPZαA**

1. Prechill the Falcon tubes on ice, and add 50 μL of *E. coli* XL1-Blue Supercompetent cells.
2. Add 1 μL of DnpI-treated pGAPZαA* (*see* Subheading 3.1).
3. Swirl the transformation reaction gently to mix and incubate on ice for 30 min. Heat pulse the mix at 42 °C for 45 s. Finally, incubate on ice for 2 min.
4. Add the mix to 0.5 mL of LB medium and incubate at 37 °C for 1 h at 225 rpm.
5. Spread the cells in LB-Zeocin-IPTG-X-gal medium for colony selection.
6. Incubate the plates during 24 h at 37 °C.
7. Select the Zeocin-resistant colonies.

3.3 Amplification and Cloning of the Genes Coding for the Ketocaroteno-genesis Synthesis in E. coli

1. Prepare the PCR mix: 50 ng of the appropriate DNA template, the oligonucleotide pair for each gene (Table 2), 2.5 U/μL of *PfuTurbo* DNA polymerase, 5 μL of 10× reaction buffer, 1 μL of dNTP mix, and the corresponding volume of double-deionized water for a final volume of 50 μL in PCR vials.
2. Do the PCR according to these parameters: 1 cycle 95 °C, 33 min; 30 cycles, 95 °C 30 s, 55–65 °C 30 s, and 72 °C 90 s; and 1 final cycle, 72 °C 7 min.
3. Run the PCR products in an agarose gel during 60 min at 70 V. For electrophoresis, the PCR products are mixed in a proportion 4:1 with loading buffer.
4. Select the PCR products according their molecular weight, after loading in the gel a 1 kb Ladder.
5. After cutting the band corresponding to the gene of interest from the agarose gel, the DNA fragment will be purified using a Illustra™ GFX™ PCR DNA and Gel Band Purification Kit. Briefly, the agarose band was cut and incubated with capture

Table 3
Mix of restriction enzymes and the different DNA fragments

Enzyme restriction	Gene	Plasmid	Final ligand	*E. coli* strain obtained
SfuI and EcoRI	*crt*E	pGAPZαA	pGAPZA-E	*E. coli crt*E
SfuI and EcoRI	*crt*B	pGAPZαA*	pGAPZA*-B	*E. coli crt*B
SfuI and EcoRI	*crt*I	pGAPZαA*	pGAPZA*-I	*E. coli crt*I
SfuI and XbaI	*crt*L	pGAPZαA*	pGAPZA*-L	*E. coli crt*L
SfuI and EcoRI	*crt*W	pGAPZαA	pGAPZA-W	*E. coli crt*W
SfuI and EcoRI	*crt*Z	pGAPZαA*	pGAPZA*-Z	*E. coli crt*Z
BamHI and BglII	pGAPZA*-B	pGAPZA*-E	pGAPZA-EB	*E. coli* EB
BamHI and BglII	pGAPZA*-I*	pGAPZA-EB	pGAPZA-EBI*	*E. coli* EBI*
BamHI and BglII	pGAPZA*-L*	pGAPZA-EBI*	pGAPZA-EBI*L*	*E. coli* EBI*L*
BamHI and BglII	pGAPZA*-Z	pGAPZA-W	pGAPZA-WZ	*E. coli* ZW

buffer 3, adding later this solution to the Microspin Column for DNA binding. After washing with the wash buffer, the DNA was recovered by elution with the elution buffer.

6. DNA digestion: Mix the right restriction enzymes with the DNA fragment. Incubate the mix during 2 h at 30 °C (Table 3). Stop the digestion reaction by loading the reaction mix in an agarose gel. Repeat the purification of the digested DNA (**Steps 4** and **5**).
7. DNA ligation: Prepare the ligation reaction mixing 3 U/μL of T4 DNA ligase, 1 μL of DNA ligase 1× buffer, 100 ng of plasmid DNA, and 1:3 ng of insert. Incubate the mix overnight at 16 °C (*see* **Note 6**).
8. The DNA constructions (Table 3) are used to transform *E. coli* TOP10 by shock thermic following the protocol described in Subheading 3.2.
9. Growth the transformed *E. coli* (Table 3) selected strain in LB-Zeocin at 37 °C and 150 rpm for 24 h.
10. Purify the plasmids from the cells using the Plasmid Maxi Kit for plasmid recovery from *E. coli*. Briefly, 500 mL overnight cultured bacteria were harvested by centrifuging at 6000 × *g*, 15 min, 4 °C. The cells were lysed by sequential addition of buffet P1 and P2 and kept at RT for 5 min after vigorously mixing. The prechilled buffer P3 was then added, and the solution was kept 20 min on ice. While samples were centrifuging at 20,000 × *g*, 30 min, and 4 °C, the columns were

equilibrated by adding 10 mL of buffer QBT. Then, the supernatant coming from the previous centrifugation was added and let it to enter by gravity flow. The samples were then washed twice with 2 × 30 mL of buffer QC. Finally, the DNA was eluted with the buffer QF and precipitated by adding RT isopropanol. After a final centrifugation at 15,000 × *g*, 30 min, 4 °C, the supernatant was discarded and the pellet resuspended in water.

11. Site-directed mutagenesis of crtI and crtL to obtain crtI* and crtL*: DNA templates, pGAPZA*-I and pGAPZA*-L; oligonucleotides, SDM-*crt*I* F and SDM-*crt*I* R; and SDM-*crt*L* F and SDM-*crt*I* R. Just follow the method described in Subheading 3.1 for site-directed mutagenesis.
12. Transform *E. coli* XL1-Blue Supercompetent with pGAPZA*-I* and pGAPZA*-L* to obtain the strains *E. coli* crtI* and *E. coli* crtL* (Table 3).
13. Purify the plasmid pGAPZA*-I* and pGAPZA*-L* from *E. coli* crtI* and *E. coli* crtL* using the Plasmid Maxi Kit for plasmid recovery from *E. coli.*
14. Construction of the plasmid pGAPZA-EB and of *E. coli*-EB: Digest the plasmid pGAPZA*-B with the restriction enzymes BamHI and BglII, as previously described and pGAPZA-E with BamHI. Ligate the digested and purified fragments as described in **Step 7,** and follow all the *E. coli* TOP10 transformation to get the strain *E. coli*-EB.
15. Construction of the plasmid pGAPZA-EBI* and *E. coli*-EBI*: Use the plasmid pGAPZA-EB and the plasmid pGAPZA*-I* as DNA donors. Digest pGAPZA*-I* with BglII and BamHI, and follow the same steps that previously described for cloning the fragment in BamHI site of pGAPZA-EB. Transform *E. coli* TOP10 by thermic shock with to get *E.coli*-EBI*.
16. Construction of the plasmid pGAPZA-EBI*L* and *E. coli*-EBI*L*: Use the plasmid pGAPZA-EBI* and the plasmid pGAPZA*-L* as DNA donors (Table 3). Digest pGAPZA*-L* with BglII and BamHI, and follow the same steps that previously described for cloning the fragment in BamHI site of pGAPZA-EBI*. Transform *E. coli* TOP10 by thermic shock with to get *E. coli*-EBI*L*.
17. Construction of the plasmid pGAPZA-WZ and *E. coli*-WZ: Use the plasmid pGAPZA-W and the plasmid pGAPZA*-Z as DNA donors (Table 3). Digest pGAPZA*-Z with BglII and BamHI and follow the same steps that previously described for cloning the fragment in BamHI site of pGAPZA-W. Transform *E. coli* TOP10 by thermic shock with to get *E. coli*-ZW.

3.4 Cloning and Expression of the Plasmids in *P. pastoris*

3.4.1 Cloning and Expression of pGAPZA-EBI* in Pichia pastoris X-33 to Obtain Pichia pastoris Pp-EBI

1. Linearization of the plasmid pGAPZA-EBI*: The plasmid must be digested with AvrII during 1 h at 37 °C. The reaction can be stopped by heat inactivation at 65 °C. The lineal fragment must be purified using agarose gels as previously described.
2. Preparation of *P. pastoris* X-33 for transformation: Grow the cells in 5 mL YPD medium overnight. Then, inoculate 500 mL of fresh medium and grow the cells until OD_{600} = 1.3–1.5. Centrifuge the cells, and resuspend them in 500 mL of ice-cold, sterile water twice. Centrifuge them again and resuspend in 20 mL ice-cold sorbitol 1 M. After a new centrifugation, resuspend the cells in 1 mL of ice-cold sorbitol 1 M (*see* **Note 7**).
3. Electro-transformation of *P. pastoris* X-33: Take 80 μL of cells in sorbitol, and add 10 μg of linearized plasmid from **Step 1**. Transfer the mix to an ice-cold electroporation cuvette 0.2 mm. Use a 1500 V, 200 Ω, and 25 μF pulse to transform the yeast. Then, resuspend the cells immediately in 1 mL of sorbitol 1 M and transfer to a sterile tube 15 mL. Incubate at 30 °C with high shaking during 2 h.
4. Plate the cells in YPDS-agar-Zeocin plates, and incubate at 30 °C during 2–3 days to let the colonies form to obtain *Pichia pastoris* Pp-EBI, lycopene producer.

3.4.2 Cloning and Expression of pGAPZA-EBI*L* in Pichia pastoris X-33 to Obtain Pichia pastoris Pp-EBIL

1. Linearization of the plasmid pGAPZA-EBI*L*: The plasmid must be digested with AvrII during 1 h at 37 °C. The reaction can be stopped by heat inactivation at 65 °C. The lineal fragment must be purified using agarose gels as previously described.
2. Preparation of *P. pastoris* X-33 for transformation: Follow the previously described protocol.
3. Follow the previously described protocol for transforming and selecting the strain *Pichia pastoris* Pp-EBIL and β-carotene producer.

3.4.3 Cloning and Expression of pGAPZA-WZ in Pichia pastoris Pp-EBIL to Obtain Pichia pastoris Pp-EBILWZ

1. Linearization of the plasmid pGAPZA-WZ: The plasmid must be digested with AvrII during 1 h at 37 °C. The reaction can be stopped by heat inactivation at 65 °C. The lineal fragment must be purified using agarose gels as previously described.
2. Preparation of *P. pastoris* Pp-EBIL for transformation: Follow the previously described protocol.
3. Follow the previously described protocol for transforming and selecting the strain *Pichia pastoris* Pp-EBILWZ and astaxanthin producer.

3.5 Carotenoids Extraction and HPLC Analysis from *P. pastoris*

3.5.1 Lycopene

1. Inoculate 5 mL of YPD medium + Zeocin 100 μg/mL with a single-colony *Pichia pastoris* Pp-EBI, and grow it overnight at 30 °C and 200 rpm. Then, inoculate 500 mL of YPD medium + Zeocin 100 μg/mL. Keep the culture during 72 h.
2. Harvest the cells and wash twice with distilled water (*see* **Note 8**).
3. Transfer the cells into 50 mL polypropylene tubes, and centrifuge for 15 min, at 3000 × *g* and 4 °C. Discard the supernatant, and freeze the cells at −80 °C.
4. Lyophilize the cells over 48 h at 0.1 mbar.
5. Weight 50 mg of lyophilized cells, and incubate with 3 mL DMSO, prewarmed at 55 °C during 30 min (1 min of strong shaking: 29 min without shaking). Residual cell suspension will be extracted with 10 mL of acetone and vortex 5 min at 4 °C.
6. Add 5 mL of hexane + 1 mL of phosphate buffer 0.1 M to the extract. Vortex 30 s and centrifuge during 10 min at 3000 × *g*. This extraction procedure should be repeated, adding fresh hexane and phosphate buffer until getting a cell pellet colorless.
7. Evaporate the crude extract under a N_2 flow.
8. Keep the pigment powder at −80 °C for HPLC analysis (*see* **Note 9**).

3.5.2 β-Carotene

Repeat the procedure of Subheading 3.5.1 using the *Pichia pastoris* Pp-EBIL.

3.5.3 Astaxanthin

Repeat the procedure of Subheading 3.5.1 using the *Pichia pastoris* Pp-EBILWZ.

3.5.4 HPCL Analysis

1. Dissolve the cryopreserved dried extracts obtained in 2 mL of chloroform/methanol/acetone (3:2:1 v/v/v), and filter them across polycarbonate 0.22 μm membranes.
2. Dissolve the carotenoid standards in the same buffer.
3. Inject 20 μL of sample in the column, and run the samples at 1 mL/min at 25 °C.
4. Compare the retention time of the standards with the peak in the extracts coming from the yeast. The majority peak in each case should match with the corresponding retention time for the standard.

The engineering of the yeast *P. pastoris* resulted in an efficient method to get a new source of carotenoids, particularly astaxanthin, a high commercial value compound, complementing the current sources. In addition, the production of astaxanthin in this yeast should be cheaper than preparation by other means.

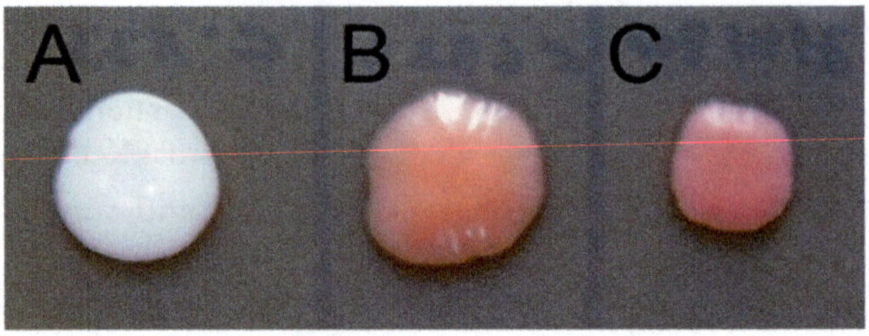

Fig. 1 Isolated colonies of recombinant *Pichia pastoris* strains harboring genes encoding different carotenoids. (**a**) Non-recombinant *Pichia pastoris* X-33 strain, recipient of carotenoid-encoding genes; (**b**) *Pichia pastoris* Pp-EBIL; (**c**) *Pichia pastoris* Pp-EBILWZ

A great advantage of the sequential introduction of the keto-carotenoids gene cluster(s) has been the construction of engineering *P. pastoris* strains, able to produce other high-value carotenoids such as lycopene or β-carotene, thus increasing the potential use of GRAS yeasts in the carotenoid-producing industries (Figs. 1 and 2, Table 4).

Future experiments will see the increase of the yield production by variations in the culture of *Pichia*, variations that include studies on physicochemical conditions of fermentations, as well as the effect of different carbon and nitrogen sources. A deep consideration in the fermentation scale-up procedures will be also taken into account.

4 Notes

1. The biggest difficulty for designing oligonucleotides to characterize unknown genes involved in the carotenogenesis is the evolutive convergence: the same gene comes from different origins, and only concrete positions of the active site show the same sequence, being more important than the 3D configuration of the site.
2. The quality of the components of the solvent phase for the HPLC should be extremely high, only HPLC-grade components, to avoid potential interferences in the spectrum.
3. The presence of organic solvents makes mandatory use of polycarbonate filters.
4. The carotenoid standards should be manipulated carefully to avoid mistakes in the identification. During the performance of these works, several blind tests were carried out to check their integrity. The standards also must be protected from the light to avoid the degradation.

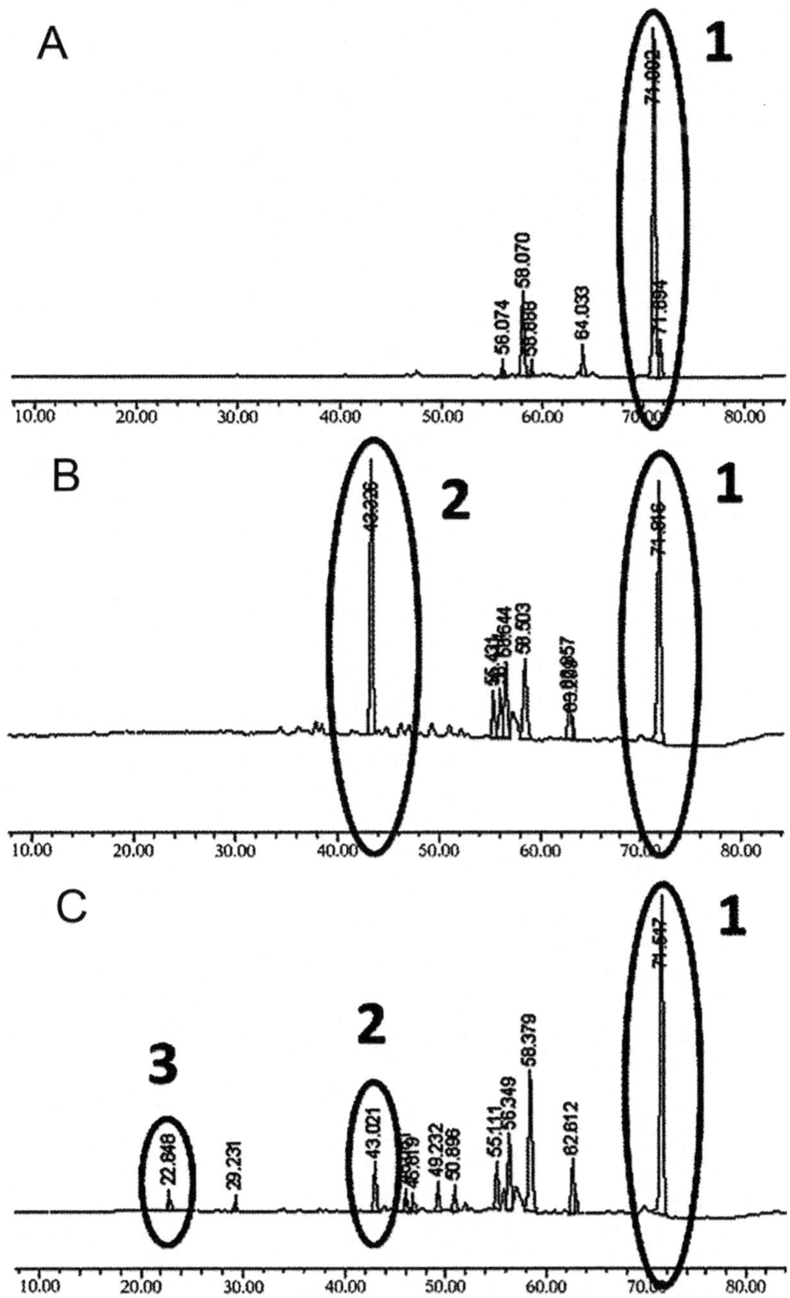

Fig. 2 HPLC analysis of different *Pichia pastoris* recombinant strains harboring carotenoid-encoding genes. (**a**) *Pichia pastoris* Pp-EBI, strain containing lycopene-synthesizing gene; (**b**) *Pichia pastoris* Pp-EBIL, strain containing lycopene and β-carotene-synthesizing genes; (**c**) strain *Pichia pastoris* Pp-EBILWZ able to synthesize lycopene, β-carotene, and astaxanthin, peaks 1, 2, and 3, respectively

Table 4
List of recombinant strains obtained in this work

Final strain	Origin strain	Plasmid	Final carotenoid
Pichia pastoris Pp-EBI	*P. pastoris* X-33	pGAPZA-EBI*	Lycopene
Pichia pastoris Pp-EBIL	*P. pastoris* X-33	pGAPZA-EBI*L*	β-Carotene
Pichia pastoris Pp-EBILWZ	*P. pastoris* Pp-EBIL	pGAPZA-WZ	Astaxanthin

5. The construction of the plasmid pGAPZαA* was one of the critical steps in this work in order to get different versions of the same plasmid with different restriction sites. This was the key for the sequential integration of the different genes described in the work. It was also necessary to modify the sequences of some of the genes to get also complementary restriction sites (see the table describing the oligonucleotides used).
6. During the manipulation of the T4 DNA ligase, the temperature is a critical step. It is highly recommended to aliquot the T4 DNA ligase buffer and to avoid repetitive cycles of freezing/thawing.
7. During the yeast transformation, the temperature of sorbitol is critical; only ice-chilled buffer should be used. Also, the obtained cells must be used during the same day. In order to keep the transformation efficiency, only fresh yeast cells must be employed.
8. The wash steps are essential to remove any rest of culture medium and possible interferences during the HPLC characterization from medium components.
9. The carotenoid powder must be stored protected from the light.

References

1. Sandmann G (2001) Carotenoid biosynthesis and biotechnological application. Arch Biochem Biophys 385(1):4–12. https://doi.org/10.1006/abbi.2000.2170
2. Giorio G et al (2013) Elevation of lutein content in tomato: a biochemical tug-of-war between lycopene cyclases. Metab Eng 20:167–176. https://doi.org/10.1016/j.ymben.2013.10.007
3. Fraser PD, Bramley PM (2004) The biosynthesis and nutritional uses of carotenoids. Prog Lipid Res 43(3):228–265. https://doi.org/10.1016/j.plipres.2003.10.002
4. Sandmann G (1994) Carotenoid biosynthesis in microorganisms and plants. Eur J Biochem 223(1):7–24
5. Araya-Garay JM et al (2011) cDNA cloning of a novel gene codifying for the enzyme lycopene beta-cyclase from *Ficus carica* and its expression in *Escherichia coli*. Appl Microbiol Biotechnol 92(4):769–777. https://doi.org/10.1007/s00253-011-3488-8
6. Seddon JM et al (1994) The use of vitamin supplements and the risk of cataract among US male physicians. Am J Public Health 84(5):788–792

7. Willett WC, Trichopoulos D (1996) Nutrition and cancer: a summary of the evidence. Cancer Causes Control 7(1):178–180
8. Murillo AG et al (2016) The potential of non-provitamin A carotenoids for the prevention and treatment of non-alcoholic fatty liver disease. Biology (Basel) 5(4):42. https://doi.org/10.3390/biology5040042
9. Yang Y et al (2016) Astaxanthin prevents and reverses the activation of mouse primary hepatic stellate cells. J Nutr Biochem 29:21–26. https://doi.org/10.1016/j.jnutbio.2015.11.005
10. Yang Y et al (2016) Histone deacetylase 9 plays a role in the antifibrogenic effect of astaxanthin in hepatic stellate cells. J Nutr Biochem 40:172–177. https://doi.org/10.1016/j.jnutbio.2016.11.003
11. Yeh PT et al (2016) Astaxanthin inhibits expression of retinal oxidative stress and inflammatory mediators in streptozotocin-induced diabetic rats. PLoS One 11(1):e0146438. https://doi.org/10.1371/journal.pone.0146438
12. Visioli F, Artaria C (2017) Astaxanthin in cardiovascular health and disease: mechanisms of action, therapeutic merits, and knowledge gaps. Food Funct 8(1):39–63. https://doi.org/10.1039/c6fo01721e
13. Guedes AC, Amaro HM, Malcata FX (2011) Microalgae as sources of carotenoids. Mar Drugs 9(4):625–644. https://doi.org/10.3390/md9040625
14. Safafar H et al (2015) Carotenoids, phenolic compounds and tocopherols contribute to the antioxidative properties of some microalgae species grown on industrial wastewater. Mar Drugs 13(12):7339–7356. https://doi.org/10.3390/md13127069
15. Poojary MM et al (2016) Innovative alternative technologies to extract carotenoids from microalgae and seaweeds. Mar Drugs 14(11):214. https://doi.org/10.3390/md14110214
16. Zhang Z et al (2017) Two-step cultivation for production of astaxanthin in Chlorella zofingiensis using a patented energy-free rotating floating photobioreactor (RFP). Bioresour Technol 224:515–522 https://doi.org/10.1016/j.biortech.2016.10.081
17. Peng L et al (2016) Cultivation of *Neochloris oleoabundans* in bubble column photobioreactor with or without localized deoxygenation. Bioresour Technol 206:255–263. https://doi.org/10.1016/j.biortech.2016.01.081
18. Gong M, Bassi A (2016) Carotenoids from microalgae: a review of recent developments. Biotechnol Adv 34(8):1396–1412. https://doi.org/10.1016/j.biotechadv.2016.10.005
19. Wade N et al (2005) Esterified astaxanthin levels in lobster epithelia correlate with shell colour intensity: potential role in crustacean shell colour formation. Comp Biochem Physiol B Biochem Mol Biol 141(3):307–313. https://doi.org/10.1016/j.cbpc.2005.04.004
20. Hix LM et al (2004) Upregulation of connexin 43 protein expression and increased gap junctional communication by water soluble disodium disuccinate astaxanthin derivatives. Cancer Lett 211(1):25–37. https://doi.org/10.1016/j.canlet.2004.01.036
21. Kurihara H et al (2002) Contribution of the antioxidative property of astaxanthin to its protective effect on the promotion of cancer metastasis in mice treated with restraint stress. Life Sci 70(21):2509–2520
22. Johnson EA et al (1980) Phaffia rhodozyma as an astaxanthin sources in salmonid diet. Aquaculture 20:123–134
23. Bubrick P (1991) Production of astaxanthin from Haematococcus. Bioresour Technol 38:237–239
24. Krubasik P, Sandmann G (2000) A carotenogenic gene cluster from *Brevibacterium linens* with novel lycopene cyclase genes involved in the synthesis of aromatic carotenoids. Mol Gen Genet 263(3):423–432
25. Lee JH et al (2004) *Paracoccus haeundaensis* sp. nov., a Gram-negative, halophilic, astaxanthin-producing bacterium. Int J Syst Evol Microbiol 54(Pt 5):1699–1702. https://doi.org/10.1099/ijs.0.63146-0
26. Misawa N et al (1995) Structure and functional analysis of a marine bacterial carotenoid biosynthesis gene cluster and astaxanthin biosynthetic pathway proposed at the gene level. J Bacteriol 177(22):6575–6584
27. Li C et al (2015) Modification of targets related to the Entner-Doudoroff/pentose phosphate pathway route for methyl-D-erythritol 4-phosphate-dependent carotenoid biosynthesis in *Escherichia coli*. Microb Cell Factories 14:117. https://doi.org/10.1186/s12934-015-0301-x
28. Shen HJ et al (2015) Engineering of *Escherichia coli* for lycopene production through promoter engineering. Curr Pharm Biotechnol 16(12):1094–1103
29. Seo YB et al (2015) Molecular cloning and co-expression of phytoene synthase gene from *Kocuria gwangalliensis* in *Escherichia coli*. J Microbiol Biotechnol 25(11):1801–1809. https://doi.org/10.4014/jmb.1505.05035

30. Li Z et al (2015) Cloning and expression of a zeta-carotene desaturase gene from *Lycium chinense*. J Genet 94(2):287–294
31. Lange N, Steinbuchel A (2011) beta-Carotene production by *Saccharomyces cerevisiae* with regard to plasmid stability and culture media. Appl Microbiol Biotechnol 91(6):1611–1622. https://doi.org/10.1007/s00253-011-3315-2
32. Miura Y, Kondo K, Shimada H, Saito T, Nakamura K, Misawa N (1998) Production of lycopene by the food yeast, *Candida utilis* that does not naturally synthesize carotenoid. Biotechnol Bioeng 58(2–3):306–308
33. Shimada H et al (1998) Increased carotenoid production by the food yeast *Candida utilis* through metabolic engineering of the isoprenoid pathway. Appl Environ Microbiol 64(7):2676–2680
34. Miura Y et al (1998) Production of the carotenoids lycopene, beta-carotene, and astaxanthin in the food yeast *Candida utilis*. Appl Environ Microbiol 64(4):1226–1229
35. Papp T et al (2006) Heterologous expression of astaxanthin biosynthesis genes in *Mucor circinelloides*. Appl Microbiol Biotechnol 69(5):526–531. https://doi.org/10.1007/s00253-005-0026-6
36. Johns AM et al (2016) Four inducible promoters for controlled gene expression in the oleaginous yeast *Rhodotorula toruloides*. Front Microbiol 7:1666. https://doi.org/10.3389/fmicb.2016.01666
37. Misawa N, Shimada H (1998) Metabolic engineering for the production of carotenoids in non-carotenogenic bacteria and yeast. J Biotechnol 59(3):169–181

Chapter 20

Isolation and Characterization of Extrachromosomal Double-Stranded RNA Elements from Carotenogenic Yeasts

Marcelo Baeza, María Fernández-Lobato, Jennifer Alcaíno, and Víctor Cifuentes

Abstract

Double-stranded RNA (dsRNA) molecules are widely found in yeasts and filamentous fungi. It has been suggested that these molecules may play an important role in the evolution of eukaryote genomes and could be a valuable tool in yeast typing. The characterization of these extrachromosomal genetic elements is usually a laborious process, especially when trying to analyze a large number of samples. In this chapter, we describe a simple method to isolate dsRNA elements from yeasts using low amounts of starting material and their application to different *Xanthophyllomyces dendrorhous* strains and other psychrotolerant carotenogenic yeasts. Furthermore, the methodologies for enzymatic and hybridization characterizations and quantification of relative dsRNA abundance are detailed.

Key words dsRNA, *Xanthophyllomyces dendrorhous*, Mycovirus, Carotenogenic yeasts

1 Introduction

Cytoplasmic inherited double-stranded RNA (dsRNA) molecules are extrachromosomal genetic elements (EGEs) encapsidated in virus-like particles (VLPs) commonly found in yeast and filamentous fungi. Their function is not well understood, and only in a few cases they have been unmistakably associated with a detectable phenotype in the host, such as mycocinogeny and hypovirulence [1]. In strains of the astaxanthin-producing yeast *Xanthophyllomyces dendrorhous*, large variations in the size and number of dsRNAs have been described [2, 3]. In addition, dsRNA patterns have been shown to be specific to the region of isolation of different strains. It has been proposed that, together with other molecular markers, the dsRNA pattern may be a valuable tool for yeast typing [4, 5]. The isolation of dsRNA from yeast may be difficult and time-consuming depending on the type of yeast and the number of samples.

Carlos Barreiro and José-Luis Barredo (eds.), *Microbial Carotenoids: Methods and Protocols*, Methods in Molecular Biology, vol. 1852, https://doi.org/10.1007/978-1-4939-8742-9_20, © Springer Science+Business Media, LLC, part of Springer Nature 2018

In this chapter, the characterization of dsRNAs in *X. dendrorhous* and other psychrotolerant carotenogenic yeasts will be used to exemplify a simple method for the isolation and determination of their chemical nature and relative abundance. Furthermore, the analysis of nucleotide similarity among dsRNAs using dot-blot hybridization is detailed.

2 Materials

2.1 Strains and Culture Media

1. *X. dendrorhous* UCD 67-385 (Phaff Yeast Culture Collection, CA, USA).
2. *X. dendrorhous* VKM Y-2266 (The All-Russian Collection of Microorganisms, Moscow, Russia).
3. *X. dendrorhous* VKM Y-2059 (The All-Russian Collection of Microorganisms, Moscow, Russia).
4. *X. dendrorhous* VKM Y-2786 (The All-Russian Collection of Microorganisms, Moscow, Russia).
5. *Saccharomyces cerevisiae* V18 [6].
6. *Mrakia psychrophila* [7].
7. *Mrakia blollopis* [7].
8. *Cryptococcus gastricus* (now *Goffeauzyma gastrica*) [7].
9. *Holtermanniella watticus* (now *Holtermanniella wattica*) [7].
10. *Rhodotorula laryngis* [7].
11. *Metschnikowia* sp. [7].
12. *Dioszegia* sp. [7].
13. *Sporidiobolus salmonicolor* [7].
14. *Mrakia gelida* [7].
15. Complete yeast malt medium (YM): 0.3% yeast extract, 0.3% malt extract, and 0.5% peptone.
16. Trace element solution: To 90 mL distilled water at room temperature, add salts in the following order: 5 g of citric acid·1H_2O, 5 g of $ZnSO_4 \cdot 7H_2O$, 1 g of $Fe(NH_4)_2(SO_4)_2 \cdot 6H_2O$,0.25 g of $CuSO_4 \cdot 5H_2O$, 0.05 g of $MnSO_4 \cdot 1H_2O$, 0.05 g of H_3BO_3, and 0.05 g of $Na_2MoO_4 \cdot 2H_2O$. Adjust the volume to 100 mL, add 1 mL of chloroform, and store at room temperature in the dark.
17. Biotin stock solution: Dissolve 5 mg of biotin in 50 mL of water or 50% of ethanol. Store aliquots of 2.5 mL at −20 °C.
18. Vogel-Bonner medium E (50× stock solution): To 755 mL distilled warm water (45 °C) in a 2-L flask placed on a magnetic stirring hot plate, add salts in the following order: 125 g of Na_3 citrate·2H_2O, 250 g of KH_2PO_4, 100 g of NH_4NO_3,

10 g of $MgSO_4 \cdot 7H_2O$, 5 g of $CaCl_2 \cdot 2H_2O$, 5 mL of trace element solution, and 2.5 mL of biotin stock solution. It is important to allow each salt to dissolve completely before adding the next. Adjust the volume to 1000 mL, add 5 mL of chloroform, and store at room temperature in the dark. Dilute to 1× for use.

2.2 Materials and Reagents for Isolation and Characterization of dsRNAs

1. Ethidium bromide (0.5 μg/mL).
2. 6 M KI.
3. Dimethylsulfoxide (DMSO).
4. Glass beads 0.5 mm diameter.
5. Chloroform/isoamyl alcohol (24:1).
6. TE buffer: 10 mM Tris-HCl, and 1 mM EDTA, pH 8.0.
7. TAE buffer: 2 M Tris base, ~1.56 M glacial acetic acid, and 0.05 M EDTA, pH 8.0.
8. Loading buffer (10×): 0.1% xylene cyanol, 0.1% bromophenol blue, 0.5% SDS, 0.1 M EDTA, and 50% glycerol.
9. 0.01× SSC buffer: 0.003 M NaCl and 0.0003 M sodium citrate, pH 7.0 [8, 9].
10. 2× SSC buffer: 0.6 M NaCl and 0.06 M sodium citrate, pH 7.0 [7, 8].
11. Wash buffer GC: 50 mM NaCl, 10 mM Tris-HCl, pH 7.5, 2.5 mM EDTA, and 50% v/v ethanol.
12. TBE buffer: 89 mM Tris base, 89 mM boric acid, and 2 mM EDTA, pH 8.0.
13. PBS buffer: 137 mM NaCl, 2.7 mM KCl, 10 mM Na_2HPO_4, and 2 mM KH_2PO_4, pH 7.4.
14. DNase I.
15. Nuclease S1.
16. RNase H.
17. RNase A.
18. DNase I buffer: 50 mM sodium acetate, 10 mM $MgCl_2$, and 2 mM $CaCl_2$, pH 6.5.
19. Nuclease S1 buffer: 30 mM potassium acetate, 0.3 M NaCl, 1 mM $ZnSO_4$, and 5% glycerol, pH 4.5.
20. RNase H buffer: 20 mM Tris-HCl; 0.1 M KCl; 10 mM $MgCl_2$; 0,1 mM dithiothreitol (DTT); and 5% sucrose, pH 7.5.
21. Saturated acid phenol: Add 500 g of phenol to an amber flask, and incubate at 60 °C until the phenol melts completely. Add 300 mL of distilled water, mix well by shaking, and incubate at room temperature for 12 h. Remove the water and add 200 mL of 50 mM sodium acetate, pH 4.0. Mix well by shaking, and

incubate at room temperature for 12 h. Store at room temperature.

22. Glass milk: Dissolve 1 g of silica in 10 mL of PBS buffer, allow the silica to settle for 2 h, remove the supernatant (containing fine particulate matter), and add 10 mL of PBS buffer. Repeat the procedure twice. Centrifuge at 2000 × *g* for 2 min, and suspend the silica in 10 mL of 3 M NaI. Store at 4 °C in the dark.
23. *Hind*III molecular weight marker.
24. 1 kb DNA ladder.
25. 1D Image Analysis Software version 2.0.1 (Kodak Scientific Image System, NY, USA).
26. Nuclease-free water.
27. Agarose.
28. KI 6 M.

2.3 Hybridization

1. Nucleotide mix: 10 mM each dTAP, dGTP, dCTP, and dTTP.
2. Random hexamers.
3. ^{32}P-dCTP.
4. Sephadex G-50 (GE Healthcare, Fairfield, CA, USA).
5. Nylon membrane.
6. Autoradiographic film.
7. M-MuLV reverse transcriptase.
8. Hybridization solution: 0.5 M Na_2HPO_4 and 7% SDS, pH 7.2.
9. Wash solution H1: 40 mM Na_2HPO_4 and 5% SDS, pH 7.2.
10. Wash solution H2: 40 mM Na_2HPO_4 and 1% SDS, pH 7.2.
11. ImageJ software [10].
12. Erlenmeyer flasks.
13. Eppendorf tubes.
14. Syringe 1-mL.
15. Glass wool.
16. Autoradiography films.

2.4 Laboratory Equipment

1. Incubator.
2. Shaker.
3. Centrifuge.
4. Microcentrifuge.
5. Vortex.
6. UV light.
7. Dry bath.

8. Fridge −80 °C.
9. Fridge −20 °C.
10. Cell disruptor.
11. Electrophoresis chamber system.
12. Power supply.

3 Methods

The methods described below are used for the characterization of dsRNA elements of carotenogenic yeasts. Yeasts were grown in complete YM medium or in minimal saline Vogel-Bonner medium E [11, 12] supplemented with glucose or succinate (2% w/v), which are used as fermentable and non-fermentable carbon sources, respectively. The method used for the isolation of RNA has been optimized for the processing of a high number of samples, obtaining a large enough quantity for their enzymatic characterization and molecular studies. The entire process is performed in Eppendorf tubes at room temperature using ~0.1 g of yeast pellet (wet weight) obtained from culture aliquots using centrifugation at 7000 × *g* for 5 min (*see* **Note 1**).

3.1 Isolation of Total RNA

1. Wash the yeast pellet twice with 1 mL of TE buffer, and suspend in 0.4 mL of TE buffer.
2. Add 0.4 mL of 0.5-mm diameter glass beads and 0.4 mL of acid phenol, and shake the tubes in a cell disruptor for 3–4 min (*see* **Note 2**). The results of RNA isolation obtained at various times of cell disruption for *X. dendrorhous* are shown in Fig. 1a.
3. Centrifuge the samples at 10,000 × *g* for 2 min, and transfer the aqueous phase to a tube containing 0.4 mL of saturated acid phenol (pH 4.0).
4. Shake in vortex for 2 min and centrifuge at 10,000 × *g* for 1 min (*see* **Note 3**).
5. Transfer the aqueous phase to a tube containing 0.4 mL of chloroform/isoamyl alcohol (24:1), vortex for 2 min, and centrifuge at 10,000 × *g* for 1 min.
6. Perform this last step twice.
7. Mix the aqueous phase with 1 volume of isopropanol and incubate at −20 °C for 2 h.
8. Centrifuge at 14,000 × *g* for 15 min, discard the supernatant, dry the pellet, and suspend it in 10–20 μL of nuclease-free water.
9. Add 1–2 μL of loading buffer (10×) to RNA samples and run using 1% agarose gel electrophoresis (AGE).

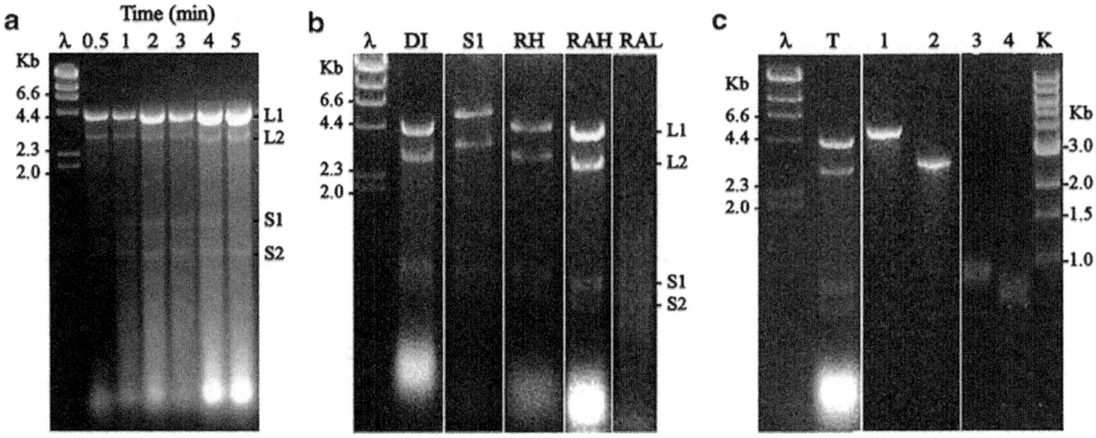

Fig. 1 Isolation and characterization of dsRNAs in *X. dendrorhous* UCD 67-385. (**a**) Total RNA obtained after different times of mechanical yeast rupture. (**b**) Determination of chemical nature using enzymatic digestions: *DI* DNase I, *S1* Nuclease S1, *RH* RNase H, *RAH* RNase A at high ionic strength, *RAL* RNase A at low ionic strength. (**c**) Purification of a particular dsRNA from agarose gels: *T* Total RNA sample, *1* Purified L1 dsRNA, *2* Purified L2 dsRNA, *3* Purified S1 dsRNA, *4* Purified S2 dsRNA. Samples were resolved in 1% agarose gels and stained with ethidium bromide. Molecular weight markers: *Hind*III (λ) and 1 kb DNA ladder (K)

10. After staining with a solution of 0.5 μg/mL ethidium bromide for 5–10 min, visualize the nucleic acid bands under UV light (*see* **Note 4**). An example of the isolation of total RNA from *X. dendrorhous* can be seen in Fig. 1a, c.

This method has been used successfully for the isolation of total RNA from different species of carotenogenic and non-carotenogenic yeasts (Table 1).

3.2 Determination of Chemical Nature

The chemical nature of EGEs can be assessed enzymatically using a set of four enzymes: DNase I for degrading the deoxyribonucleic acids, nuclease S1 for degrading single-stranded nucleic acids, RNase H for degrading RNA from a DNA/RNA hybrid, and RNase A at low ionic strength (0.01× SSC buffer) for degrading all ribonucleic acids and RNase A at high ionic strength (2× SSC buffer) for degrading only single-stranded RNAs (*see* **Note 5**).

1. Digestion with DNase I: Mix 0.5 μg of total RNA, 0.5 U of DNase I, and 2 μL of DNase I buffer; add nuclease-free water to reach 20 μL final volume and incubate for 1 h at 37 °C.
2. Digestion with nuclease S1: Mix 0.5 μg of total RNA, 0.5 U of nuclease S1, and 2 μL of nuclease S1 buffer; add nuclease-free water to reach 20 μL final volume and incubate for 1 h at 37 °C.
3. Digestion with RNase H: Mix 0.5 μg of total RNA, 0.5 U of RNase H, and 2 μL of RNase H buffer; add nuclease-free water 20 μL final volume and incubate for 1 h at 37 °C.
4. Digestion with RNase A at low ionic strength: Mix 0.5 μg of total RNA, 0.5 μg of RNase A, and 2 μL 0.01× SSC buffer;

Table 1
Analysis of dsRNAs in carotenogenic yeast species

Yeast	Main carotenoid[a]	dsRNAs (kpb)
M. psychrophila	Nd	+ (5)
M. blollopis	Nd	+ (5)
Cr. gastricus	2-γ-Carotene	–
H. watticus	Nd	+ (5)
R. laryngis	Torulene	–
Metschnikowia sp.	Nd	–
Dioszegia sp.	OHK torulene	–
S. salmonicolor	γ-Carotene	+ (5)
M. gelida	Nd	+ (5)

[a]References [16, 17]. *Nd* not detected, + dsRNAs detected, – no dsRNAs detected

add nuclease-free water to reach 20 μL final volume and incubate for 1 h at 37 °C.

5. Digestion with RNase A at high ionic strength: Mix 0.5 μg of total RNA, 0.5 μg of RNase A, and 2 μL 2× SSC buffer; add nuclease-free water to reach 20 μL final volume, and incubate for 1 h at 37 °C.
6. Analyze the samples by AGE, as previously indicated.
7. As shown in Fig. 1b, the EGEs of *X. dendrorhous* are resistant to DNase I, nuclease S1, RNase H, and RNase A at high ionic strength but sensitive to RNase A at low ionic strength. With these results, it can be concluded that these four EGEs correspond to dsRNAs. In the same way, the presence of dsRNAs was determined in yeasts belonging to different genera (Table 1).

3.3 Purification of dsRNAs

An inexpensive method for the isolation of a specific dsRNA element with a high degree of purity was previously described [13].

1. Add loading buffer (10×) to samples of total RNA, and subject them to 1% AGE in TAE buffer (*see* **Note 6**).
2. Stain with ethidium bromide and visualize under UV light (Fig. 1c).
3. Excise the bands corresponding to each dsRNA molecule from the gel (*see* **Note 7**), place in an Eppendorf tube, and add 3 volumes of 6 M KI (*see* **Note 8**).

4. Incubate tubes at 55 °C until the gel fragments completely dissolve.
5. Cool to room temperature and mix with 10 μL of glass milk.
6. Incubate on ice for 10 min and centrifuge at 10,000 × *g* for 30 s.
7. Discard the supernatant, and wash the pellet three times with 500 μL of wash buffer GC, centrifuging at 10,000 × *g* for 30 s between washing steps.
8. Remove the wash buffer and suspend the pellet in 15–20 μL of nuclease-free water.
9. Incubate at 55 °C for 10 min.
10. Centrifuge at 10,000 × *g* for 20 s, and transfer the supernatant containing the nucleic acids to a new tube.

The results of purification of each type of dsRNA from *X. dendrorhous* UCD 67-385 are shown in Fig. 1c (*see* **Note 9**). Samples of dsRNAs purified from gels were used for the estimation of molecular size (*see* **Note 10**) and hybridization experiments.

3.4 Estimation of the Length of dsRNAs

1. Resolve the dsRNA samples in 1% and 2% agarose gels in TAE buffer.
2. Stain with ethidium bromide and photograph under UV light.
3. Estimate the molecular size values by comparison against *Hind*III and 1 kb DNA markers using 1D Image Analysis Software version 2.0.1.
4. Correct the values obtained by the difference in electrophoretic mobility between dsDNA and dsRNA [14].

Using this method, the following lengths were obtained for dsRNAs: L1, 5.0 kb; L2, 3.7 kb; S1, 0.9 kb; and S2, 0.8 kb (Fig. 1c). These calculated lengths are in accord with the lengths obtained in molecular characterization and identification of dsRNAs of *X. dendrorhous* [15]. The lengths estimated for dsRNAs of other carotenogenic yeasts are indicated in Table 1.

3.5 Fixation of dsRNAs to Membrane

1. Incubate dsRNA samples at 95 °C for 10 min in the presence of 18% DMSO.
2. Immediately cool in an ice/water bath.
3. Dot 10 μL (~200 ng) onto a nylon membrane.
4. Incubate at 80 °C for 15 min.

3.6 Synthesis and Cleaning of cDNA Probes

1. Incubate samples of 400 ng of dsRNA at 95 °C for 15 min, and quickly cool in an ice/water bath.
2. Add 200 U of M-MuLV reverse transcriptase, 1 μL of 25 mM random hexamers, 1 μL of 10 mM nucleotide mix, and 3 μCi of ^{32}P-dCTP.

3. Incubate at 42 °C for 90 min and stop the reaction by incubation at 70 °C for 10 min.
4. Prepare a Sephadex G-50 column: Weigh 10 g of Sephadex G-50 powder, transfer to a 200-mL Erlenmeyer flask, and add 100 mL of TE buffer. Incubate for 1 h at room temperature. Fill a 1-mL syringe barrel, plugged at the bottom with glass wool, with the equilibrated Sephadex G50, place into 15-mL centrifuge tube, and spin at 1000 × *g* for 2 min. Then refill with equilibrated Sephadex G50 and spin at 1000 × *g* for 2 min. Repeat the last step until the 90% volume of the syringe is filled. For the collection of the eluate, place the bottom on the syringe into a microfuge tube, and then both into a 15-mL centrifuge tube. In this way, after the centrifugation step, the eluate is collected in the microfuge tube.
5. Load the sample onto Sephadex G-50 columns and centrifuge at 1000 × *g* for 2 min.
6. Collect the eluate, incubate at 95 °C for 5 min, and quickly cool in an ice/water bath (denatured cDNA probe).

3.7 Hybridization

1. Incubate membranes at 65 °C for 10 min in 10 mL of hybridization buffer.
2. Add the denatured cDNA probe and incubate at 65 °C for 16 h (*see* **Note 11**).
3. Wash the membranes twice at 65 °C for 30 min with wash buffer H1.
4. Wash the membranes twice at 65 °C for 30 min with wash buffer H2.
5. Dry membranes and expose to autoradiography films at −80 °C (*see* **Note 12**).
6. Develop the films (Fig. 2).

3.8 Determination of the Relative Abundance of dsRNAs

The relative abundance between two dsRNAs of *X. dendrorhous* was analyzed at different times of growth in complete or minimal medium supplemented with glucose or succinate.

1. Inoculate 1000 mL of medium (YM or Vogel-Bonner medium E) with 10 mL of a late exponential phase culture of *X. dendrorhous*.
2. Incubate at 22 °C with constant agitation, and harvest 10 mL of samples at several time points.
3. Make tenfold serial dilutions from 1 mL of culture sample.
4. Seed onto YM agar plates and incubate at 22 °C for 3–5 days.
5. Calculate the colony-forming units (CFUs) and the CFU/mL for each dilution.

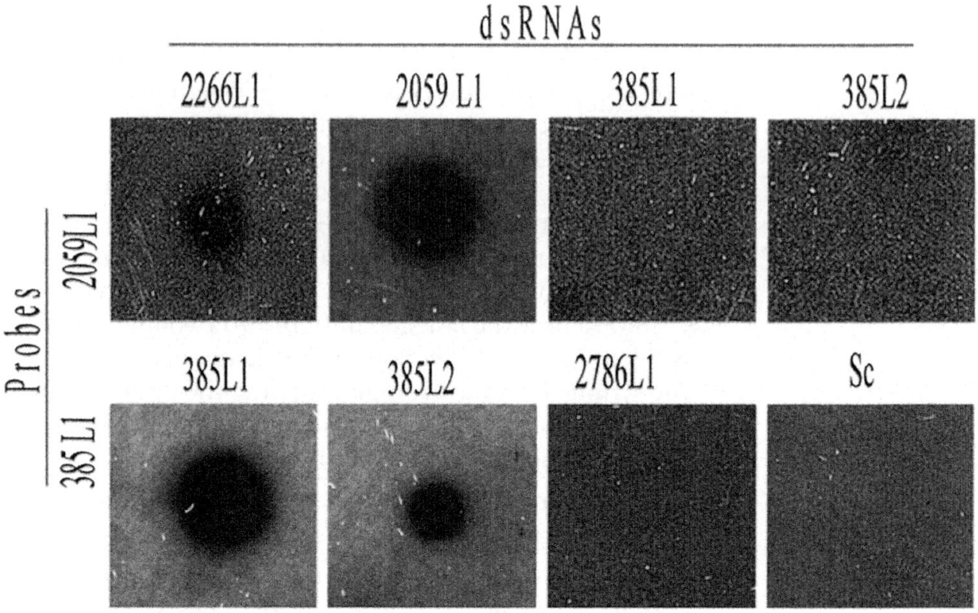

Fig. 2 Dot-blot hybridization. dsRNAs were purified from *X. dendrorhous* strains VKM Y-2266 (2266L1), VKM Y-2059 (2059L1), VKM Y-2786 (2786L1), and UCD 67-385 (385L1 and 385L2) and from *S. cerevisiae* strain V18 (Sc) [18]. dsRNA samples were heat denatured and fixed onto nylon membranes. The radioactive cDNA probes were synthesized from L1dsRNA of *X. dendrorhous* UCD 67-385 (p385L1, upper panels) and *X. dendrorhous* VKM Y2059 (p2059L1, lower panels) strains

6. Isolate the total RNA from each culture sample, analyze by 1% AGE, stain with ethidium bromide, and photograph under UV light.
7. Perform the analysis in triplicate using different volumes (2–10 μL) of each RNA sample to obtain more information.
8. Compare intensities of dsRNA bands separated in the same gel using ImageJ software [10].

The relative abundance of L2 versus L1 dsRNAs during different time points of *X. dendrorhous* growth in different media is shown in Fig. 3. The estimation of relative abundance by this simple method was in accord to estimation using two-step RT-qPCR [15].

4 Notes

1. The volume of yeast culture to be centrifuged is variable and will depend on the growth phase of harvesting. Typically, 0.1 g of moist pellet is obtained from 7 to 10 mL of stationary-phase culture. Alternatively, yeast can be obtained from cultures on semisolid media.

2. The results shown here were obtained using the Mini-Beadbeater cell disruptor. The time of rupture must be defined for other cell disruptor equipment. An inexpensive alternative is to perform cell rupture by vigorously shaking in a vortex for 30–40 min.
3. When extraction with acid phenol is used, the DNA molecules solubilize in the organic phase, producing an aqueous phase enriched mainly in RNA molecules.
4. There are alternatives to ethidium bromide for nucleic staining that are safe and eco-friendly and stain well the dsRNAs.
5. For a rapid analysis of a high number of samples, it is sufficient the enzymatic analysis with DNase I and Nuclease S1. The samples in which nucleic acid bands are detected in AGE after the enzymatic treatments are selected for a complete enzymatic analysis.
6. The purification of fragments from TBE-based gels is more complicated than from TAE-based gels.

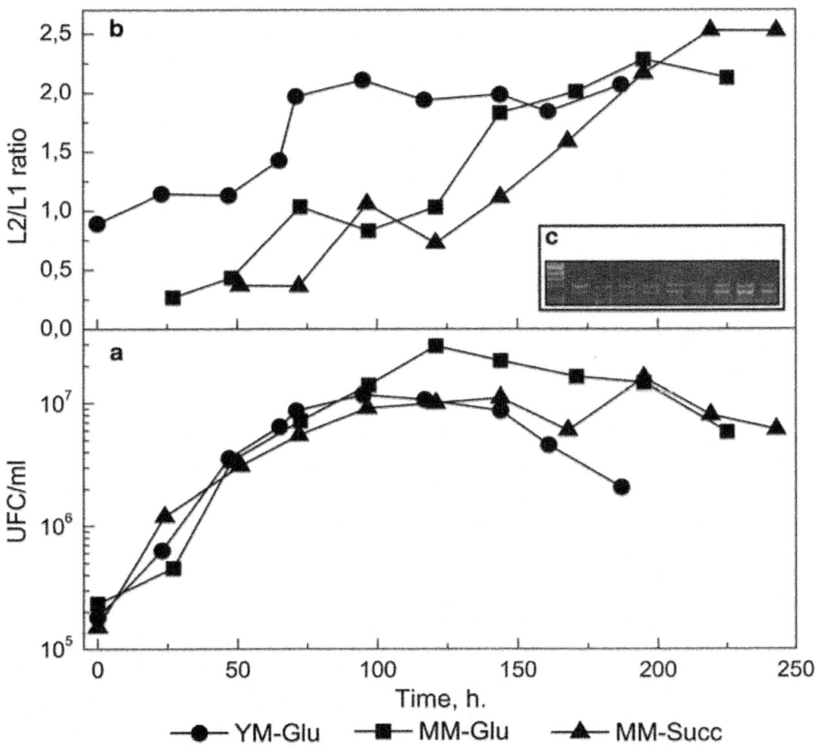

Fig. 3 The relative abundance of dsRNA elements in *X. dendrorhous* at different growth times. (**a**) *X. dendrorhous* UCD 67-385 was grown in YM medium supplemented with glucose (YM-Glu) and in Vogel-Bonner medium E supplemented with glucose (MM-Glu) or succinate (MM-Succ). Total RNA was extracted from culture samples which were collected at each time point, and the L2/L1 dsRNA ratios (**b**) were calculated from the band intensities of each dsRNA elements separated in agarose gels, as shown in (**c**) for MM-Glu

7. Only the section of gel that contains the nucleic acids should be excised. This facilitates the dissolving of the gel and allows the use of several gel pieces in a single tube, increasing the amount of dsRNA recovered.
8. The reagent NaI is used in the original published protocol. However, we use the cheaper reagent KI, obtaining similar results.
9. Several pieces of agarose gel are put in a single tube to obtain a high concentration of dsRNA.
10. The electrophoretic mobility of the dsRNA molecules in agarose gels is changed by the presence of small molecules, which generates error in the calculations of molecular sizes. These molecules include other dsRNAs or ribosomal RNA, as can be seen in Fig. 1b (compare lanes DI and S1).
11. The incubation temperature must be adapted to the degree of stringency desired.
12. The time of exposure varies due to a wide range of possible identities among the cDNA probes and dsRNA targets. Because of this variance, at least three autoradiography films are placed in each cassette and exposed at different times. In this way, overexposure of the strong signals is avoided.

Acknowledgments

This work was supported by grants FONDECYT 11060157 and 1130333, Innova CORFO 07CN13PZT-17, and INACH RT_07-13.

References

1. Tavantzis SM (2001) dsRNA genetic elements: concepts and applications in agriculture, forestry, and medicine. CRC Press, Boca Raton
2. Baeza M, Sanhueza M, Flores O, Oviedo V, Libkind D, Cifuentes V (2009) Polymorphism of viral dsRNA in *Xanthophyllomyces dendrorhous* strains isolated from different geographic areas. Virol J 6:160
3. Pfeiffer I, Kucsera J, Varga J, Párducz A, Ferenczy L (1996) Variability and inheritance of double-stranded RNA viruses in *Phaffia rhodozyma*. Curr Genet 30:294–297
4. Maqueda M, Zamora E, Rodríguez-Cousiño N, Ramírez M (2010) Wine yeast molecular typing using a simplified method for simultaneously extracting mtDNA, nuclear DNA and virus dsRNA. Food Microbiol 27:205–209
5. Cubeta MA, Vilgalys R (1997) Population biology of the *Rhizoctonia solani* complex. Phytopathology 87:480–484
6. Baeza ME, Sanhueza MA, Cifuentes VH (2008) Occurrence of killer yeast strains in industrial and clinical yeast isolates. Biol Res 41:173–182
7. Carrasco M, Rozas JM, Barahona S, Alcaíno J, Cifuentes V, Baeza M (2012) Diversity and extracellular enzymatic activities of yeasts isolated from King George Island, the sub-Antarctic region. BMC Microbiol 12:251
8. Pryor A, Boelen MG (1987) A double-stranded RNA mycovirus from the maize rust *Puccinia sorghi*. Can J Bot 65:2380–2383
9. Castillo A, Cifuentes V (1994) Presence of double-stranded RNA and virus-like particles

in *Phaffia rhodozyma*. Curr Genet 26:364–368

10. Rasband, W.S U.S. National Institutes of Health, Bethesda, Maryland, USA, http://imagej.nih.gov/ij/, 1997–2016
11. Retamales P, Hermosilla G, Leon R, Martınez C, Jimenez A, Cifuentes V (2002) Development of the sexual reproductive cycle of *Xanthophyllomyces dendrorhous*. J Microbiol Methods 48:87–93
12. Vogels HJ (1956) A convenient growth medium for *Neurospora*. Microbiol Genet Bull 13:42–43
13. Boyle JS, Lew AM (1995) An inexpensive alternative to glassmilk for DNA purification. Trends Genet 11:8
14. Livshits MA, Amosova OA, Lyubchenko YL (1990) Flexibility difference between double-stranded RNA and DNA as revealed by gel electrophoresis. J Biomol Struct Dyn 7:1237–1249
15. Baeza M, Bravo N, Sanhueza M, Flores O, Villarreal P, Cifuentes V (2012) Molecular characterization of totiviruses in *Xanthophyllomyces dendrorhous*. Virol J 9:140
16. Villarreal P, Carrasco M, Barahona S, Alcaíno J, Cifuentes V, Baeza M (2016) Tolerance to ultraviolet radiation of psychrotolerant yeasts and analysis of their carotenoid, mycosporine, and ergosterol content. Curr Microbiol 72:94–101
17. Barahona S, Yuivar Y, Socias G, Alcaíno J, Cifuentes V, Baeza M (2016) Identification and characterization of yeasts isolated from sedimentary rocks of Union Glacier at the Antarctica. Extremophiles 20:479–491
18. Reyes E, Barahona S, Fischman O, Niklitschek M, Baeza M, Cifuentes V (2004) Genetic polymorphism of clinical and environmental strains of *Pichia anomala*. Biol Res 37:747–757

Index

Carlos Barreiro and José-Luis Barredo (eds.), *Microbial Carotenoids: Methods and Protocols*, Methods in Molecular Biology, vol. 1852, https://doi.org/10.1007/978-1-4939-8742-9, © Springer Science+Business Media, LLC, part of Springer Nature 2018

Printed by Printforce, the Netherlands